TODAY'S TECHNICIAN

Classroom Manual for

Manual Transmissions and Transaxles

TODAY'S TECHNICIAN

Classroom Manual for

Manual Transmissions and Transaxles

Jack Erjavec

Columbus State Community College
Columbus, Ohio

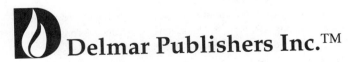

Delmar Publishers Inc.™

I(T)P™

NOTICE TO THE READER

Publisher does not warrant or guarantee any of the products described herein or perform any independent analysis in connection with any of the product information contained herein. Publisher does not assume, and expressly disclaims, any obligation to obtain and include information other than that provided to it by the manufacturer.

The reader is expressly warned to consider and adopt all safety precautions that might be indicated by the activities described herein and to avoid all potential hazards. By following the instructions contained herein, the reader willingly assumes all risks in connection with such instructions.

The publisher makes no representations or warranties of any kind, including but not limited to, the warranties of fitness for particular purpose or merchantability, nor are any such representations implied with respect to the material set forth herein, and the publisher takes no responsibility with respect to such material. The publisher shall not be liable for any special, consequential or exemplary damages resulting, in whole or in part, from the readers' use of, or reliance upon, this material.

Portions of materials contained herein have been reprinted with permission of General Motors Corporation, Service Technology Group.

COVER PHOTO: Courtesy of Chevrolet Motor Division

DELMAR STAFF

Senior Administrative Editor: Vernon Anthony
Developmental Editor: Catherine Eads
Project Editor: Eleanor Isenhart
Production Coordinator: Karen Smith
Art/Design Coordinator: Heather Brown

For information address Delmar Publishers Inc.
3 Columbia Circle, Box 15015
Albany, New York 12212-5015

Printed in the United States of America
Published simultaneously in Canada
by Nelson Canada
a division of the Thomson Corporation

10 9 8 7 6 5 4 3 2 XXX 00 99 98 97 96 95 94

Library of Congress Cataloging-in-Publication Data

Erjavec, Jack.
Manual transmissions and transaxles / Jack Erjavec.
p. cm. — (Today's technician)
Includes index.
Contents: v. 1. Classroom manual — v. 2. Shop manual.
ISBN 0-8273-6181-5 (set)
1. Motor vehicles—Transmission devices—Maintenance and repair—Handbooks, manuals, etc. 2. Motor Vehicles—Transmission devices—Maintenance and repair—Problems, exercises, etc. I. Title.
II. Series.
TL262.E75 1994
629.24'4'0288—dc20

93-33478
CIP

CONTENTS

Preface

Unlike yesterday's mechanic, the technician of today and for the future must know the underlying theory of all automotive systems and be able to service and maintain those systems. Today's technician must also know how these individual systems interact with each other. Standards and expectations have been set for today's technician, and these must be met in order to keep the world's automobiles running efficiently and safely.

The *Today's Technician* series, by Delmar Publishers, features textbooks that cover all mechanical and electrical systems of automobiles and light trucks. Principal titles correspond with the eight major areas of ASE (National Institute for Automotive Service Excellence) certification. Additional titles include remedial skills and theories common to all of the certification areas and advanced or specialized subject areas that reflect the latest technological trends.

Each title is divided into two manuals: a Classroom Manual and a Shop Manual. Dividing the material into two manuals provides the reader with the information needed to begin a successful career as an automotive technician without interrupting the learning process by mixing cognitive and performance-based learning objectives.

Each Classroom Manual contains the principles of operation for each system and subsystem. It also discusses the design variations used by different manufacturers. The Classroom Manual is organized to build upon basic facts and theories. The primary objective of this manual is to allow the reader to gain an understanding of how each system and subsystem operates. This understanding is necessary to diagnose the complex automobile systems.

The understanding acquired by using the Classroom Manual is required for competence in the skill areas covered in the Shop Manual. All of the high priority skills, as identified by ASE, are explained in the Shop Manual. The Shop Manual also includes step-by-step instructions for diagnostic and repair procedures. Photo Sequences are used to illustrate many of the common service procedures. Other common procedures are listed and are accompanied with fine-line drawings and photographs that allow the reader to visualize and conceptualize the finest details of the procedure. The Shop Manual also contains the reasons for performing the procedures, as well as when that particular service is appropriate.

The two manuals are designed to be used together and are arranged in corresponding chapters. Not only are the chapters in the manuals linked together, the contents of the chapters are also linked. Both manuals contain clear and thoughtfully selected illustrations. Many of the illustrations are original drawings or photos prepared for inclusion in this series. This means that the art is a vital part of each manual.

The page layout is designed to include information that would otherwise break up the flow of information presented to the reader. The main body of the text includes all of the "need-to-know" information and illustrations. In the side margins are many of the special features of the series. Items such as definition of new terms, common trade jargon, tools list, and cross-referencing are placed in the margin, out of the normal flow of information so as not to interrupt the thought process of the reader.

The content of each chapter has been verified by recognized experts from industry and from schools of education. Their sole purpose is to check the accuracy, completeness, and timeliness of the material. Each manual in this series is organized in a like manner and contains the same features.

Classroom Manual

To stress the importance of safe work habits, the Classroom Manual dedicates one full chapter to safety. Included in this chapter are common safety practices, safety equipment, and safe handling of hazardous materials and wastes. This includes information on MSDS sheets and OSHA regulations. Other features of this manual include:

Cognitive Objectives

These objectives define the contents of the chapter and define what the student should have learned upon completion of the chapter.
Each topic is divided into small units to promote easier understanding and learning.

Marginal Notes

Page numbers for cross-referencing appear in the margin. Some of the common terms used for components, and other bits of information, also appear in the margin. This provides an understanding of the language of the trade and helps when conversing with an experienced technician.

Cautions and Warnings

Throughout the text, cautions are given to alert the reader of potential hazardous materials or unsafe conditions. Warnings are also given to advise the student of things that can go wrong if instructions are not followed or if a nonacceptable part or tool is used.

References to the Shop Manual

Reference to the appropriate page in the Shop Manual is given whenever necessary. Although the chapters of the two manuals are synchronized, material covered in other chapters of the Shop Manual may be fundamental to the topic discussed in the Classroom Manual.

A Bit of History

This feature gives the student a sense of the evolution of the automobile. This feature not only contains nice-to-know information, but also should spark some interest in the subject matter.

Summaries

Each chapter concludes with summary statements that contain the important topics of the chapter. These are designed to help the reader review the contents.

Terms to Know

A list of new terms appears next to the Summary. Definitions for these terms can be found in the Glossary at the end of the manual.

Review Questions

Short answer essay, fill-in-the-blank, and multiple-choice type questions follow each chapter. These questions are designed to accurately assess the student's competence in the stated objectives at the beginning of the chapter.

Shop Manual

To stress the importance of safe work habits, the Shop Manual also dedicates one full chapter to safety. Other important features of this manual include:

Performance Objectives

These objectives define the contents of the chapter and define what the student should have learned upon completion of the chapter. These objectives also correspond with the list of required tasks for ASE certification. *Each ASE task is addressed.*

Although this textbook is not designed to simply prepare someone for the certification exams, it is organized around the ASE task list. These tasks are defined generically when the procedure is commonly followed and specifically when the procedure is unique for specific vehicle models. Imported and domestic model automobiles and light trucks are included in the procedures.

Photo Sequences

Many procedures are illustrated in detailed photo sequences. These detailed photographs show the students what to expect when they perform particular procedures. They also can provide a student a familiarity with a system or type of equipment, which the school may not have.

Marginal Notes

New terms are pulled out and defined. Common trade jargon also appears in the margin and gives some of the common terms used for components. This allows the reader to speak and understand the language of the trade, especially when conversing with an experienced technician.

Cautions and Warnings

Throughout the text, cautions are given to alert the reader of potential hazardous materials or unsafe conditions. Warnings are also given to advise the student of things that can go wrong if instructions are not followed or if a nonacceptable part or tool is used.

References to the Classroom Manual

Reference to the appropriate page in the Classroom Manual is given whenever necessary. Although the chapters of the two manuals are synchronized, material covered in other chapters of the Classroom Manual may be fundamental to the topic discussed in the Shop Manual.

Customer Care

This feature highlights those little things a technician can do or say to enhance customer relations.

Tools Lists

Each chapter begins with a list of the Basic Tools needed to perform the tasks included in the chapter. Whenever a Special Tool is required to complete a task, it is listed in the margin next to the procedure.

Service Tips

Whenever a short-cut or special procedure is appropriate, these are described in the text. These tips are generally those things commonly done by experienced technicians.

Case Studies

Case Studies concentrate on the ability to properly diagnosis the systems. Each chapter ends with a case study, in which a vehicle has a problem and the logic used by a technician to solve the problem is explained.

Terms to Know

Terms in this list can be found in the Glossary at the end of the manual.

Diagnostic Chart

Chapters include detailed diagnostic charts linked with the appropriate ASE task. These charts list common problems and most probable causes. They also list a page reference in the Classroom Manual for better understanding of the system's operation and a page reference in the Shop Manual for details on the procedure necessary for correcting the problem.

ASE Style Review Questions

Each chapter contains ASE style review questions that reflect the performance objectives listed at the beginning of the chapter. These questions can be used to review the chapter as well as to prepare for the ASE certification exam.

Reviewers

Michael Moore
Attawaba Technical Institute
Baxley, GA

Charles B. Weir
Black River Technical College
Pochontas, AK

Dennis Wierima
Northwest Technical College—East Grand Folks
East Grand Folks, MN

Lawrence F. LeGree
Montcalm Community College
Sidney, MI

Paul D. Hope
Indian Hills Community College
Ottumwa, IA

John Sharpe
Mott Community College
Linden, MI

Jack Erjavec

CHAPTER 1

Safety

Upon completion and review of this chapter, you should be able to:

- ❑ Explain how safety practices are part of professional behavior.
- ❑ Dress safely and professionally.
- ❑ Recognize fire hazards.
- ❑ Inspect equipment and tools for unsafe conditions.
- ❑ Properly work around batteries.
- ❑ Explain the procedures for responding to an accident.
- ❑ Identify substances that could be regarded as hazardous materials.

Safety is everyone's job. You should work safely to protect yourself and the people around you. Perhaps the best single safety rule is "Think before you act." Too often people working in shops gamble by working in an unsafe way. Often they gamble and win and no one gets hurt, but it takes only one accident to lose all those past winnings. The gamble may save five minutes, but cost you an eye or a hand. By acting first and thinking last, you can ruin your back and lose your career. Accidents in a shop can be prevented by others and by YOU. Safe work habits also prevent damage to the vehicles and equipment in the shop.

It has been said that 50% of all shop accidents could have been prevented by a single individual, the technician.

Personal Safety

When you have neat work habits, you display a professional attitude. Actually, neat habits are also safe habits. Cleaning up spills and keeping equipment and tools out of the path of others prevents accidents. This is only common sense, but too often, time is not taken to remove these safety hazards. It is the rush to complete a job that usually results in unsafe conditions. True professionals take time to clean their tools and work areas. A professional does a better job, makes more money, and has fewer accidents than others who don't take the time to be neat and safe.

With a professional attitude, you do not clown around in the shop, you do not throw items in the shop, and you do not create an unsafe condition for the sake of saving time. Instead, a professional saves time by effective diagnostics and proper procedures.

Shop Manual
Chapter 1, page 8

An accident is something that happens unintentionally and is a consequence of doing something else.

Dress and Appearance

How you dress and appear to others says something about your personality and your attitude, including your attitude about safety. Clothing that hangs out freely, such as shirttails, can create a safety hazard and cause serious injury. Nothing you wear should be allowed to dangle in the engine compartment or around equipment. Long hair should be tied up or tucked under a hat. Shirts should be tucked in and buttoned and long sleeves buttoned or carefully rolled up (Figure 1-1).

Rotating Pulleys and Belts

Be very careful around belts, pulleys, wheels, chains, or any other rotating mechanism. Be especially careful while leaning against a belt or pulley when it's not moving. Make sure no one accidentally activates the machine. When working around an engine's drive belts and pulleys, make sure your hands, shop towels, or loose clothing do not come in contact with the moving parts. It may not seem that these parts are rotating or traveling at high rates of speed, but they are. Hands and fingers can be quickly pulled into a revolving belt or pulley even at engine idle speeds.

Figure 1-1 Proper dress prevents injuries.

The thermostatic switch for the electric cooling fans may be disconnected to prevent the fan from coming on.

CAUTION: Be careful when working around electric engine cooling fans. These fans are controlled by a thermostat and can come on without warning, even when the engine is not running. Whenever you must work around these fans, disconnect the electrical connector to the fan motor before reaching into the area around the fan.

Jewelry

Rings, necklaces, bracelets, and watches should not be worn while working. A ring can rip your finger off, a watch or bracelet can cut your wrist, and a necklace can choke you. This is especially true when working with or around electrical wires. The metal used to make jewelry conducts electricity very well and can easily cause a short, through you, if it touches a bare wire.

An electrical short is basically an alternative path for the flow of electricity.

Foot Protection

You should also protect your feet. Tennis and jogging shoes will provide little protection if something falls on your foot. Boots or shoes made of leather, or a material that approaches the strength of leather, offer much better protection from falling objects. There are many designs of safety shoes and boots made of leather or similar material with steel plates built into the toe and shank to protect your feet. Many also have soles that are designed to resist slipping on wet surfaces. Foot injuries are not only quite painful but can also put you out of work for some time.

The shank of a shoe is that portion of the shoe that protects the ball of your foot.

Keep your clothing clean. If you spill gasoline or oil on yourself, change that item of clothing immediately. Oil against your skin for a prolonged period of time can produce rashes or other allergic reactions. Gasoline can irritate cuts and sores. As well as being bad for your health, greasy uniforms do not look professional.

Safety Glasses

Brake fluid is not only used in brake systems but is also commonly used as the hydraulic fluid for hydraulic clutch assemblies.

You should wear some sort of eye protection whenever and wherever there is a possibility of dirt or metal particles flowing in the air (Figure 1-2). Dirt, grease, or rust can get into your eyes and cause serious damage. Safety glasses or goggles should be worn when you are working under a vehicle and when you are using any machining equipment, grinding wheels, chemicals, compressed air, or fuels. Chemicals, such as brake fluid, can cause serious eye irritations, which can

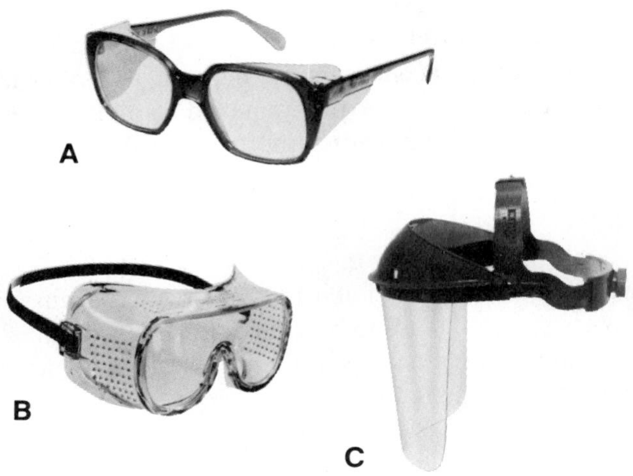

Figure 1-2 Different types of eye protection worn by automotive technicians: (A) safety glasses; (B) goggles; and (C) face shield. (Courtesy of Goodson Shop Supplies)

lead to blindness. The lenses of safety glasses should be made of tempered glass or safety plastic. Common sense should tell you to wear safety glasses nearly all of the time you are working in the shop. Some schools and repair shops require that you do.

Hand Protection

Good hand protection is often overlooked. A scrape, cut, or burn can seriously impair your ability to work for many days. A well-fitting pair of heavy work gloves should be worn while grinding, welding, or when handling chemicals or high temperature components. Special rubber gloves are recommended for handling caustic chemicals.

A caustic material has the ability to destroy or eat through something. Caustic materials are considered extremely corrosive.

Fire Hazards and Prevention

There are many items around a typical shop that pose a fire hazard. These include gasoline, diesel fuel, cleaning solvents, and dirty rags. Each of these should be treated as potential fire bombs.

Gasoline

Gasoline is present in shops so often that its dangers are often forgotten. A slight spark or an increase in heat can cause a fire or explosion. Gasoline fumes are heavier than air. Therefore when an open container of gasoline is sitting about, the fumes spill out over the sides of the container onto the floor. These fumes are more flammable than liquid gasoline and can easily explode.

CAUTION: Never siphon gasoline or diesel fuel with your mouth. These liquids are poisonous and can make you sick or fatally ill.

Never smoke around gasoline or in a shop filled with gasoline fumes because even the droppings of hot ashes can ignite the gasoline. If the engine has a gasoline leak or you have caused a leak by disconnecting a fuel line, wipe it up immediately and stop the leak. While stopping the leak, be extra careful not to cause any sparks. Make sure that any grinding or welding that may be taking place in the area is stopped until the spill is totally cleaned up and the floor has been flushed with water. The rags used to wipe up the gasoline should be taken outside to dry. Immediately wipe up any gasoline that has spilled on the floor. If vapors are present in the shop, have the doors open and the ventilating system turned on to get rid of those fumes. Remember, it takes only a small amount of fuel mixed with air to cause combustion.

Gasoline should always be stored in approved containers and never in a glass bottle or jar. If the glass jar was knocked over or dropped, a terrible explosion could take place. Never use gasoline to clean parts. Never pour gasoline into a carburetor air horn to start the car. Repair any fuel leak immediately.

Diesel Fuel

The *volatility* of a substance is a statement of how easily the substance vaporizes or explodes.

Diesel fuel is not as volatile as gasoline, but it should be stored and handled in the same manner as gasoline. It is also not as refined as gasoline and tends to be a very dirty fuel. It normally contains many impurities, including active microscopic organisms that can be highly infectious. If diesel fuel happens to get on an open cut or sore, thoroughly wash it immediately.

Solvents

The *flammability* of a substance is a statement of how well the substance supports combustion.

Cleaning solvents are not as volatile as gasoline, but they are still flammable. They should be stored and treated in the same way as gasoline (Figure 1-3).

Figure 1-3 Combustible materials should be stored in approved safety containers and cabinets. (Courtesy of Sherwin-Williams Company)

Rags

A *hydrocarbon* is a substance composed of hydrogen and carbon molecules.

Oily or greasy rags can also be a source for fires. These rags should be stored in an approved container and never thrown out with normal trash. Like gasoline, oil is a hydrocarbon and can ignite with or without a spark or flame.

Safe Tools and Equipment

When you work with any equipment, make sure you use it properly and that it is set up according to the manufacturer's instructions. All equipment should be properly maintained and periodically inspected for unsafe conditions. Frayed electrical cords or loose mountings can cause serious injuries. All electrical outlets should be equipped to allow for the use of three-pronged electrical cords. The third prong allows for a safety ground connection. All equipment with rotating parts should be equipped with safety guards that reduce the possibility of the parts coming loose and injuring someone (Figure 1-4). Do not depend on someone else to inspect and maintain equipment. Check it out before you use it! If you find the equipment unsafe, put a sign on it to warn others and notify the person in charge.

Figure 1-4 Equipment such as grinding wheels should be equipped with safety shields and tool guards. (Courtesy of Snap-On Tools Corporation)

Tools and equipment should never be used for purposes other than those for which they are designed. Using the proper tool in the correct way will not only be safer, but will also allow you to do a better job. Safety is everyone's job! Take it upon yourself to ensure your safety and the safety of others.

Batteries

CAUTION: The active chemical in a battery (Figure 1-5), the electrolyte, is basically sulfuric acid. Sulfuric acid can cause severe skin burns and permanent eye damage, including blindness, if it gets in your eye. If some battery acid gets on your skin, wash it off immediately and flush your skin with water for at least 5 minutes. If the electrolyte gets into your eyes, immediately flush them with water and promply see a doctor. NEVER rub your eyes; just flush them well and go to a doctor. Working with and around batteries is a commonsense time to wear safety glasses or goggles.

The hydrogen gases that form in the top of a battery while it is being charged are very explosive. Never smoke or introduce any form of heat around a charging battery. An explosion will not only destroy the battery but may also spray sulfuric acid all over you, the car, and the shop. When connecting a battery charger to a battery, leave the charger off until all of the charger's leads are connected. This will prevent electrical sparks, thereby preventing a possible explosion.

CAUTION: Always double-check the polarity of the battery charger's connections and leads before turning the charger on. Incorrect polarity can damage the battery or cause it to explode.

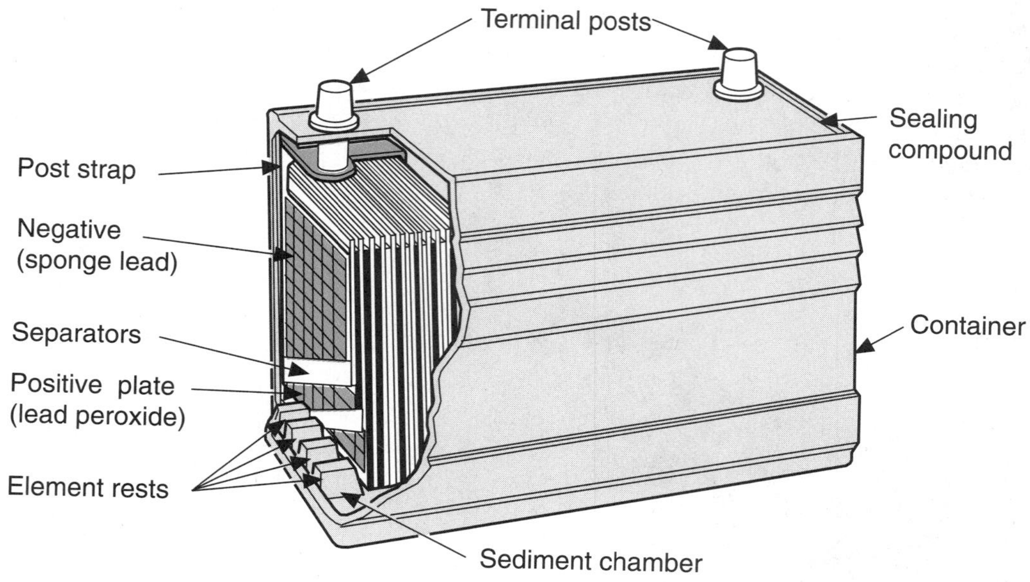

Figure 1-5 The construction of a typical battery.

The most dangerous battery is one that has been overcharged. It is hot and has been, or still may be, producing large amounts of hydrogen. Allow the battery to cool before working with or around it. Also never use or charge a battery that has frozen electrolyte. Extreme amounts of hydrogen may be released from the electrolyte as it thaws. If a battery does explode and battery acid gets on you, immediately flood the acid off, then get medical help.

Accidents

If there is a fire, the quicker it is controlled, the less damage it will do. The same is true of accidents—that is, the quicker you respond, the less damage there will be. For example, if battery acid gets in your eye, the sooner your eye is flushed and examined by a doctor, the better chance you have of not being blinded. Never pass off an eye injury as something unimportant. Often, when a chemical gets in your eye, it takes a few hours to cause great discomfort and by that time the damage may already be done.

Your supervisor should be immediately informed of all accidents that occur in the shop. It's a good idea to keep a list of up-to-date emergency telephone numbers posted next to the telephone. The numbers should include a doctor, hospital, and the fire and police departments. The work area should also have a first-aid kit (Figure 1-6) for treating minor injuries. Facilities for flushing eyes should also be near or in the shop area. Know where they are.

First Aid

It's also a good idea for you to know first aid. The knowledge of how to treat certain injuries can save someone's life. The American Red Cross offers many low cost but thorough courses on first aid. You will realize the importance of these classes the first time you have to give first aid to someone or when someone must give it to you.

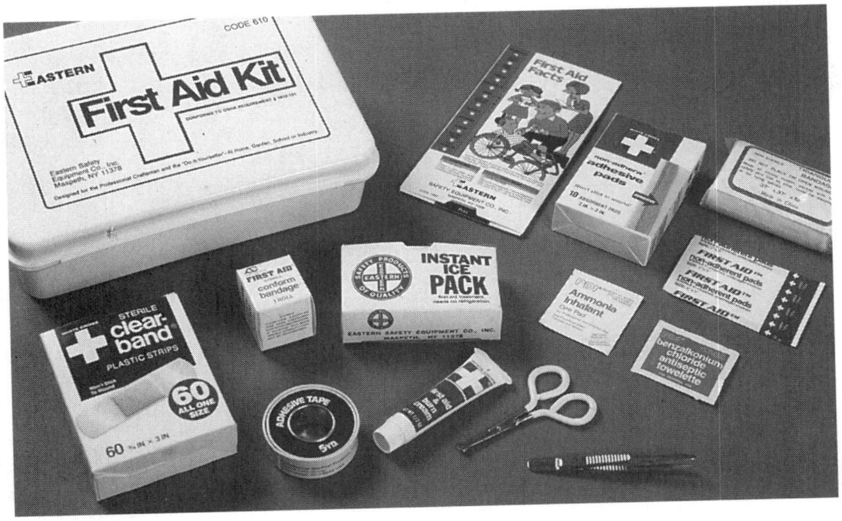

Figure 1-6 A typical first aid kit and its contents.

Hazardous Materials

Many solvents and other chemicals used in an auto shop have warning and caution labels that should be read and understood by everyone that uses them. These are typically considered hazardous materials. Also many service procedures generate what are known as hazardous wastes. Dirty solvents and liquid cleaners are good examples.

Right-to-Know Law

Every employee in a shop is protected by "Right-to-Know" laws concerning hazardous materials and wastes. The general intent of these laws is for employers to provide a safe working place, as it relates to hazardous materials. All employees and students must be trained about their rights under the legislation, the nature of the hazardous chemicals in their workplace, the labeling of chemicals, and the information about each chemical listed and described on Material Safety Data Sheets (MSDS). These sheets (Figure 1-7) are available from the manufacturers and suppliers of the chemicals. They detail the chemical composition and precautionary information for all products that can present health or safety hazards.

Employees must be familiar with the intended purposes of the substance, the recommended protective equipment, accident and spill procedures, and any other information regarding the safe handling of hazardous materials. This training must be given annually to employees and provided to new employees as part of their job orientation. The Canadian equivalents to the MSDS are called Workplace Hazardous Materials Information Systems (WHMIS).

CAUTION: When handling any hazardous material, always wear the appropriate safety protection. Always follow the correct procedures while using the material and be familiar with the information given on the MSDS for that material.

All hazardous materials should be properly labeled, indicating what health, fire, or reactivity hazard it poses and what protective equipment is necessary when handling each chemical. The manufacturer of the hazardous material must provide all warnings and precautionary information, which must be read and understood by all users before they use it. You should pay great attention to the label information. By doing so you will use the substance in the proper and safe way, thereby preventing hazardous conditions.

Reactivity is a statement of how easily a substance can cause or be part of a chemical reaction.

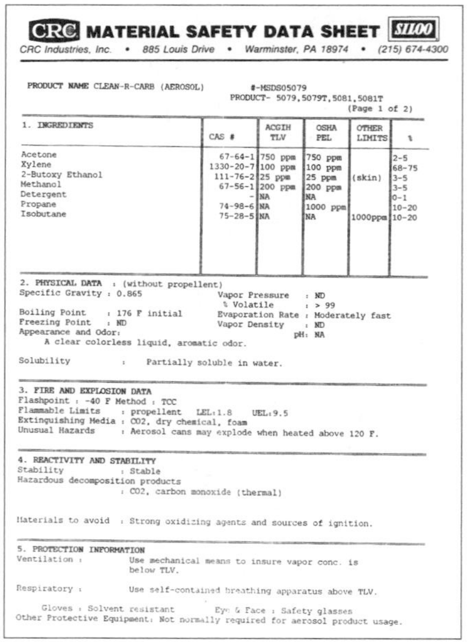

CRC MATERIAL SAFETY DATA SHEET SILOO

CRC Industries, Inc. • 885 Louis Drive • Warminster, PA 18974 • (215) 674-4300

PRODUCT NAME CLEAN-R-CARB (AEROSOL) #-MSDS05079
PRODUCT- 5079,5079T,5081,5081T
(Page 1 of 2)

1. INGREDIENTS	CAS #	ACGIH TLV	OSHA PEL	OTHER LIMITS	%
Acetone	67-64-1	750 ppm	750 ppm		2-5
Xylene	1330-20-7	100 ppm	100 ppm		68-75
2-Butoxy Ethanol	111-76-2	25 ppm	25 ppm	(skin)	3-5
Methanol	67-56-1	200 ppm	200 ppm		3-5
Detergent	-	NA	NA		0-1
Propane	74-98-6	NA	1000 ppm		10-20
Isobutane	75-28-5	NA	NA	1000ppm	10-20

2. PHYSICAL DATA : (without propellent)
Specific Gravity : 0.865
Vapor Pressure : ND
% Volatile : > 99
Boiling Point : 176 F initial
Evaporation Rate : Moderately fast
Freezing Point : ND
Vapor Density : ND
Appearance and Odor:
A clear colorless liquid, aromatic odor.
pH: NA
Solubility : Partially soluble in water.

3. FIRE AND EXPLOSION DATA
Flashpoint : -40 F Method : TCC
Flammable Limits : propellent LEL:1.8 UEL:9.5
Extinguishing Media : CO2, dry chemical, foam
Unusual Hazards : Aerosol cans may explode when heated above 120 F.

4. REACTIVITY AND STABILITY
Stability : Stable
Hazardous decomposition products
: CO2, carbon monoxide (thermal)

Materials to avoid : Strong oxidizing agents and sources of ignition.

5. PROTECTION INFORMATION
Ventilation : Use mechanical means to insure vapor conc. is below TLV.

Respiratory : Use self-contained breathing apparatus above TLV.

Gloves : Solvent resistant Eye & Face : Safety glasses
Other Protective Equipment: Not normally required for aerosol product usage.

Figure 1-7 A sample of a Material Safety Data Sheet. (Courtesy of CRC Industries, Inc.)

A list of all hazardous materials used in the shop should be posted for the employees to see. Shops must maintain documentation on the hazardous chemicals in the workplace, proof of training programs, records of accidents or spill incidents, satisfaction of employee requests for specific chemical information via the MSDS, and a general right-to-know compliance procedure manual utilized within the shop.

Hazardous Wastes

Waste is considered hazardous if it is on the EPA list of known harmful materials or if it has one or more of the following characteristics: ignitability, corrosivity, reactivity, or EP toxicity. A complete list of EPA hazardous wastes can be found in the Code of Federal Regulations. It should be noted that no material is considered hazardous waste until the shop is finished using it and is ready to dispose of it. All hazardous wastes must be disposed of properly (Figure 1-8).

A substance that has high *ignitability* is one that can catch fire quickly.

Corrosivity is a statement that defines how likely it is that a substance will destroy or eat away at other substances.

A substance with high *toxicity* is very poisonous.

Figure 1-8 Most shops hire full-service and licensed waste management firms to remove and dispose of hazardous wastes. (Courtesy of DuPont Company)

Summary

- ❑ You have an important role in creating an accident-free work environment. Your appearance and work habits will go a long way toward preventing accidents. Common sense and concern for others will result in fewer accidents.
- ❑ Accidents can be prevented by not having anything dangle near rotating equipment and parts.
- ❑ Jewelry should never be worn when working on a car. Any type of jewelry can get caught in rotating parts or can short out an electrical circuit.
- ❑ Safety glasses should be worn whenever you are working under a vehicle or when you are using machining equipment, grinding wheels, chemicals, compressed air, or fuels.
- ❑ Gasoline spills should be immediately cleaned up and gasoline should only be stored in approved containers.
- ❑ Dirty and oily rags should be stored in approved containers.
- ❑ All equipment should be inspected for safety hazards before being used.
- ❑ Tools and equipment should only be used for the purposes for which they were designed.
- ❑ Fire, heat, and sparks should be kept away from a battery. These could cause the battery to explode.
- ❑ Accidents can happen and when they do, you should respond immediately to prevent further injury or damage.
- ❑ All employees have the right to know what hazardous materials they are using to perform their job. Most of the needed information is contained on a MSDS.
- ❑ Hazardous wastes must be properly disposed of. Typically, shops hire full-service waste management firms to remove and dispose of these wastes.

Terms To Know

Caustic
Corrosivity
Electrolyte
Ignitability
Reactivity
Shank
Sulfuric acid
Toxicity
Volatile

Review Questions

Short Answer Essays

1. Why can being careful around the shop be considered as having a professional attitude?
2. How does the way you dress affect your personal safety?
3. When should you wear safety glasses?
4. How should gasoline be stored?
5. Name five basic rules for using tools and equipment safely.
6. What should you do in the case of an accident in the shop?
7. If a chemical gets into someone's eye, what should you do?
8. What sort of information is given on a MSDS?
9. How should you dispose of hazardous wastes?
10. What do all employees of a shop have the right to know?

Fill-in-the-Blanks

1. Basic safety rules center around two general behaviors: ___________________ ___________________ and ___________________ for others.

2. Safety glasses should be worn whenever you are working under a vehicle or when you are using ___________________ ___________________ , ___________________ ___________________ , ___________________ , ___________________ ___________________ , and ___________________ .

3. ___________________ , ___________________ , and ___________________ should be kept away from a battery. These could cause the battery to explode.

4. Most of the information needed to know about the hazardousness of a chemical is found on a ___________________ .

5. Safe ___________________ habits can prevent ___________________ .

6. The ___________________ of a substance is a statement of how easily the substance vaporizes or explodes.

7. There are many items around a typical shop that pose a fire hazard. These include: ___________________ , ___________________ ___________________ , ___________________ ___________________ , and ___________________ ___________________.

8. The ___________________ of a substance is a statement of how well the substance supports combustion.

9. ___________________ and ___________________ shoes provide little protection if something falls on your foot. Boots or shoes made of ___________________ or a material that approaches the strength of ___________________ offers much better protection from falling objects.

10. The general intent of ___________________ ___________________ ___________________ laws is for employers to provide a safe working place, as it relates to hazardous materials.

ASE Style Review Questions

1. While discussing ways to prevent fires:
   Technician A says all dirty and oily rags should be stored in approved containers.
   Technician B says all dirty and oily rags should be kept in a pile until the end of the day, then they should be moved to a suitable container.
   Who is correct?
   A. A only
   B. B only
   C. Both A and B
   D. Neither A nor B

2. *Technician A* says accidents can be prevented by not having anything dangle near rotating equipment and parts.
   Technician B says accidents can be prevented by having common sense.
   Who is correct?
   A. A only
   B. B only
   C. Both A and B
   D. Neither A nor B

3. While discussing what to do when an accident does occur:
   Technician A says technicians should immediately determine the cause of the accident.
   Technician B says technicians should respond immediately to prevent further injury or damage.
   Who is correct?
   A. A only
   B. B only
   C. Both A and B
   D. Neither A nor B

4. *Technician A* says gasoline spills should be immediately cleaned up.
   Technician B says gasoline should only be stored in approved containers.
   Who is correct?
   A. A only
   B. B only
   C. Both A and B
   D. Neither A nor B

5. *Technician A* says unsafe equipment should have its power disconnected and marked with a sign to warn others.
   Technician B says that all equipment should be inspected for safety hazards before being used.
   Who is correct?
   A. A only
   B. B only
   C. Both A and B
   D. Neither A nor B

6. *Technician A* says hazardous wastes must be properly disposed of.
   Technician B says shops hire full-service waste management firms to remove and dispose of hazardous wastes.
   Who is correct?
   A. A only
   B. B only
   C. Both A and B
   D. Neither A nor B

7. While discussing ways to create an accident-free work environment:
   Technician A says everyone in the shop should take full responsibility for ensuring safe work areas.
   Technician B says the appearance and work habits of technicians can help prevent accidents.
   Who is correct?
   A. A only
   B. B only
   C. Both A and B
   D. Neither A nor B

8. *Technician A* says that safe hand tools are clean.
   Technician B says all tools and equipment should only be used for the purposes for which they were designed.
   Who is correct?
   A. A only
   B. B only
   C. Both A and B
   D. Neither A nor B

9. While discussing hazardous materials:
Technician A says the employer must inform the employees, after they quit or get fired, of the hazardous materials they used on a particular job.
Technician B says nearly all of the information about the hazardousness of a substance can be found in a service manual.
Who is correct?
A. A only
B. B only
C. Both A and B
D. Neither A nor B

10. *Technician A* says jewelry should never be worn when working on a car.
Technician B says jewelry can short out an electrical circuit.
Who is correct?
A. A only
B. B only
C. Both A and B
D. Neither A nor B

CHAPTER 2

Drive Train Theory

Upon completion and review of this chapter, you should be able to:

- ❑ Identify the major components of a vehicle's drive train.
- ❑ Explain how a set of gears can increase torque.
- ❑ Define and determine the ratio between two meshed gears.
- ❑ State the purpose of a transmission.
- ❑ Describe the difference between a transmission and a transaxle.
- ❑ State the purpose of a clutch assembly.
- ❑ Describe the differences between a typical FWD and RWD car.
- ❑ Describe the construction of a drive shaft.
- ❑ State the purpose of U- and CV-joints.
- ❑ State the purpose of a differential.
- ❑ Identify and describe the various gears used in modern drive trains.
- ❑ Identify and describe the various bearings used in modern drive trains.

Introduction

An automobile can be divided into four major systems or basic components: (1) the engine, which serves as a source of power; (2) the power train, or drive train, which transmits the engine's power to the car's wheels; (3) the chassis, which supports the engine and body and includes the brake, steering, and suspension systems; and (4) the car's body, interior, and accessories, which include the seats, heater and air conditioner, lights, windshield wipers, and other comfort and safety features.

About 50% of the questions on the ASE Manual Transmission and Transaxles Certification Test are based on transmissions. The remaining questions are related to the other drive train components.

Basically, the drive train has four primary purposes: to connect and disconnect the engine's power to the wheels, to select different speed ratios, to provide a way to move the car in reverse, and to control the power to the drive wheels for safe turning of the automobile. The main components of the drive train are the clutch, transmission, and drive axles (Figure 2-1).

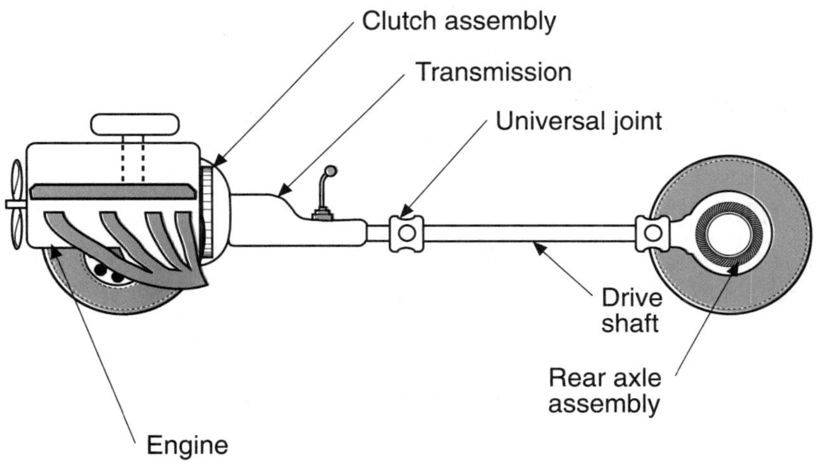

Figure 2-1 Typical drive train components for a rear-wheel-drive car.

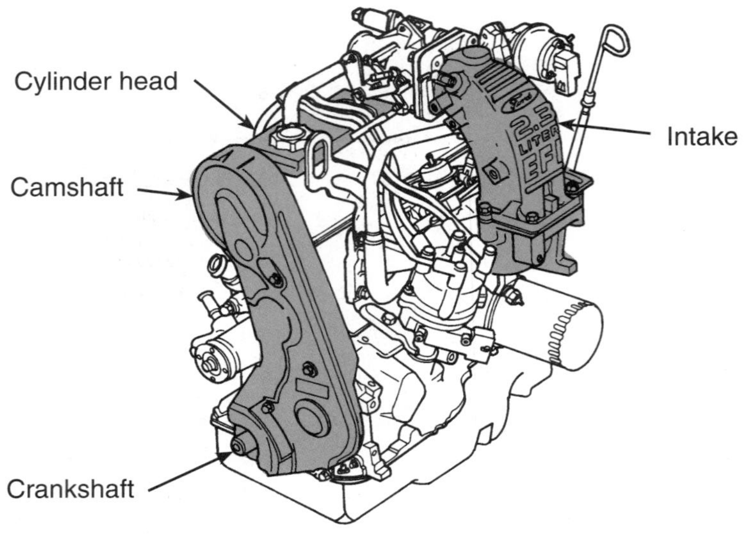

Figure 2-2 Major components of a eight-cylinder engine. (Reprinted with the permission of Ford Motor Company)

Engine

Although the engine is a major system by itself (Figure 2-2), its output should be considered a component of the drive train. An engine develops a rotary motion or torque that, when multiplied by the transmission gears, will move the car under a variety of conditions. The engine produces power by burning a mixture of fuel and air in its combustion chambers. Combustion causes a high pressure in the cylinders, which forces the pistons downward. Connecting rods transfer the downward movement of the pistons to the crankshaft, and supplies power to the drive train.

Most automotive engines are four-stroke cycle engines. The opening and closing of the intake and exhaust valves are timed to the movement of the piston. As a result, the engine passes through four different events or strokes during one combustion cycle. The four strokes are called the intake, compression, power, and exhaust strokes. As long as the engine is running, this cycle of events repeats itself, resulting in the production of engine torque.

Engine Torque

To convert foot-pounds to Newton-meters, multiply the number of foot-pounds by 1.355.

The rotating or twisting force of the engine's crankshaft is called engine torque. Engine torque is measured in foot-pounds (ft.-lb.) or in the metric measurement Newton-meters (Nm). Most engines produce a maximum amount of torque while operating within a range of engine speeds. When an engine reaches the maximum speed of that range, torque is no longer increased. This range of engine speeds is normally referred to as the engine's torque curve. Ideally, the engine should operate within its torque curve at all times (Figure 2-3).

As a car is climbing up a steep hill, its driving wheels slow down, which causes engine speed to decrease as well as reduce the engine's torque. The driver must downshift the transmission, which increases engine speed and allows the engine to produce more torque (Figure 2-4)

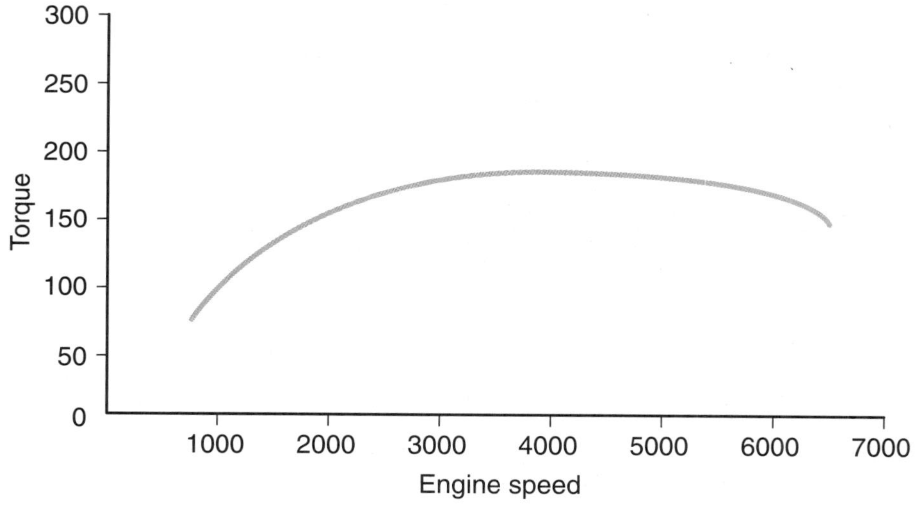

Figure 2-3 The amount of torque produced by an engine varies with the speed of the engine.

When the car reaches the top of the hill and begins to go down, its speed and the speed of the engine rapidly increase. The driver can now upshift, which allows the engine's speed to decrease and places it back in the torque curve.

Downshifting is the shifting to a lower gear.

Torque Multiplication

Measurements of horsepower indicate the amount of work being performed and the rate at which it is being done. The term *power* actually means a force at work, that is, doing work. The

One horsepower is the equivalent of moving 33,000 pounds 1 foot in 1 minute.

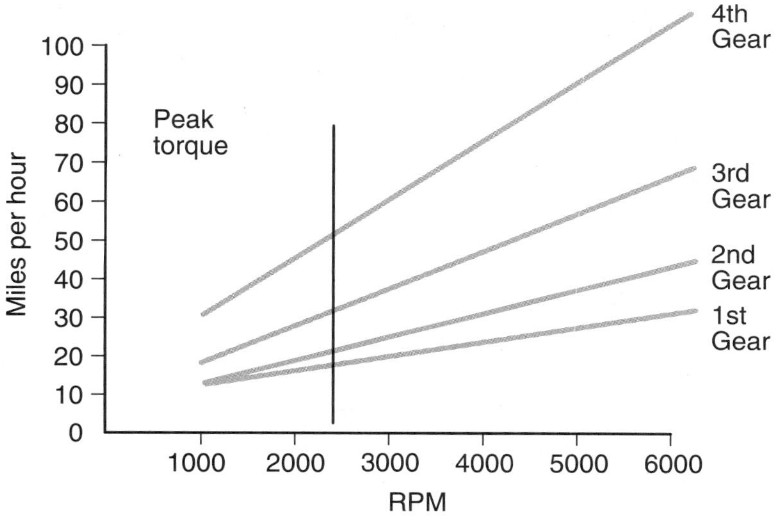

Figure 2-4 If the car represented by this chart were going up a hill in fourth gear and the hill caused the car's speed to drop to 30 MPH, the engine would labor heavily because it would be operating below its peak torque. To overcome the hill and allow the car to increase speed, the driver should place the car into third gear.

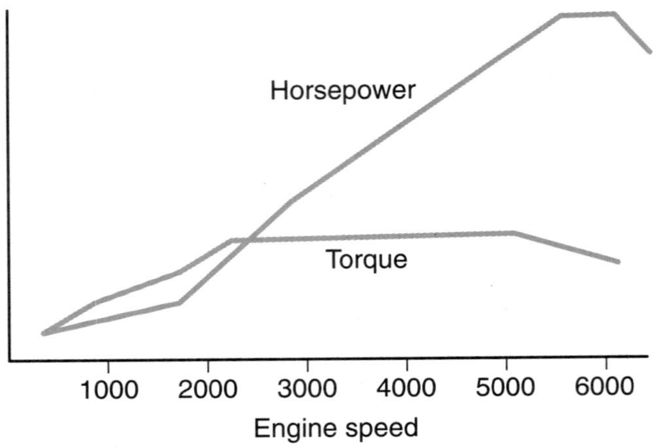

Figure 2-5 After an engine has reached its peak torque, its horsepower output increases with an increase in engine speed.

drive line can transmit power and multiply torque, but it cannot multiply power. When power flows through one gear to another, the torque is multiplied in proportion to the different gear sizes. Torque is multiplied, but the power remains the same, because the torque is multiplied at the expense of rotational speed (Figure 2-5).

Basic Gear Theory

The primary components of the drive train are gears. Gears apply torque to other rotating parts of the drive train and are used to multiply torque. As gears with different numbers of teeth mesh, each rotates at a different speed and torque. Torque is calculated by multiplying the force by the distance from the center of the shaft to the point where the force is exerted (Figure 2-6).

For example, if you tighten a bolt with a wrench that is one foot long and apply a force of 10 pounds to the wrench, you are applying 10 ft.-lb. of torque to the bolt. Likewise, if you apply a force of 20 pounds to the wrench, you are applying 20 ft.-lb. of torque. You could also apply 20 ft.-lb. of torque by applying only 10 pounds of force if the wrench were 2 feet long (Figure 2-7).

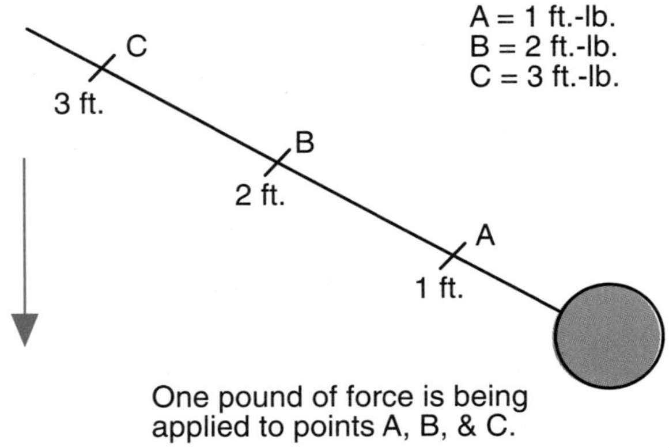

Figure 2-6 Torque is calculated by multiplying the force (1 pound) by the distance from the center of the shaft to the point (Points A, B, and C) where force is exerted.

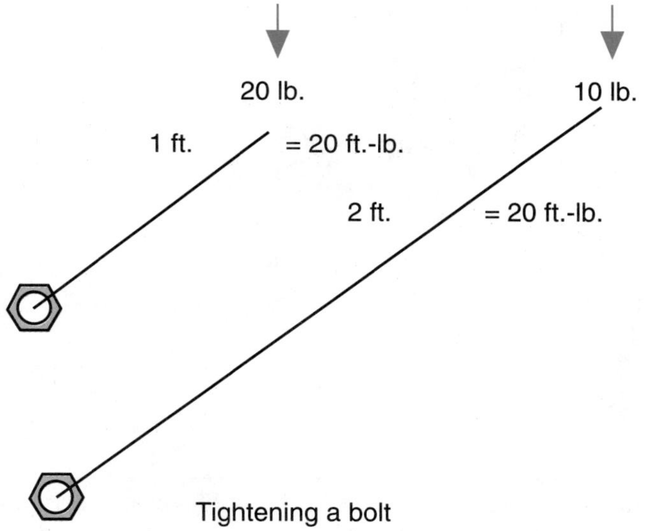

Figure 2-7 The torque applied to both bolts is 20 ft-lb.

The distance from the center of a circle to its outside edge is called the *radius.* On a gear, the radius is the distance from the center of the gear to the point on its teeth where force is applied.

If a tooth on the driving gear is pushing against a tooth on the driven gear with a force of 25 pounds and the force is applied at a distance of 1 foot, which is the radius of the driving gear, a torque of 25 ft.-lb. is applied to the driven gear. The 25 pounds of force from the teeth of the smaller (driving) gear is applied to the teeth of the larger (driven) gear. If that same force were applied at a distance of 2 feet from the center, the torque on the shaft at the center of the driven gear would be 50 ft.-lb. The same force is acting at twice the distance from the shaft center (Figure 2-8).

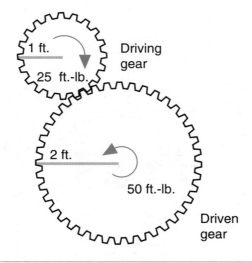

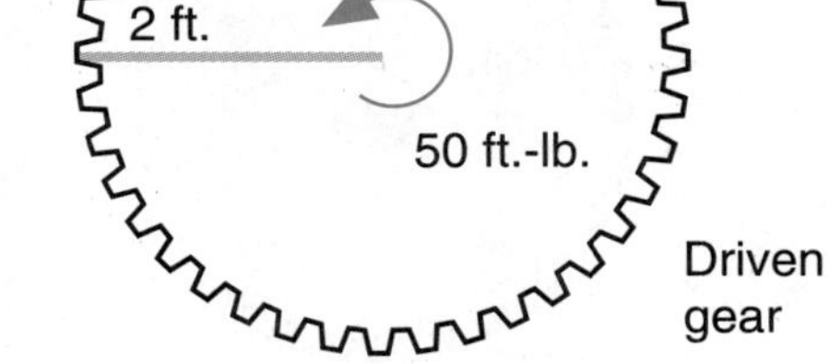

Figure 2-8 The driven gear will turn at half the speed but twice the torque because it is two times larger than the driving gear.

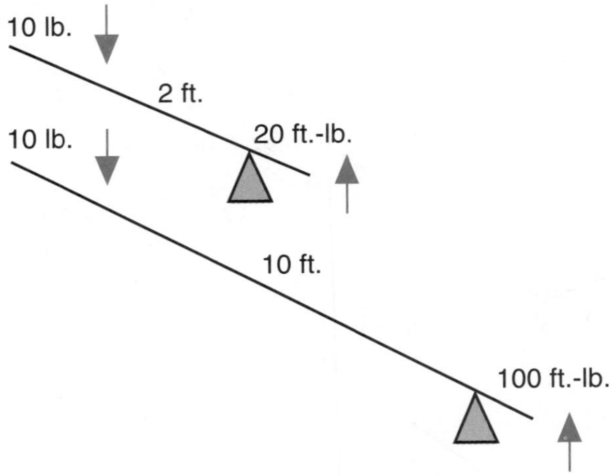

Figure 2-9 The principle of lever action is that a long lever is able to perform more work with less force than a short lever.

The meshing of gears describes the fit of one tooth of one gear between two teeth of the other gear.

Gear ratios are normally expressed in terms of some number to one (1) and use a colon (:) to show the numerical comparison, e.g., 3.5:1, 1:1, 0.85:1.

The amount of torque that can be applied from a power source is proportional to the distance from the center at which it is applied. If a fulcrum or pivot point is placed closer to the object being moved, more torque is available to move the object, but the lever must move farther than if the fulcrum was farther away from the object (Figure 2-9). The same principle is used for gears in mesh: A small gear will drive a large gear more slowly, but with greater torque.

A drive train consisting of a driving gear with 24 teeth and a radius of 1 inch and a driven gear with 48 teeth and a radius of 2 inches will have a torque multiplication factor of 2 and a speed reduction of 1/2. Thus, it doubles the amount of torque applied to it at half the speed (Figure 2-10). The radii between the teeth of a gear act as levers; therefore, a gear that is twice the size of another has twice the lever arm length of the other.

Gear ratios express the mathematical relationship of one gear to another. Gear ratios can be varied by changing the diameter and number of teeth of the gears in mesh. A gear ratio also expresses the amount of torque multiplication between two gears. The ratio is obtained by dividing the diameter or number of teeth of the driven gear by the diameter or teeth of the drive gear. If the

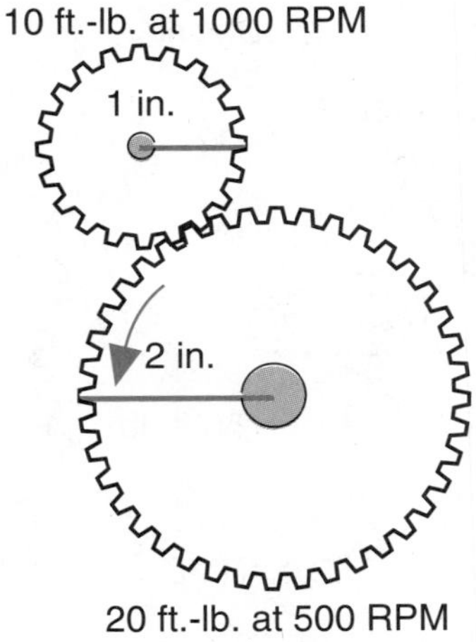

Figure 2-10 The one-inch gear will turn the two-inch gear at half its speed but twice the torque.

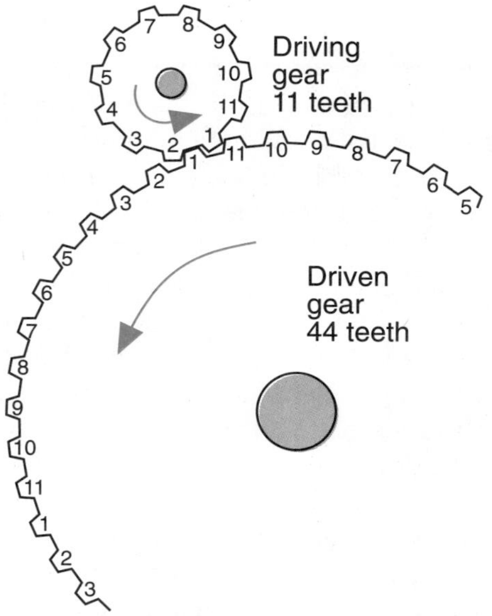

Figure 2-11 The driving gear must rotate four times to rotate the driven gear once. The ratio of the gear set is 4:1.

smaller driving gear had 10 teeth and the larger gear had 40 teeth, the ratio would be 4:1 (Figure 2-11). The gear ratio tells you how many times the driving gear has to turn to rotate the driven gear once. With a 4:1 ratio, the smaller gear must turn four times to rotate the larger gear once.

The larger gear turns at one-fourth the speed of the smaller gear, but has four times the torque of the smaller gear. In gear systems, speed reduction means torque increase. For example, when a typical four-speed transmission is in first gear, there is a speed reduction of 12:1 from the engine to the drive wheels, which means that the crankshaft turns 12 times to turn the wheels once. The resulting torque is 12 times the engine's output: Therefore, if the engine produces 100 ft.-lb. of torque, a torque of 1,200 ft.-lb. is applied to the drive wheels (Figure 2-12).

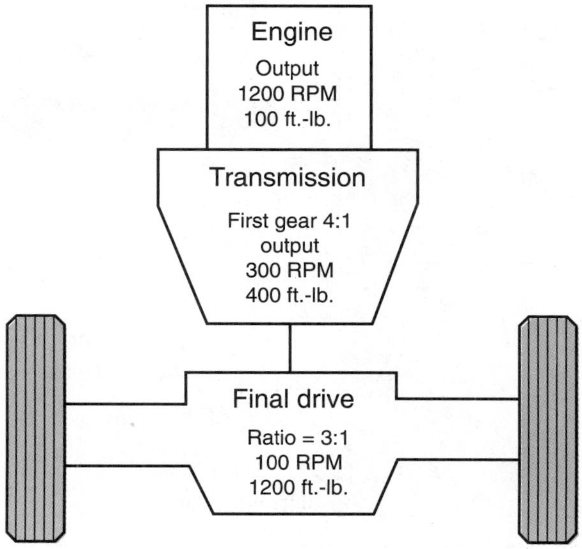

Figure 2-12 The torque and subsequent speed reduction of a typical vehicle's drive train.

Transmission

Transmissions are often called gearboxes.

Torque converters use fluid flow to connect and disconnect the engine's power to the transmission.

Manual transmissions are commonly called standard shift or "stick-shift" transmissions.

High gear is a common term used for the top gear in each type of transmission.

The transmission is mounted to the rear of the engine and is designed to allow the car to move forward and in reverse. It also has a neutral position. In this position, the engine can run without applying power to the drive wheels. Therefore, although there is input to the transmission when the vehicle is in neutral, there is no output from the transmission because the driving gears are not engaged to the output shaft.

There are two basic types of transmissions: automatic and manual transmissions. Automatic transmissions use a combination of a torque converter and a planetary gear system to change gear ratios automatically. A manual transmission is an assembly of gears and shafts that transmits power from the engine to the drive axle and changes in gear ratios is controlled by the driver.

By moving the shift lever, various gear and speed ratios can be selected. The gears in a transmission are selected to give the driver a choice of both speed and torque. Lower gears allow for lower vehicle speeds but more torque. Higher gears provide less torque but higher vehicle speeds. Gear ratios state the ratio of the number of teeth on the driven gear to the number of teeth on the drive gear. There is often much confusion about the terms high and low gear ratios. A gear ratio of 4:1 is lower than a ratio of 2:1. Although numerically the 4 is higher than the 2, the 4:1 gear ratio allows for lower speeds and hence is termed a low gear ratio.

Automotive manual transmissions are constant mesh, fully synchronized transmissions. Constant mesh means that the transmission gears are constantly in mesh, regardless of whether the car is stationary or moving. "Fully synchronized" refers to a mechanism of brass rings and synchronizers used to bring the rotating shafts and gears of the transmission to one speed for smooth up and down shifting.

Today, most manual transmissions have four or five forward speeds. These speeds or gears are identified as first, second, third, fourth, and fifth (Figure 2-13). Different gear ratios are necessary because an engine develops relatively little power at low engine speeds. The engine must be turning at a fairly high speed before it can deliver enough power to get the car moving. Through selection of the proper gear ratio, torque applied to the drive wheels can be multiplied.

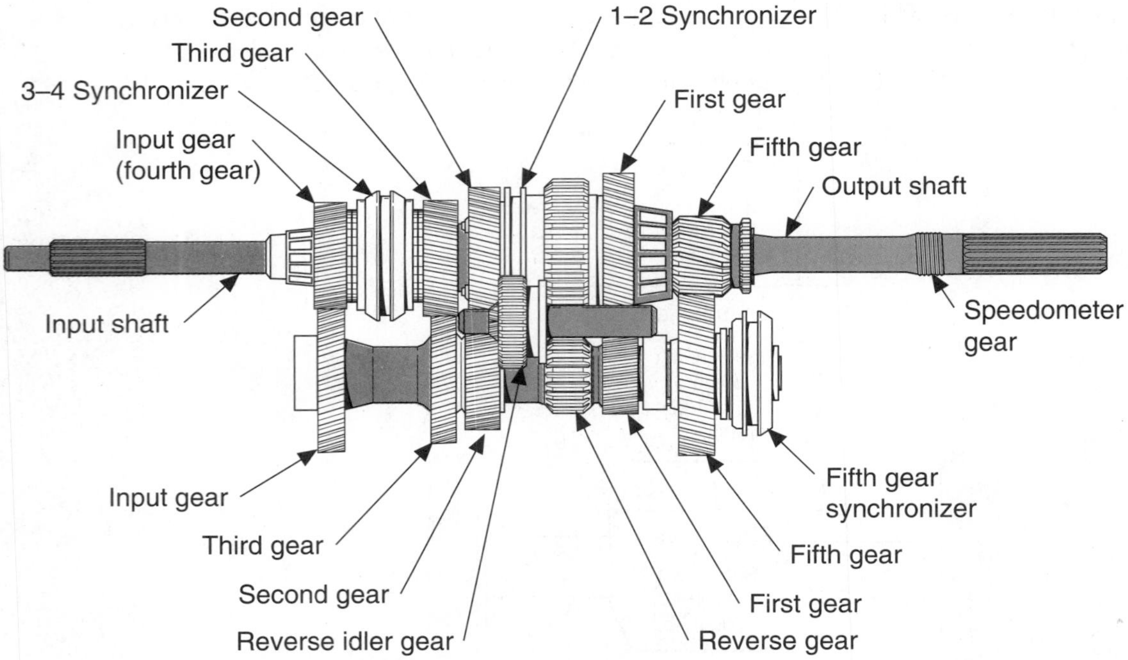

Figure 2-13 The arrangement of the gears and shafts in a typical five-speed transmission. (Reprinted with the permission of Ford Motor Company)

Transmission Gears

Transmissions contain several combinations of large and small gears. In low or first gear, a small gear on the input shaft drives a large gear on another shaft. This reduces the speed of the larger gear but increases its turning force or torque. Connected to the second shaft is a small gear that drives a larger gear that is connected to the drive shaft, which in turn is connected to the driving axle. This reduces the speed and increases the torque still more, giving a higher gear ratio for starting movement or pulling heavy loads. First gear is primarily used to initiate movement. It has the lowest gear ratio of any gear in a transmission. It also allows for the most torque multiplication (Figure 2-14).

First and second gears are the low gears in a typical transmission, whereas third, fourth, and fifth are the high gears.

Shift levers are typically called "shifters."

Synchronizers serve as clutches for the brass rings and the gears.

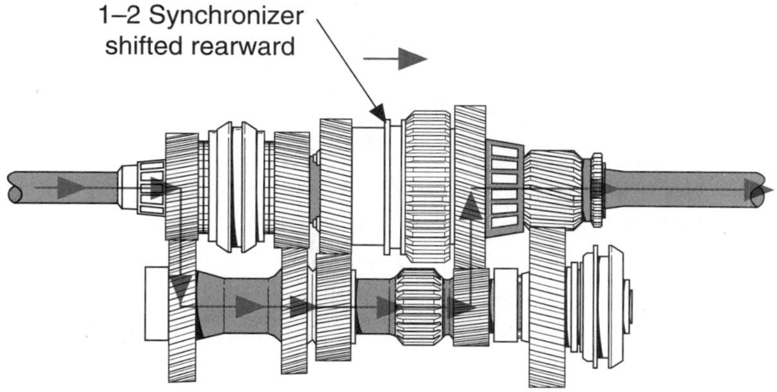

Figure 2-14 Power flow through first gear. (Reprinted with the permission of Ford Motor Company)

Second gear uses the same first pair of gears as low. However, the second pair of gears is disconnected and the power flows through another gearset. A larger gear on the second shaft drives a smaller gear connected to the drive shaft, resulting in less overall speed reduction than in first gear. Second gear is used when the need for torque multiplication is less than the need for vehicle speed and acceleration. Because the car is already in motion, less torque is needed to move the car (Figure 2-15).

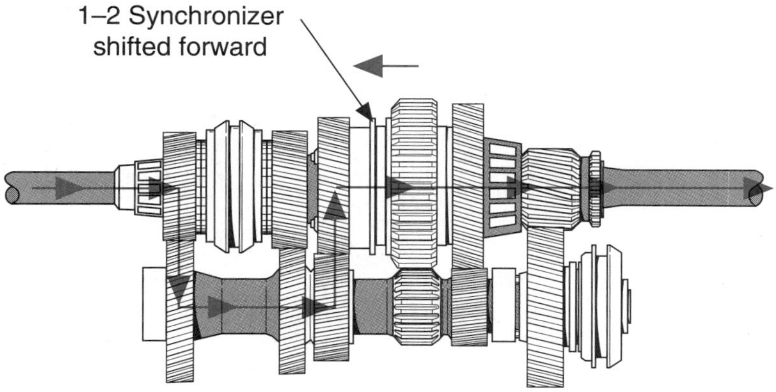

Figure 2-15 Power flow through second gear. (Reprinted with the permission of Ford Motor Company)

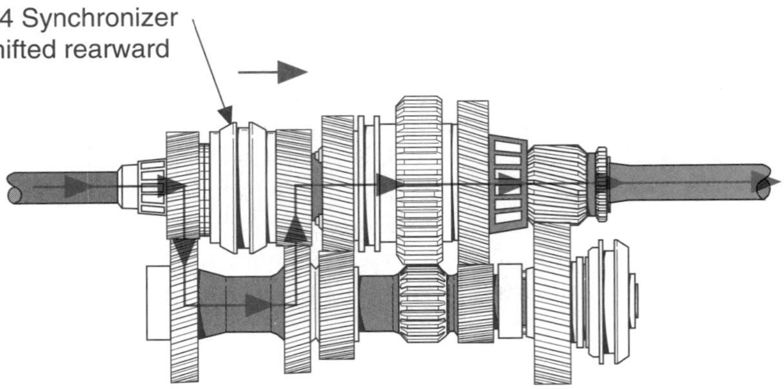

Figure 2-16 Power flow through third gear. (Reprinted with the permission of Ford Motor Company)

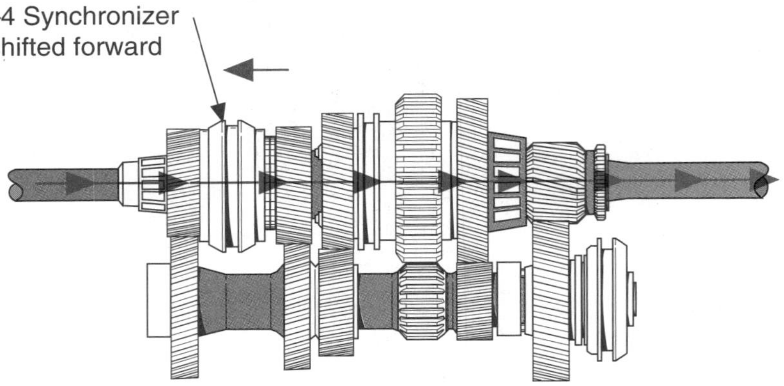

Figure 2-17 Power flow through fourth gear. (Reprinted with the permission of Ford Motor Company)

Direct drive is characterized by the transmission's output shaft rotating at the same speed as its input shaft.

Overdrive causes the output shaft of the transmission or driveshaft to rotate faster than the input shaft or engine.

Third gear allows for a further decrease in engine speed and torque multiplication, while increasing vehicle speed and encouraging fuel economy (Figure 2-16).

Fourth gear typically provides a direct drive (1:1) ratio, so that the amount of torque entering the transmission is also the amount of torque that passes through and out of the transmission output shaft (Figure 2-17). This gear is used at cruising speeds and promotes fuel economy. While the car is in fourth gear, it lacks the performance characteristics of the second and third gears. To pass slower moving vehicles, the transmission often must be downshifted to third to take advantage of third gear's slight torque multiplication, resulting with improved acceleration.

Many of today's transmissions have a fifth gear, called an overdrive gear. Overdrive gears have ratios of less than 1:1. These ratios are achieved by using a small driving gear meshed with a smaller driven gear (Figure 2-18). Output speed is increased and torque is reduced. The purpose of overdrive is to promote fuel economy and reduce operating noise while maintaining highway cruising speed.

Through the use of a reverse idler gear, the direction of the incoming torque is reversed and the transmission output shaft rotates in the opposite direction of the forward gears (Figure 2-19).

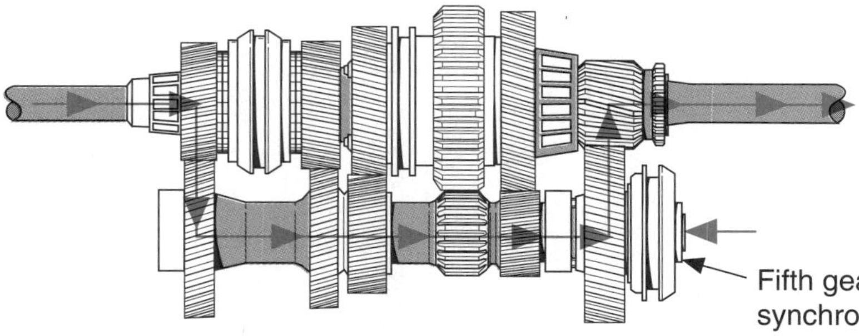

Figure 2-18 Power flow through fifth gear. (Reprinted with the permission of Ford Motor Company)

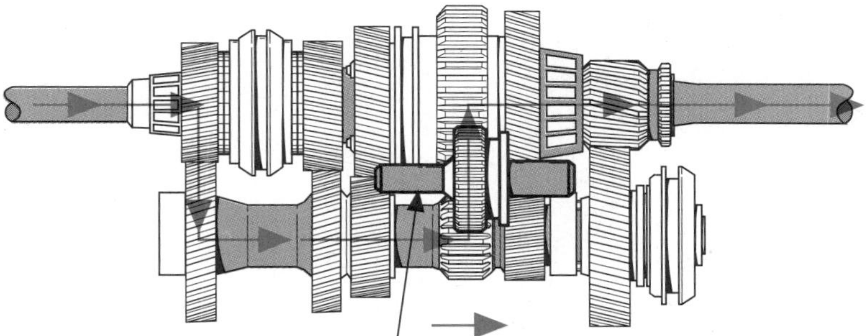

Figure 2-19 Power flow through reverse gear. (Reprinted with the permission of Ford Motor Company)

Normally, reverse uses the same gears as first gear with the addition of the reverse idler gear. Therefore, only low speeds can be obtained in reverse.

The transmission's gear ratios are further increased by the gear ratio of the ring and pinion gears in the drive axle assembly. Typical axle ratios are between 2.5 and 4.5:1. The final (overall) drive gear ratio is calculated by multiplying the transmission gear ratio by the final drive ratio. If a transmission is in first gear with a ratio of 3.63:1 and has a final drive ratio of 3.52:1, the overall gear ratio is 12.87:1 (Figure 2-20). If fourth gear has a ratio of 1:1, using the same final drive ratio, the overall gear ratio is 3.52:1.

Final drive ratio is also called the overall gear ratio.

Another aspect of transmission gear ratios is their spacing. Spacing is the "distance" between gear ratios of the various gears. For example, in a four-speed transmission, first gear may have a gear ratio of 3.63:1, a second gear a 2.37:1 ratio, third gear a 1.41:1 ratio, and fourth gear a 1:1 ratio, this is a wide ratio transmission. In a close ratio transmission for the same car, the ratios could be 2.57:1 in first, 1.72:1 in second, 1.26:1 in third, and 1:1 in fourth. Fourth gear is the same in both transmissions, but the close ratio gearbox moves the three lower gears closer to fourth. This makes the car more difficult to start from a dead stop, but allows for faster acceleration once the car is rolling. Quicker acceleration is possible because there is less loss of engine speed between gear changes.

Both RWD and FWD have a final drive ratio.

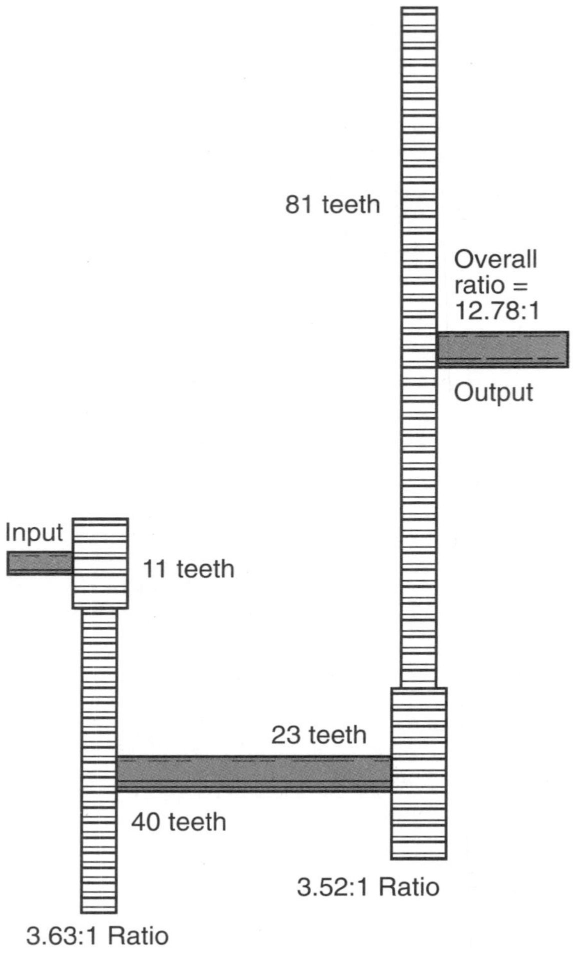

Figure 2-20 The overall gear ratio is calculated by multiplying the ratio of the first set of gears by the ratio of the second (3.63 times 3.52 equals 12.78).

Clutch

The flywheel also serves to dampen crankshaft vibrations, add inertia to the rotation of the crankshaft, and as a large gear for the starter motor.

A flywheel, which is a large and heavy disc, is attached to the rear end of the crankshaft (Figure 2-21). The rear face of the flywheel is machined flat and smooth to serve as the driving member of the clutch. When the clutch is engaged, the rotary motion of the engine's crankshaft is transferred from the flywheel, through the clutch, to the transmission. The rotary motion is then delivered by the transmission to the differential, then transferred by axle shafts to the tires, which push against the ground to move the car.

The clutch assembly is located immediately to the rear of the engine and is directly connected to the engine's crankshaft by the flywheel. Its purpose is to smoothly engage and disengage the engine's power to the rest of the drive train. When the driver depresses the clutch pedal, the clutch assembly is disengaged and the engine is disconnected from the rest of the drive train. This allows the engine to run while the car is standing still. The clutch is said to be engaged when the clutch pedal is up from the floor and the engine is connected to the drive train.

If two discs are attached to a shaft and are not touching each other, we can spin one disc as fast as we want to without affecting the other disc. As the discs move together, however, the

Figure 2-21 A typical flywheel mounted to the rear of an engine's crankshaft.

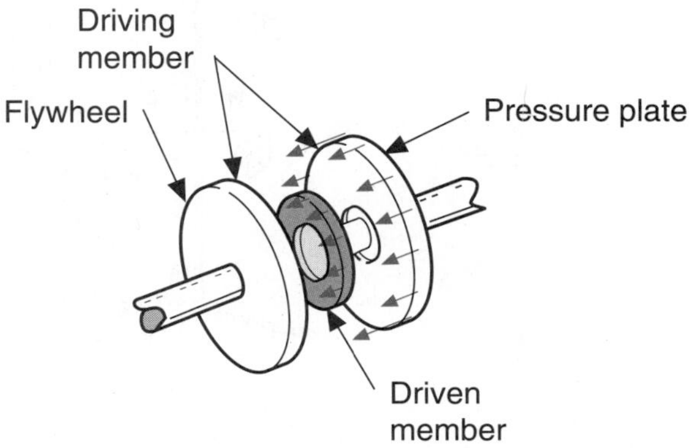

Figure 2-22 When the clutch is engaged, the driven member (clutch disc) is squeezed between the two driving members (pressure plate and flywheel). The transmission is connected to the driven member. (Courtesy of General Motors Corporation)

spinning disc will cause the other to begin to turn. When the discs are held tightly together, both discs will turn as if they were one. Automotive clutches use this two-disc system. The flywheel serves as one of the discs and the pressure plate of the clutch assembly serves as the other. A third disc, the clutch disc, is used to connect these two discs (Figure 2-22).

The clutch disc is made of a high frictional material that provides for a solid connection between the two discs. The discs are forced together by strong springs, which clamp the clutch disc between the pressure plate and flywheel. The spring pressure is released by pushing down on the clutch pedal, which allows the discs to separate (Figure 2-23).

Today's cars are designed to transfer the engine's power to either the front or rear wheels. In a FWD car, the transmission and driving axle are both located in one cast aluminum housing called a transaxle assembly. All of the driving components are located compactly at the front of the vehicle (Figure 2-24). One of the major advantages of front wheel drive is that the weight of the power train components is placed over the driving wheels, which provides for improved traction on slippery road surfaces.

RWD cars have the power train, with the exception of the engine, located beneath the body. The engine is mounted at the front of the chassis and the related power train components extend to the rear driving wheels. The transmission's internal parts are located within an aluminum or cast iron housing called the transmission case assembly. The driving axle is located at the rear of the vehicle, in a separate housing called the rear axle assembly. A drive shaft connects the output of the transmission to the rear axle.

FWD is the standard abbreviation for front-wheel-drive.

RWD is the standard abbreviation for rear-wheel-drive.

A rear-wheel-drive car is best described as a car that moves by the power exerted by its rear wheels.

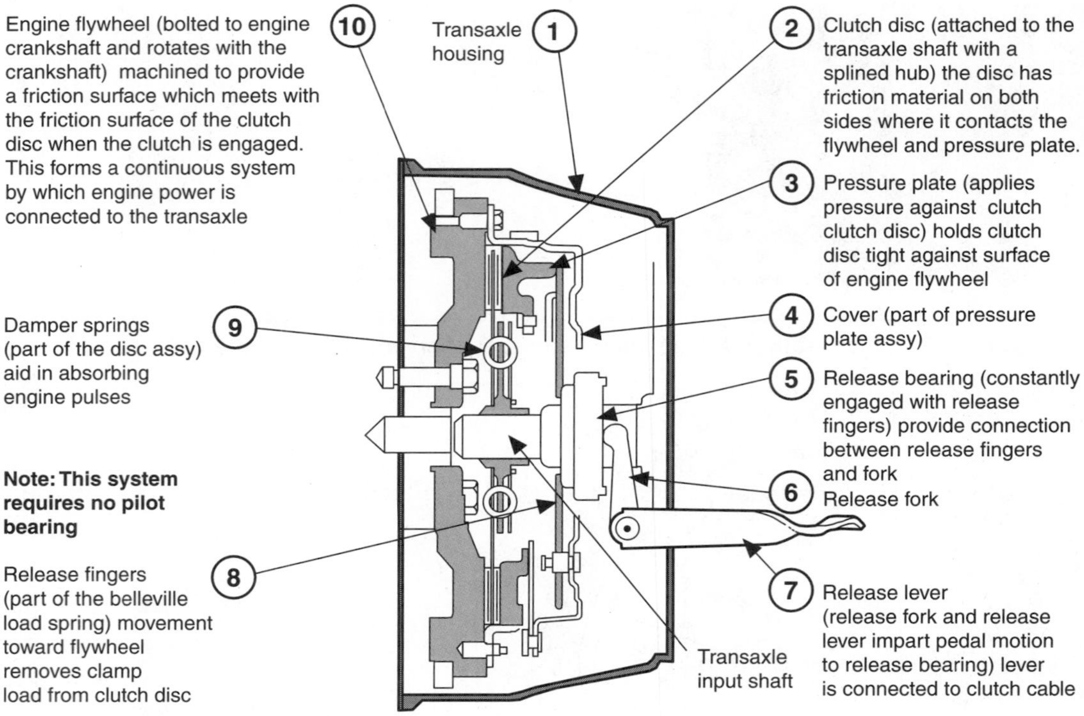
Engine flywheel (bolted to engine crankshaft and rotates with the crankshaft) machined to provide a friction surface which meets with the friction surface of the clutch disc when the clutch is engaged. This forms a continuous system by which engine power is connected to the transaxle
10
Transaxle housing
1
2
Clutch disc (attached to the transaxle shaft with a splined hub) the disc has friction material on both sides where it contacts the flywheel and pressure plate.
3
Pressure plate (applies pressure against clutch clutch disc) holds clutch disc tight against surface of engine flywheel
4
Cover (part of pressure plate assy)
Damper springs (part of the disc assy) aid in absorbing engine pulses
9
5
Release bearing (constantly engaged with release fingers) provide connection between release fingers and fork
6
Release fork
Note: This system requires no pilot bearing
Release fingers (part of the belleville load spring) movement toward flywheel removes clamp load from clutch disc
8
7
Release lever (release fork and release lever impart pedal motion to release bearing) lever is connected to clutch cable
Transaxle input shaft

Figure 2-23 The major components of a clutch assembly and the operation of each. (Reprinted with the permission of Ford Motor Company)

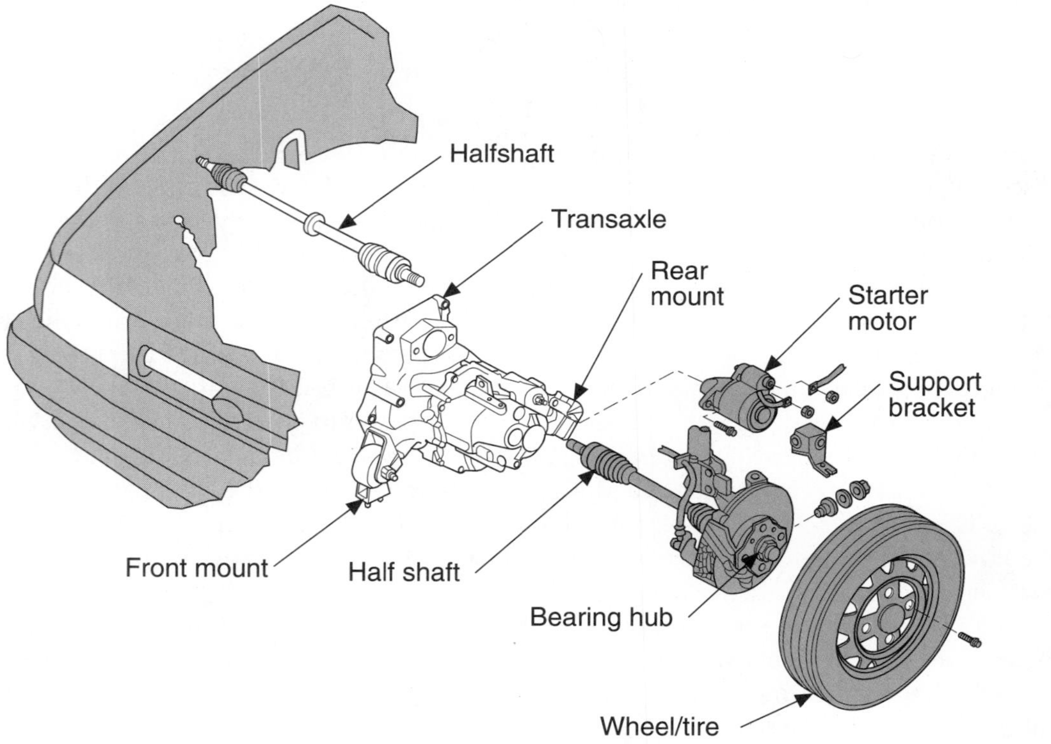
Halfshaft
Transaxle
Rear mount
Starter motor
Support bracket
Front mount
Half shaft
Bearing hub
Wheel/tire

Figure 2-24 A typical transaxle assembly. (Reprinted with the permission of Ford Motor Company)

A BIT OF HISTORY

The French built Panhard, 1892, was the first vehicle to have its power generated by a front-mounted, liquid fueled, internal combustion engine and transmitted to the rear driving wheels by a clutch, transmission, differential, and drive shaft.

Drive Line

The car's drive shaft and its joints are often called the drive line. The drive line transmits torque from the transmission to the driving wheels. RWD cars use a long drive shaft that connects the transmission to the rear axle. The engine and drive line of FWD cars are located between the front driving wheels.

Drive Line for RWD cars

A drive shaft is a steel tube normally consisting of two universal joints and a slip joint (Figure 2-25). The drive shaft transfers power from the transmission output shaft to the rear drive axle. A differential in the axle housing transmits the power to the rear wheels, which then move the car forward or backward.

A sleeve yoke is commonly called a slip yoke.

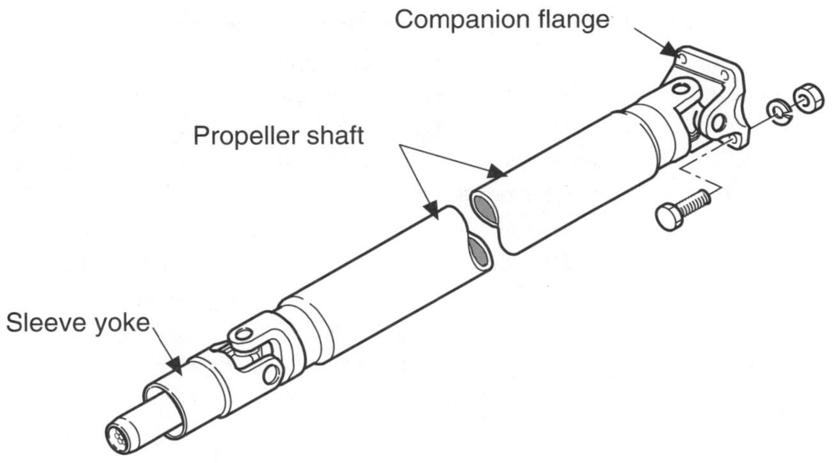

Figure 2-25 A typical drive shaft. (Courtesy of Nissan Motor Company)

Drive shafts differ in construction, lining, length, diameter, and type of slip joint. Typically, the drive shaft is connected at one end to the transmission and at the other end to the rear axle, which moves up and down with wheel and spring movement.

Drive shafts are typically made of hollow, thin-walled steel tubing with the universal joint yokes welded at either end. Universal joints allow the drive shaft to change angles in response to the movements of the rear axle assembly (Figure 2-26). As the angle of the drive shaft changes, its length must also change. The slip yoke normally fitted to the front universal joint allows the shaft to remain in place as its length requirements change.

Universal joints are most often called U-joints.

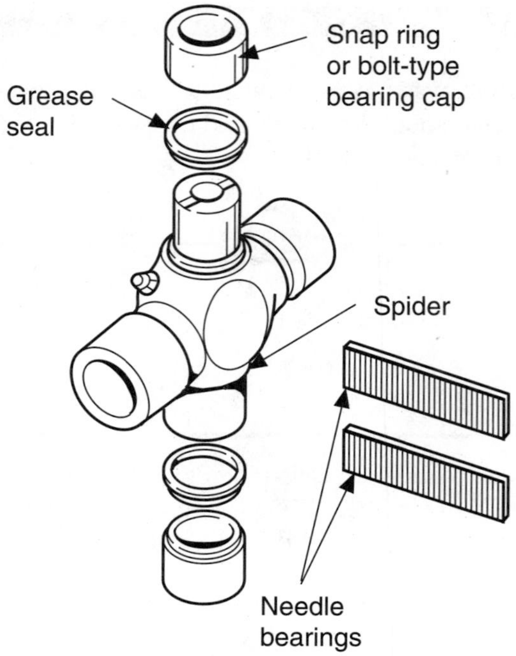

Figure 2-26 An exploded view of a universal joint. (Reprinted with the permission of Ford Motor Company)

Rear Axle Assembly

The rear axle housing encloses the complete rear-wheel driving axle assembly. In addition to housing the parts, the axle housing also serves as a place to mount the vehicle's rear suspension and braking system. The rear axle assembly serves three other major functions: It changes the direction of the power flow 90 degrees, allows for the drive wheels to rotate at different speeds while cornering, and acts as the final gear reduction unit.

The rear axle consists of two sets of gears: the ring and pinion gear set and the differential gears. When torque leaves the transmission, it flows through the drive shaft to the ring and pinion gears, where it is further multiplied (Figure 2-27). By considering the engine's torque curve, the car's weight, and tire size, manufacturers are able to determine the best rear axle gear ratios for proper acceleration, hill-climbing ability, fuel economy, and noise level limits

The primary purpose of the differential gear set is to allow a difference in driving wheel speed when the vehicle is rounding a corner or curve. The differential also transfers torque equally to both driving wheels when the vehicle is traveling in a straight line (Figure 2-28).

The torque on the ring gear is transmitted to the differential, where it is split and sent to the two driving wheels. When the car is traveling in a straight line, both driving wheels travel the same distance at the same speed. However, when the car is making a turn, the outer wheel must travel farther and faster than the inner wheel. When the car is steered into a 90-degree turn to the left and the inner wheel turns on a 20-foot radius, the inner wheel travels about 31 feet. The outer wheel, being nearly 5 feet from the inner wheel, turns on a 24 2/3-foot radius and travels nearly 39 feet.

Without some means for allowing the drive wheels to rotate at different speeds, the wheels would skid when the car was turning. This would result in little control during turns and in excessive tire wear. The differential eliminates these troubles by allowing the outer wheel to rotate faster as turns are made.

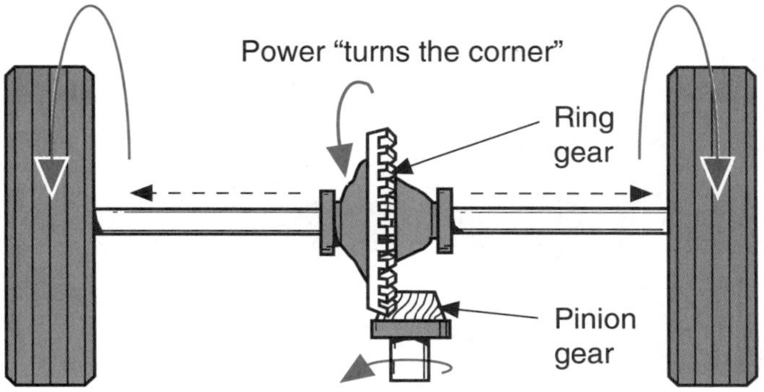

Figure 2-27 The gears of a differential unit not only multiply torque, but also transmit power "around the corner" to the drive wheels.

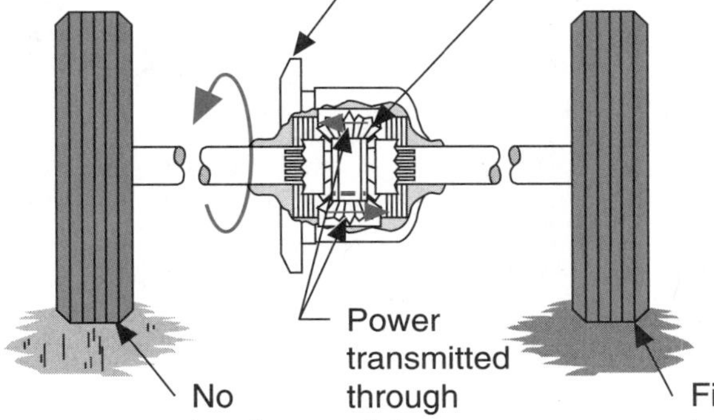

Figure 2-28 The differential allows for one drive wheel to turn at a different speed than the other while the car is turning and allows for more power to the wheel with less traction.

Differential Design

On FWD cars, the differential unit is normally part of the transaxle assembly (Figure 2-29). On RWD cars, it is part of the rear axle assembly. A differential unit is located in a cast iron or aluminum casting, the differential case, and is attached to the center of the ring gear. Located inside the case are the differential pinion shafts and gears and the axle side gears.

The differential assembly revolves with the ring gear (Figure 2-30). Axle side gears are splined to the rear axle or front axle drive shafts.

When an automobile is moving straight ahead, both wheels are free to rotate. Engine power is applied to the pinion gear, which rotates the ring gear. Beveled pinion gears are carried around by the ring gear and rotate as one unit. Each axle receives the same power, so each wheel turns at the same speed.

Pinion gears are typically mounted or attached to a shaft and supply the input to a gear set.

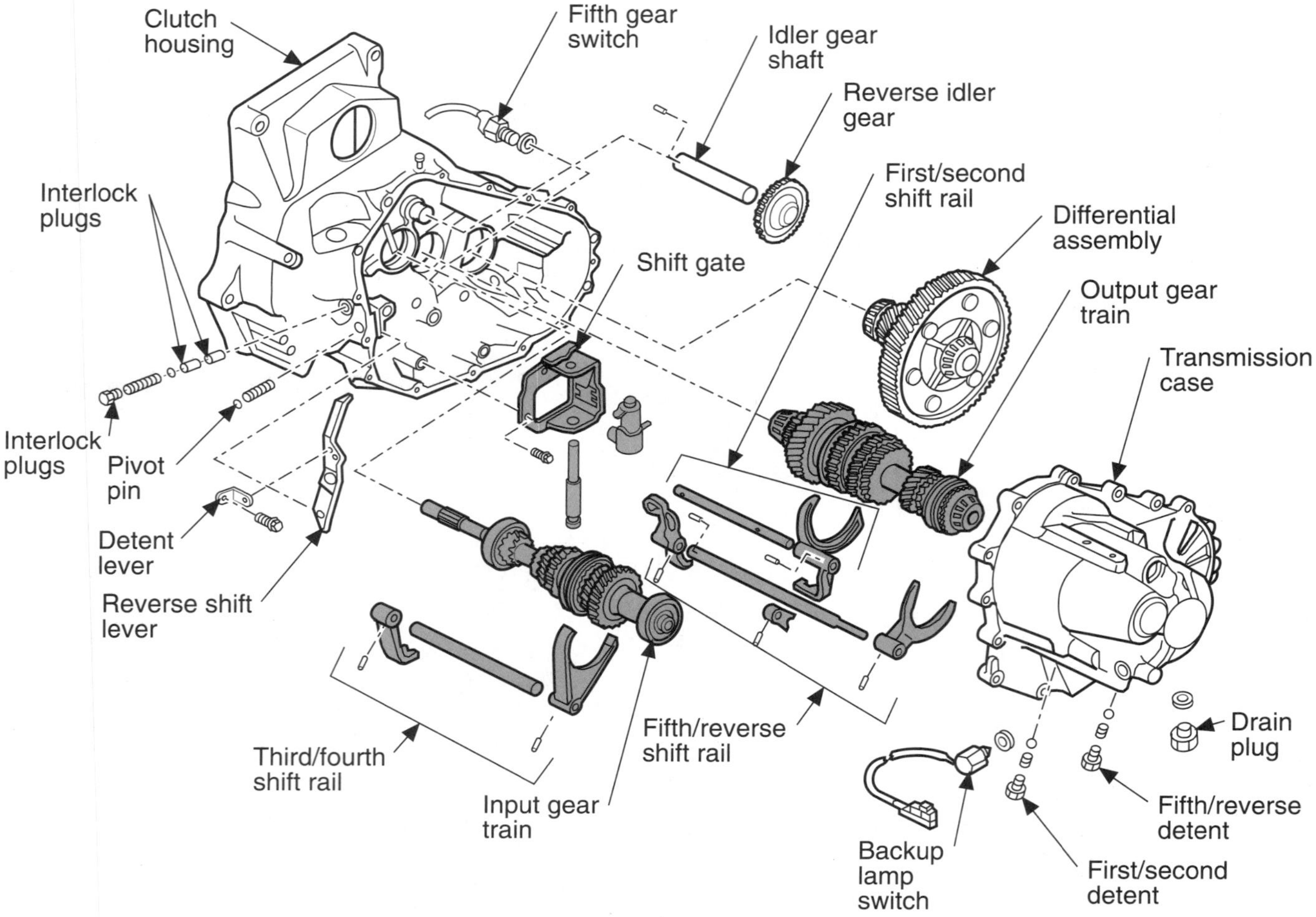

Figure 2-29 An exploded view of a transaxle and its differential. (Reprinted with the permission of Ford Motor Company)

When the car turns a sharp corner, only one wheel rotates freely. Torque still comes in on the pinion gear and rotates the ring gear, carrying the beveled pinions around with it. However, one axle is held stationary and the beveled pinions are forced to rotate on their own axis and "walk around" their gear. The other side is forced to rotate because it is subjected to the turning force of the ring gear, which is transmitted through the pinions.

During one revolution of the ring gear, one gear makes two revolutions, one with the ring gear and another as the pinions "walk around" the other gear. As a result, when the drive wheels have unequal resistance applied to them, the wheel with the least resistance turns more revolutions. As one wheel turns faster, the other turns proportionally slower.

To prevent a loss of power on slippery surfaces, a differential lock is often used to lock the two axles together until the slippery spot is passed, at which point they are released. These differentials are referred to as *limited-slip* or *traction-lock* differentials. When the car is proceeding in a straight line, the differential gears are locked against rotation due to gear reaction. When the vehicle turns a corner or a curve, the differential pinion gears rotate around the differential pinion shaft. The differential pinion gears allow the inside axle shaft and driving wheels to slow down. On the opposite side, the pinion gears allow the outside wheels to accelerate. Both

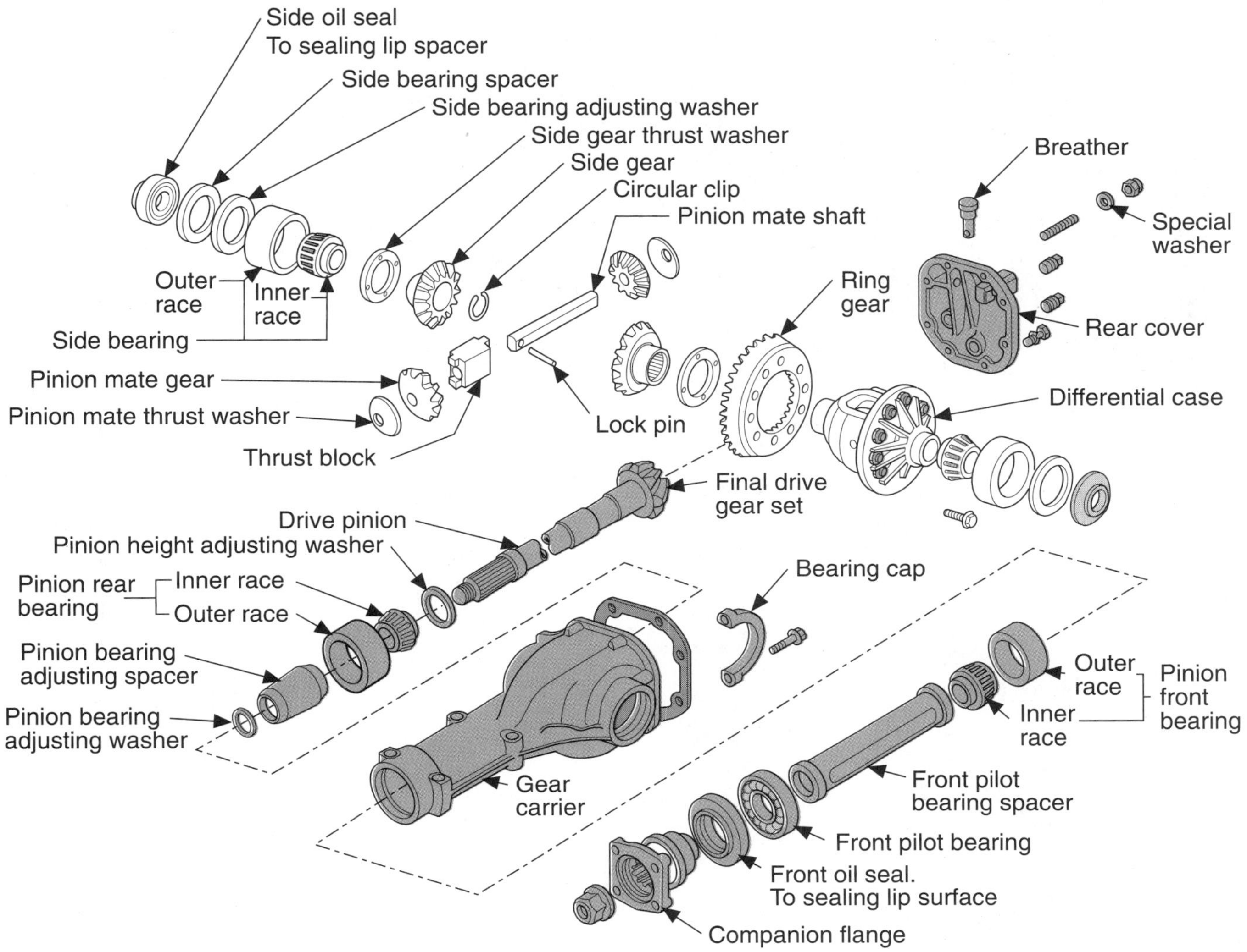

Figure 2-30 A cross-sectional view of a rear-wheel-drive differential. (Courtesy of Nissan Motor Company)

driving wheels resume equal speeds when the vehicle completes the corner or curve. This differential action improves vehicle handling and reduces driving wheel tire wear.

Driving Axles

On RWD automobiles, the axle shafts or drive axles are located within the hollow horizontal tubes of the axle housing. The purpose of an axle shaft is to transmit the torque from the differential's side gears to the driving wheels. Axle shafts are heavy steel bars splined at the inner end to mesh with the axle side gear in the differential. The driving wheel is bolted to the wheel flange at the outer end of the axle shaft (Figure 2-31). The car's wheels rotate with the axles, which allow the car to move.

The drive axles of a FWD car extend from the sides of the transaxle to the drive wheels. CV-joints are fitted to the axles to allow the axles to move with the car's suspension and steering systems.

CV-joints are constant velocity joints, which allow the angle of the axle shafts to change with no loss in rotational speed.

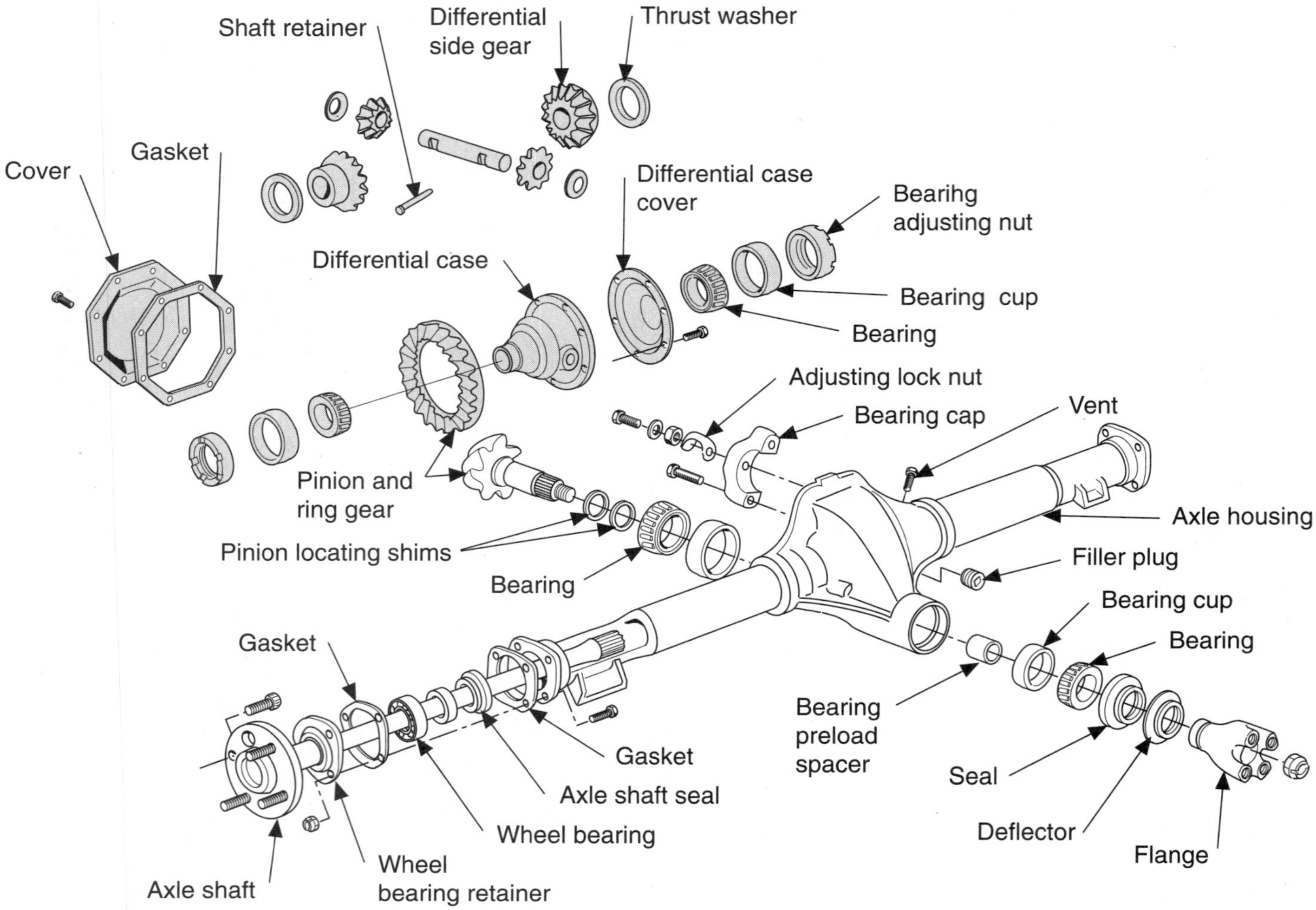

Figure 2-31 Axle shafts are supported by a bearing in the axle housing and are driven by the differential side gears. The vehicle's wheels are bolted to the axle shaft's outer flange. (Reprinted with the permission of Ford Motor Company)

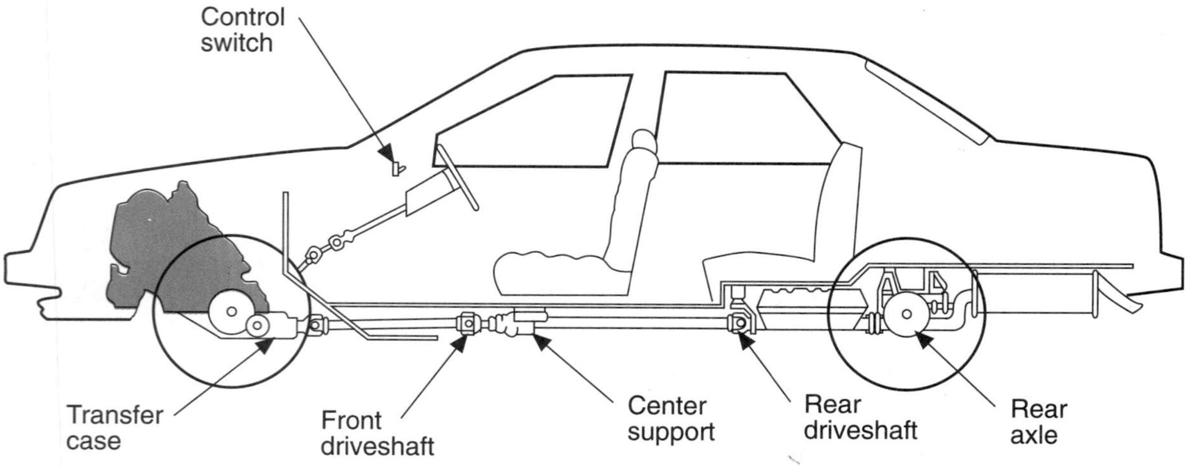

Figure 2-32 The arrangement of the drive train components in a typical 4WD car. (Reprinted with the permission of Ford Motor Company)

Four-wheel-drive (4WD) vehicles, especially those that are used off the road, can deliver power to all four wheels. The driver can select two-wheel or four-wheel drive. Some 4WD vehicles are always engaged in four wheel drive; these vehicles are commonly said to have "full-time" four-wheel drive. Each drive shaft to the front and rear axles has universal joints and a slip joint. Torque from the transmission enters a transfer case (Figure 2-32). The transfer case has gears that can be engaged to send power to only the rear wheels, or to both the front and the rear wheels.

4WD is the standard abbreviation for four-wheel-drive

Full-time 4WD is also called "all-wheel drive."

Types of Gears

Gears are normally used to transmit torque from one shaft to another. These shafts may operate in line, parallel to each other, or at an angle to each other. These different applications require a variety of gear designs that vary primarily in the size and shape of the teeth.

In order for gears to mesh, they must have teeth of the same size and design. Meshed gears have at least one pair of teeth engaged at all times. Some gear designs allow for contact between more than one pair of teeth. Gears are normally classified by the type of teeth they have and by the surface on which the teeth are cut.

Automobiles use a variety of gear types to meet the demands of speed and torque. The most basic type of gear is the spur gear, which has its teeth parallel to and in alignment with the center of the gear (Figure 2-33). Early transmissions used straight-cut spur gears, which were easier to machine but were noisy and difficult to shift. Today these gears are used mainly for slow speeds to avoid excessive noise and vibration. They are commonly used in simple devices such as hand or powered winches.

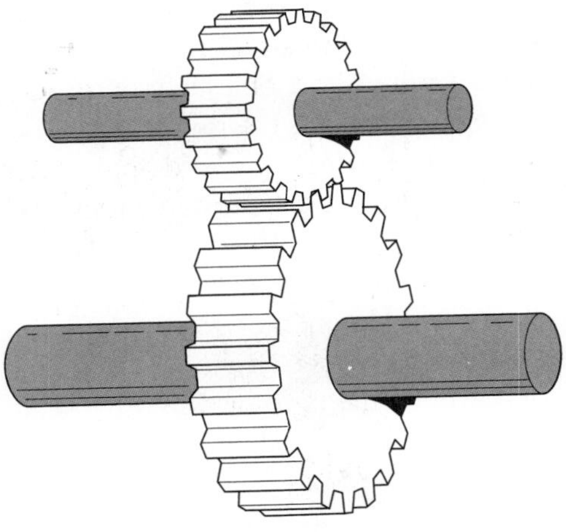

Figure 2-33 Spur gears have teeth cut straight across the gear's edge and parallel to the shaft.

Helical gears are like spur gears except that their teeth have been twisted at an angle from the gear center line (Figure 2-34). These gears get their name from being cut in a helix, which is a form of curve. This curve is more difficult to machine but is used because it reduces gear noise. Engagement of these gears begins at the tooth tip of one gear and rolls down the trailing edge of the teeth. This angular contact tends to cause side thrusts that are absorbed by a bearing. However, helical spur gears are quieter in operation and have greater strength and durability than straight spur gears, simply because the contacting teeth are longer. Helical spur gears are widely used in transmissions today because they are quieter at high speeds and are durable.

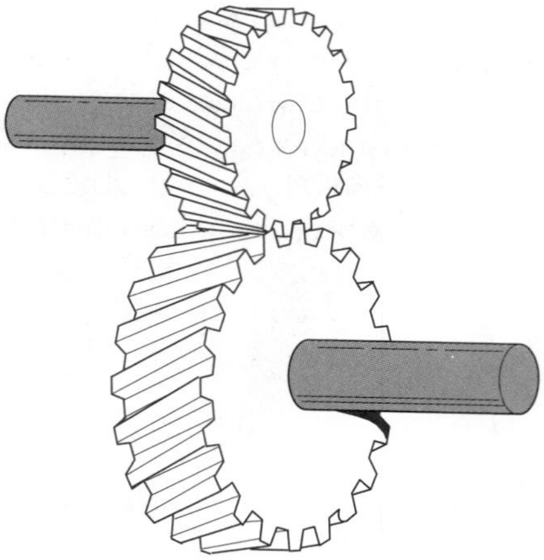

Figure 2-34 Helical gears have teeth cut at an angle to the gear's axis of rotation.

Herringbone gears are actually double helical gears with teeth angles reversed on opposite sides. This causes the thrust produced by one side to be counterbalanced by the thrust produced by the other side. The two sets of teeth are often separated at the center by a narrow gap for better alignment and to prevent oil from being trapped at the apex. Herringbone gears are best suited for quiet, high-speed, low-thrust applications where heavy loads are applied. Large turbines and generators frequently use herringbone gears because of their durability.

Bevel gears are shaped like a cone with its top cut off. The teeth point inward toward the peak of the cone. These gears permit the power flow to "turn a corner." Spiral bevel gears were developed for use where higher speed and strength as well as a change in the angle of the power flow were required. Their teeth are cut obliquely on the angular faces of the gears. The most commonly used spiral beveled gear set is the ring and pinion type used in heavy truck differentials. Bevel-type gears are also used for slow-speed applications that are not subject to high impact forces. Handwheel controls that must operate some remote device at an angle use straight bevel gears.

The hypoid gear resembles the spiral bevel gear but the pinion drive gear is located below the center of the ring gear. Its teeth and general construction are the same as the spiral bevel gear. The most common use for hypoid gears is in modern differentials (Figure 2-35). Here, they allow for lower body styles by lowering the transmission drive shaft.

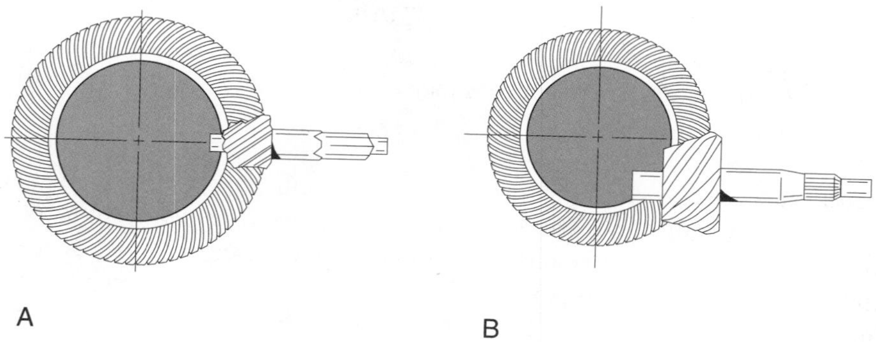

Figure 2-35 (A) Spiral bevel differential gears and (B) hypoid gears. (Reprinted with the permission of Ford Motor Company of Canada, Ltd.)

The worm gear is actually a screw capable of high speed reductions in a compact space. Its mating gear has teeth that are curved at the tips to permit a greater contact area. Power is supplied to the worm gear, which drives the mating gear. Worm gears usually provide right-angle power flows. The most common use for the worm gear is in applications where the power source operates at high speed and the output is at slow speed with high torque. Many steering mechanisms use a worm gear connected to the steering shaft and wheel and a partial (sector) gear connected to the steering linkage. Small power hand tools frequently use a high-speed motor with a worm gear drive.

Some tractors and off-road vehicles use worm gears as final drive gears.

Rack and pinion gears convert straight-line motion into rotary motion, and vice versa. Rack and pinion gears also change the angle of power flow with some degree of speed change. The teeth on the rack are cut straight across the shaft, and those on the pinion are cut like a spur gear. These gear sets can provide control of arbor presses and other devices where slow speed is involved. Rack and pinion gears also are commonly used in automotive steering boxes.

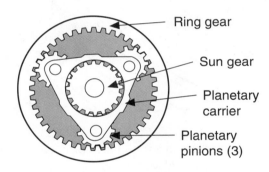

Internal gears have their teeth pointing inward and are commonly used in the planetary gear set used in automatic transmissions and transfer cases. These are gear sets in which an outer ring gear has internal teeth that mate with teeth on smaller planetary gears. These gears, in turn, mesh with a center or sun gear (Figure 2-36). Many changes in speed and torque are possible, depending on which parts are held stationary and which are driven. In a planetary gearset, one gear is normally the input, another is prevented from moving or held, and the third gear is the output gear. Planetary gears are widely used because each set is capable of more than one speed change. The gear load is spread over several gears, reducing stress and wear on any one gear.

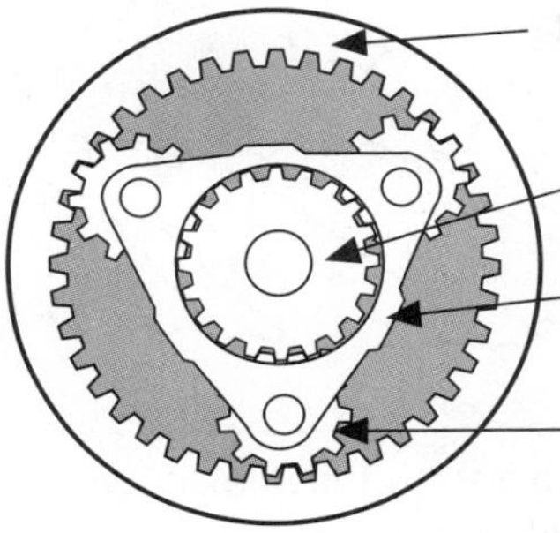

Figure 2-36 A simple planetary gear set.

Gear Wear

All new gear teeth have slight imperfections, but these normally disappear during break-in as the teeth are oiled and become polished. When a lack of lubrication or other factors cause a gear to fail, a thorough visual inspection can normally determine the cause of failure.

The teeth of a normally worn gear will have a polished surface that extends the full length of the tooth from the bottom to the tip of the tooth. Gears that are manufactured properly, well lubricated, and not overloaded or improperly installed will look this way after many hours of service.

Excessive wear and/or grooves on the teeth are normally caused by fine particles carried in the lubricant or embedded in the tooth surfaces. The usual sources of these fine particles are metal particles from gear teeth, abrasives left in the gear case, or sand and scale from the housing casting.

Scratching on the surface of the teeth can also be caused by suspended or embedded particles. However, scratching of these surfaces normally occurs when the particles are larger than the fine abrasives that cause excessive wear. These particles are normally metal pieces from the gears themselves, resulting from the gear being subjected to loads for which it was not designed.

If the contact surface of the teeth is worn smoothly, the gears have been overloaded and metal has been removed by the sliding pressure of the meshing gears. This causes a worn depression along the length of the teeth. Continuous use under this condition will result in backlash and severe peening, which can be misleading as to the real cause of the wear.

Rolling is the result of overload and sliding, which leaves a burr on the tooth edge. Insufficient bearing support results in rolling of the metal due to the sliding pressure.

Peening is the result of backlash, which causes the teeth to hammer on others with tremendous impact so that they look as if they were beaten with a hammer. The impact forces the lubricant out of the teeth so that the gears mesh without a protective layer of oil, causing heavy metal-to-metal contact.

A wavy surface or "fish scales" on the teeth at right angles to the direction of the sliding action is referred to as *rippling*. This may be caused by a lack of lubrication, heavy loads, or vibration.

Scoring is caused by a temperature rise and thinning or rupture of the lubricant's film due to extreme loads. Pressure and sliding action heats the gear and permits the metal surface to transfer from one tooth to the face of another. As the process continues, chunks of metal loosen and gouge the teeth in the direction of the sliding motion (Figure 2-37).

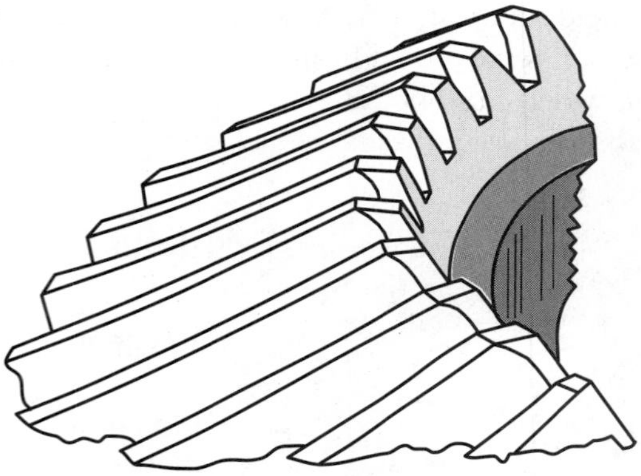

Figure 2-37 Contaminated or old lubricant is the primary cause for scored gear teeth. (Reprinted with the permission of Ford Motor Company)

Pitting is a condition normally associated with a thin oil film, possibly due to high oil temperatures. A very small amount of pitting gives the surface of the gear teeth a gray appearance.

Spalling is a common wear condition that starts with fine surface cracks and eventually results in large flakes or chips coming off the tooth face. Improperly hardened teeth are most often subject to this kind of damage due to the brittle nature of the metal. Spalling may occur on only one or two teeth, but the chips may then damage the other teeth.

Corrosive wear results in an erosion of the tooth surfaces by acid. The acid is the result of moisture combining with impurities in the lubricant and contaminants in the air. Normally, the surfaces initially become pitted and then become chipped or spalled.

Evidence of high heat or burning is usually caused by failure of the lubricant or by a lack of lubrication. During high stress and sliding motion, friction causes excessive heat and the temperature limits of the metal are exceeded. Burned gear teeth become extremely brittle and are easily broken.

A misalignment of the gears that places heavy loads on small portions of a gear will result in interference wear. It can also be the result of meshed gears with teeth that are not designed to work together.

Ridges are scratches appearing near one end of a tooth, especially on a hypoid pinion gear. These are caused by excessive loads or a lack of lubrication.

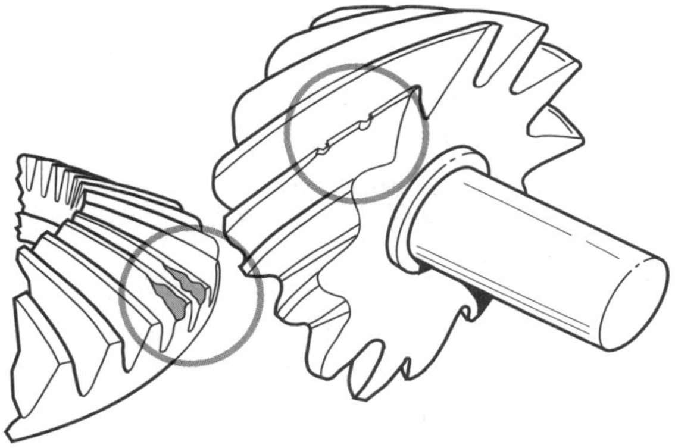

Figure 2-38 Damaged or broken gear teeth are common causes of gear-related noises. (Reprinted with the permission of Ford Motor Company)

Broken teeth can be the result of many defects: high impact forces, overloading, fatigue, or defective manufacturing processes. Examine the break carefully. If the break shows fresh metal all over the break, an impact overload was probably the cause. If the break shows an area in the center of fresh metal with the edges dark and old looking, the breakage was due to fatigue that started with a fine surface crack (Figure 2-38).

Cracks are normally caused by improper heat treating during manufacturing or improperly machined tooth dimensions. Most cracks resulting from improper heat treating are extremely fine and won't show up until a gear has been used for some time.

Bearings

Gears are either securely attached to a shaft or designed to move freely on the shaft. The ease with which the gears rotate on the shaft or the shaft rotates with the gears partially determines the amount of power needed to rotate them. If they rotate with great difficulty due to high friction, much power is lost. High friction will also cause excessive wear to the gears and shaft. To reduce the friction, bearings are fitted to the shaft and/or gears.

The simplest type of bearing is a cylindrical hole formed in a piece of material, into which the shaft fits freely. The hole is usually lined with a brass or bronze lining, or *bushing,* which not only reduces the friction but also allows for easy replacement when wear occurs. Bushings usually have a tight fit in the hole (Figure 2-39).

Bushings are often referred to as plain bearings.

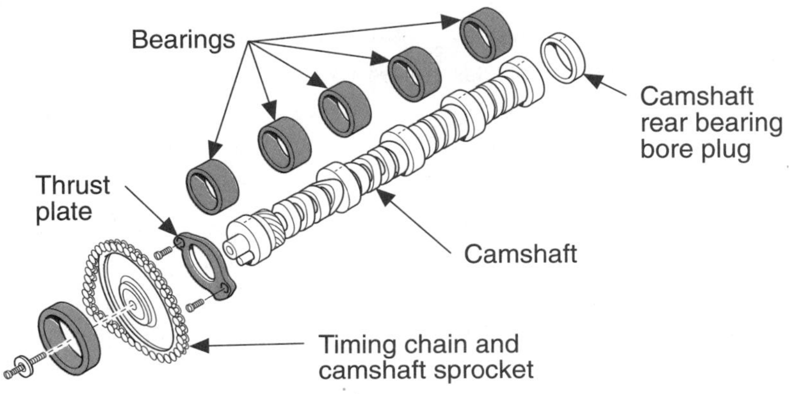

Figure 2-39 Plain bearings are used as engine camshaft bearings. (Reprinted with the permission of Ford Motor Company)

Ball- or roller-type bearings are used wherever friction must be minimized. With these types of bearings, rolling friction replaces the sliding friction that occurs in plain bearings. Typically two bearings are used to support a shaft instead of a single long bushing. Bearings have three purposes: They support a load, maintain alignment of a shaft, and reduce rotating friction.

A journal is the area on a shaft that rides on the bearing.

Most bearings are capable of withstanding only loads that are perpendicular to the axis of the shaft. Such loads are called *radial loads* and bearings that carry them are called *radial* or *journal bearings.*

To prevent the shaft from moving in the axial direction, *shoulders* or *collars* may be formed on it or secured to it. If both collars are made integral with the shaft, the bearings must be split or made into halves, and the top half or cap bolted in place after the shaft has been put in place. The collars or shoulders withstand any end thrusts, and bearings designed this way are termed *thrust bearings.*

A single row journal or radial ball bearing has an inner race made of a ring of case-hardened steel with a groove or track formed on its outer circumference for a number of hardened steel balls to run upon (Figure 2-40). The outer race is another ring with a track on its inner circumference. The balls fit between the two tracks and roll around in the tracks as either race turns. The balls are kept from rubbing against each other by some form of cage. These bearings can withstand radial loads and can also withstand a considerable amount of axial thrust. Therefore, they are often used as combined journal and thrust bearings.

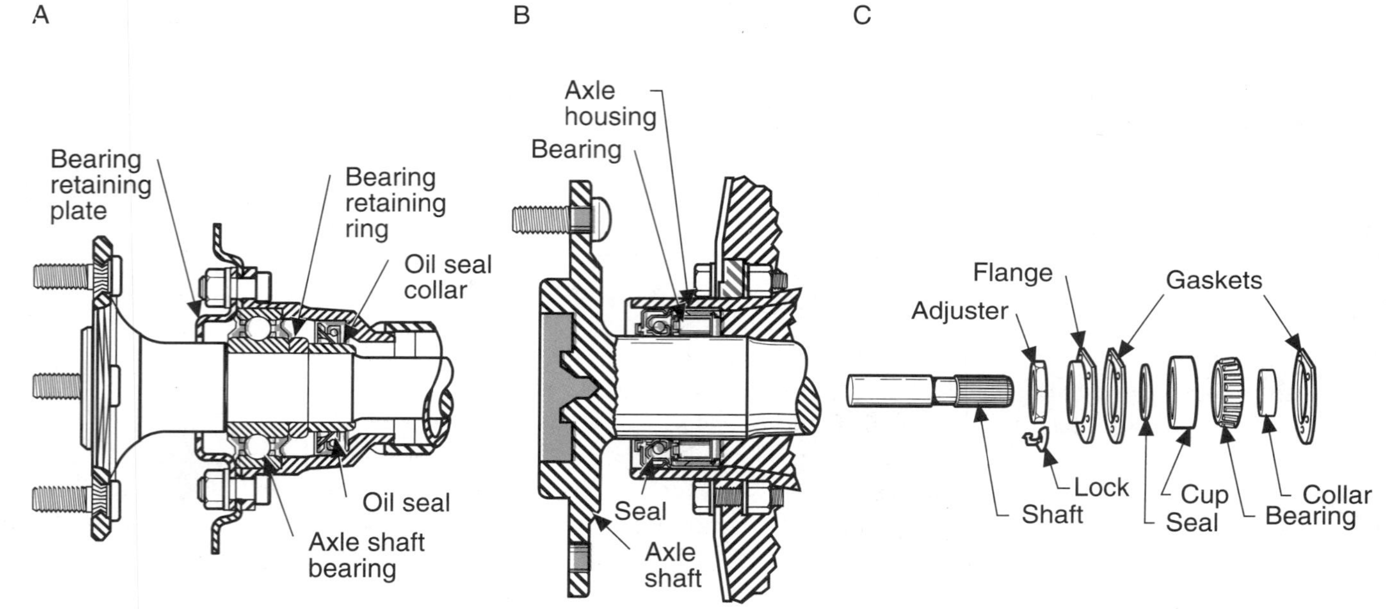

Figure 2-40 (A) Ball-type bearings supporting an axle shaft. (B) Straight roller bearing installed in an axle housing. (Courtesy of Oldsmobile Div.—GMC) (C) Tapered roller drive axle bearing. (Courtesy of Chrysler Corporation)

A bearing designed to take only radial loads has only one of its races machined with a track for the balls. Other bearings are designed to take thrust loads in only one direction. If this type of bearing is installed wrong, the slightest amount of thrust will cause the bearing to come apart.

Another type of ball bearing uses two rows of balls. These are designed to withstand considerable amounts of radial and axial loads. Constructed like two single-row ball bearings joined together, these bearings are often used in rear axle assemblies.

Roller bearings are used wherever it is desirable to have a large bearing surface and low amounts of friction. Large bearing surfaces are needed in areas of extremely heavy loads. The rollers are usually fitted between a journal of a shaft and an outer race. As the shaft rotates, the rollers turn and rotate in the race.

General Maintenance

Most of the drive line needs little maintenance other than periodic oil changes. However, when you repair a transmission, examine the entire gear train to locate worn or faulty parts and repair or replace them at that time. In this way you may prevent a breakdown and the need to disassemble the transmission once again.

Summary

- ❑ The drive train has four primary purposes: to connect the engine's power to the drive wheels, to select different speed ratios, to provide a way to move the vehicle in reverse, and to control the power to the drive wheels for safe turning of the vehicle.
- ❑ The main components of the drive train are the clutch, transmission, and drive axles.
- ❑ The rotating or twisting force of the engine's crankshaft is called engine torque.
- ❑ Gears are used to apply torque to other rotating parts of the drive train and to multiply torque.
- ❑ Torque is calculated by multiplying the applied force by the distance from the center of the shaft to the point where the force is exerted. Torque is measured in foot-pounds and Newton-meters.
- ❑ Gear ratios express the mathematical relationship, in size and number of teeth, of one gear to another.
- ❑ Gear ratios are determined by dividing the number of teeth on the driven gear by the number of teeth on the driving gear.
- ❑ Transmissions offer various gear ratios through the use of various sized gears.
- ❑ Reverse gear is accomplished by adding a third gear to a two gear set. This gear, the reverse idler gear, causes the driven gear to rotate in reverse.
- ❑ Connected to the rear of the crankshaft is the flywheel, which serves many functions, including acting as the driving member of the clutch assembly.
- ❑ The clutch assembly is comprised of another driving disc, the pressure plate, and a driven disc, the clutch disc.
- ❑ The clutch disc is mounted to the input shaft of the transmission and carries the engine's torque to the transmission when the clutch assembly is engaged.
- ❑ In FWD cars, the transmission and drive axle is located in a single assembly called a transaxle. In RWD cars, the drive axle is connected to the transmission through a drive shaft.
- ❑ The drive shaft and its joints are called the drive line of the car.
- ❑ Universal joints allow the drive shaft to change angles in response to movements of the car's suspension and rear axle assembly.
- ❑ The rear axle housing encloses the entire rear-wheel driving axle assembly.

Terms to Know

Ball bearing
Bevel gear
Bushing
Close ratio
Clutch disc
Constant velocity joint
Differential
Direct drive
Downshifting
Drive shaft
Engine torque
Flywheel
Foot-pounds
Gear ratios
Helical gear
Herringbone gear
Horsepower
Hypoid gear
Limited-slip
Meshing
Neutral
Newton-meters
Over drive
Overall gear ratio
Pitting

Terms to Know (continued)

Planetary gear
Pressure plate
Reverse idler gear
Roller bearing
Scoring
Slip yoke
Spalling
Speed reduction
Spur gear
Thrust bearing
Torque
Traction-lock
Transfer case
Universal joint
Wide ratio

- ❑ The primary purpose of the differential is to allow a difference in driving wheel speed when the vehicle is rounding a corner or curve. The ring and pinion in the drive axle also multiplies the torque it receives from the transmission.
- ❑ On FWD cars, the differential is part of the transaxle assembly.
- ❑ The drive axles on FWD cars extend from the sides of the transaxle to the drive wheels. CV-joints are fitted to the axles to allow the axles to move with the car's suspension.
- ❑ 4WD vehicles typically use a transfer case that relays engine torque to both a front and rear driving axle.
- ❑ An understanding of gears and bearings is the key to effective troubleshooting and repair of drive line components.

Review Questions

Short Answer Essays

1. What are primary purposes of a vehicle's drive train?
2. Why does torque increase when a smaller gear drives a larger gear?
3. How are gear ratios calculated?
4. Why are transmissions equipped with many different forward gear ratios?
5. What is the primary difference between a transaxle and a transmission?
6. Why are U- and CV-joints used in the drive line?
7. What does a differential unit do to the torque it receives?
8. What is the purpose of the clutch assembly? How does it work?
9. What kind of gears are commonly used in today's automotive drive trains?
10. When are ball- or roller-type bearings used?

Fill-in-the-Blanks

1. The main components of the drive train are the ____________________ , ____________________ , and ____________________ ____________________ .
2. The rotating or turning effort of the engine's crankshaft is called ____________________ ____________________ .
3. Gears are used to apply torque to other rotating parts of the drive train and to ____________________ torque.
4. Torque is calculated by multiplying the applied force by the ____________________ from the center of the ____________________ to the point where the force is exerted.

5. Torque is measured in ____________________ - ____________________ and

____________________ - ____________________ .

6. Gear ratios are determined by dividing the number of teeth on the

____________________ gear by the number of teeth on the ____________________

gear.

7. Reverse gear is accomplished by adding a ____________________

____________________ to a two gear set. This gear, the reverse idler gear, causes the driven gear to rotate in reverse.

8. The clutch assembly is comprised of a driving disc, called the

____________________ ____________________ , and a driven disc, called the

____________________ ____________________ .

9. In FWD cars, the transmission and drive axle are located in a single assembly called a ____________________ .

10. In RWD cars, the drive axle is connected to the transmission through a

____________________ ____________________ .

ASE Style Review Questions

1. While discussing the purposes of a drive train:
Technician A says that it connects the engine's power to the drive wheels.
Technician B says that it controls the power to the drive wheels for safe turning of the vehicle.
Who is correct?
A. A only
B. B only
C. Both A and B
D. Neither A nor B

2. *Technician A* says gears are used to apply torque to other rotating parts of the drive train.
Technician B says gears are used to multiply torque.
Who is correct?
A. A only
B. B only
C. Both A and B
D. Neither A nor B

3. While discussing gear ratios:
Technician A says they express the mathematical relationship, according to the number of teeth, of one gear to another.
Technician B says they express the size difference of two gears by stating the ratio of the smaller gear to the larger gear.
Who is correct?
A. A only
B. B only
C. Both A and B
D. Neither A nor B

4. While discussing reverse gear:
Technician A says reverse is accomplished by adding a third gear.
Technician B says the reverse idler gear causes the driven gear to rotate in reverse.
Who is correct?
A. A only
B. B only
C. Both A and B
D. Neither A nor B

5. While discussing the purpose of a flywheel:
Technician A says it increases engine torque.
Technician B says it acts as the driving member of the clutch assembly.
Who is correct?
A. A only
B. B only
C. Both A and B
D. Neither A nor B

6. *Technician A* says the clutch disc is splined to the input shaft of the transmission.
Technician B says the clutch disc carries the engine's torque to the transmission when the clutch assembly is engaged.
Who is correct?
A. A only
B. B only
C. Both A and B
D. Neither A nor B

7. While discussing universal joints:
Technician A says they eliminate vibrations caused by the power pulses of the engine.
Technician B says they allow the drive shaft to change angles in response to movements of the car's suspension and rear axle assembly.
Who is correct?
A. A only
B. B only
C. Both A and B
D. Neither A nor B

8. While discussing the purpose of a differential:
Technician A says it allows for equal wheel speed while the vehicle is rounding a corner or curve.
Technician B says the ring and pinion in the drive axle multiplies the torque it receives from the transmission.
Who is correct?
A. A only
B. B only
C. Both A and B
D. Neither A nor B

9. While discussing FWD vehicles:
Technician A says the differential is normally part of the transaxle assembly.
Technician B says the drive axles extend from the sides of the transaxle to the drive wheels.
Who is correct?
A. A only
B. B only
C. Both A and B
D. Neither A nor B

10. *Technician A* says 4WD vehicles typically use a transfer case to transfer engine torque to both a front and a rear driving axle.
Technician B says 4WD vehicles normally have two clutches, two drive shafts, and two differentials.
Who is correct?
A. A only
B. B only
C. Both A and B
D. Neither A nor B

CHAPTER 3

Clutches

Upon completion and review of this chapter, you should be able to:

- ❑ Define the purpose of a clutch assembly.
- ❑ Describe the major components of a clutch assembly.
- ❑ Describe the operation of a clutch.
- ❑ Define the role of each major component in a clutch.
- ❑ Describe the operation of the various mechanical and cable-type clutch linkages.
- ❑ Describe the operation of a hydraulic clutch linkage.
- ❑ Diagnose clutch-related problems by analyzing the symptoms.

The manual transmission clutch is a device used to connect and disconnect engine power flow to the transmission at the will of the driver. A driver operates the clutch with a clutch pedal inside the vehicle. This pedal allows engine power flow to be gradually applied when the vehicle is starting out from rest and interrupts power flow to avoid gear clashing when shifting gears. Engagement of the clutch allows for power transfer from the engine to the transmission and eventually to the drive wheels. Disengagement of the clutch provides the necessary halt of power transfer that allows the engine to continue running while no power is supplied to the drive wheels. Engagement and disengagement of the clutch is controlled by a pedal and clutch linkage that must be properly adjusted. The machined surfaces of the flywheel and pressure plate must be flat and free of cracks and scores in order to adequately clamp the clutch disc. Clutch slippage, vibration, and noise is minimized by the proper alignment of engine and transmission/transaxle and of the clutch components.

Approximately 35% of all cars sold in North America are equipped with a manually operated clutch and transmission.

Normally, a clutch will last 50 to 100 thousand miles. However the life of the clutch depends heavily on the driver, the loads the vehicle carries, and how well the vehicle is maintained.

Clutch Location

Although the exact physical location of the clutch assembly varies with vehicle design, the clutch is placed between the engine and the transmission. This is true regardless of engine, transmission, or drive axle location. With few exceptions, the clutch assembly and flywheel are bolted to the rear of the engine's crankshaft.

Clutch Design

The main parts of the clutch assembly are the clutch housing, flywheel, clutch shaft, clutch disc, pressure plate assembly, release bearing, and clutch linkage (Figure 3-1). The clutch housing is a large bell-shaped metal casting that connects the engine and the transmission/transaxle. It houses the clutch assembly and supports the transmission/transaxle.

A clutch housing is often called a bell housing.

Flywheel

A flywheel is bolted to the engine's crankshaft to serve many purposes. It acts as a balancer for the engine and it smoothens out, or dampens, engine vibrations caused by firing pulses. It adds inertia to the rotating crankshaft. It provides a machined surface from which the clutch can contact and pick up engine torque and transfer it to the transmission. The flywheel also acts as a friction surface and heat sink for one side of the clutch disc.

Shop Manual
Chapter 3, page 62

A clutch shaft is another name for the transmission's input shaft.

A flywheel is a large diameter heavy disc, usually made of nodular cast iron with a high graphite content.

A heat sink is a piece of material that absorbs heat to prevent the heat from settling on another component.

On some flywheels, a special plate is bolted onto the flywheel to provide a frictional surface for the clutch.

A flex-plate is also referred to as a drive-plate.

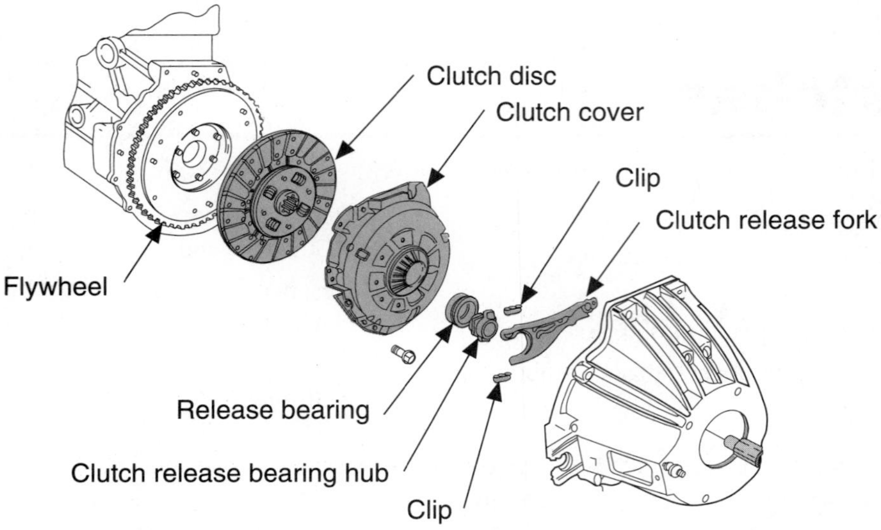

Figure 3-1 Basic clutch components. (Courtesy of Toyota Motor Manufacturing, U.S.A., Inc.)

The flywheel is bolted to the engine's crankshaft and the clutch assembly's pressure plate is bolted to the flywheel. In the center of the flywheel is the bore for the pilot bearing or bushing. The teeth around the circumference of the flywheel form a ring gear for the engine starting motor to contact. The ring gear is not actually a part of the flywheel, rather it is pressed around the outside of the flywheel (Figure 3-2).

Vehicles with automatic transmissions do not have a flywheel. Instead, they use a flex-plate and the weight of a torque converter to dampen the engine vibrations and provide inertia for the

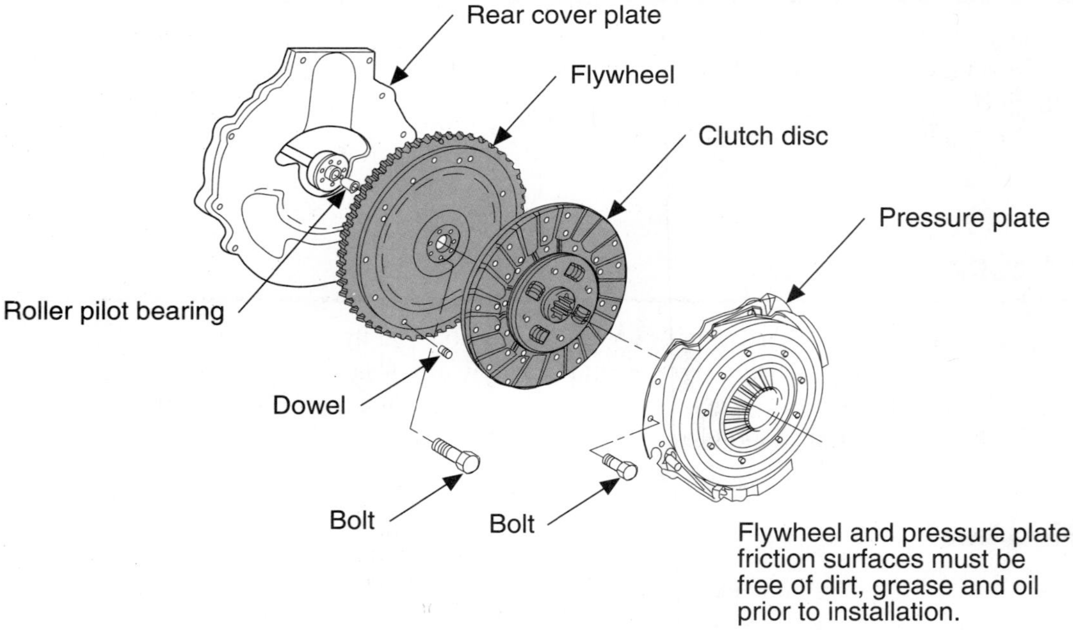

Figure 3-2 Location and mounting of typical flywheel. (Reprinted with the permission of Ford Motor Company)

crankshaft. Flex-plates are lightweight, stamped steel discs and are used as the attaching point for the torque converter to the engine's crankshaft. They have no clutch friction surface and will not interchange with manual transmission flywheels. They do have a ring gear for the engine's starter motor. However, the ring gear is normally part of the plate and is not replaceable except on some Chrysler products.

Clutch Shaft

The clutch shaft (Figure 3-3) projects from the front of the transmission. Most clutch shafts have a smaller shaft or pilot that projects from its outer end. This pilot rides in the pilot bearing in the engine crankshaft flange. The pilot bearing or bushing serves as a support for the outer end of the input shaft and it maintains proper alignment of the shaft with the crankshaft. A pilot bushing is pressed into the bore of the flywheel. Some flywheels are fitted with a ball or needle-type pilot bearing in place of a bushing. Most transaxles are not equipped with a pilot bearing or bushing because their input shaft is short and does not require outward support. The splined area of the shaft allows the clutch disc to move laterally a small amount along the splines, while preventing the disc from rocking on the shaft. When the clutch is engaged, the clutch disc drives the transmission's input shaft through these splines (Figure 3-4).

A clutch shaft is also called the input shaft of the transmission.

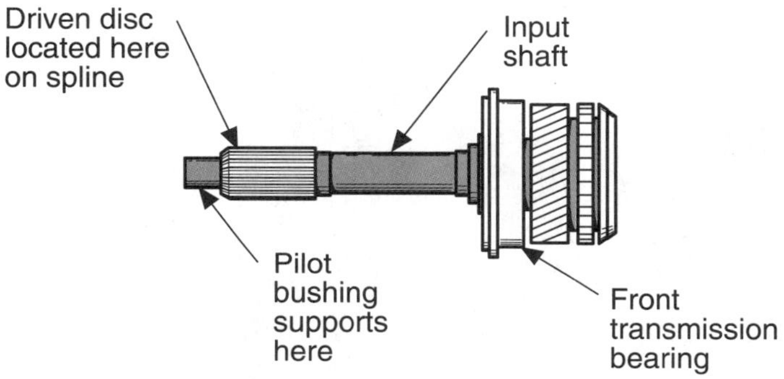

Figure 3-3 A typical clutch shaft.

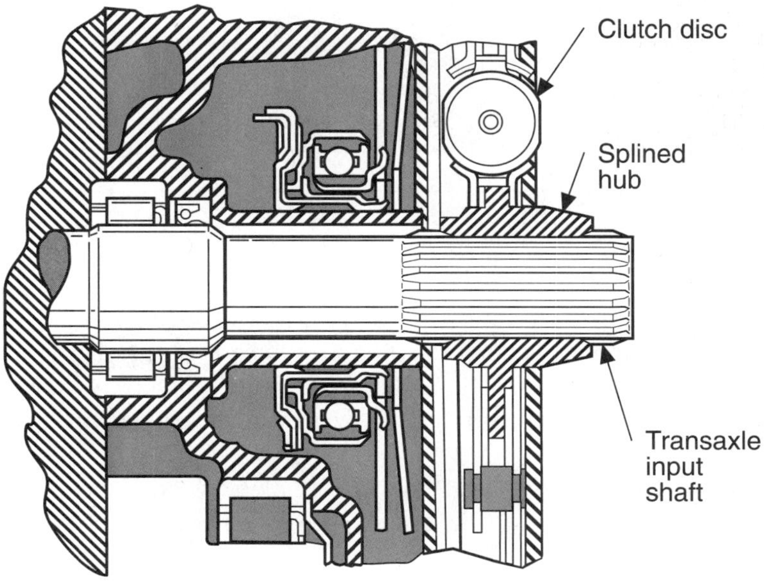

Figure 3-4 The clutch disc is splined to the input shaft.

Shop Manual
Chapter 3, page 60

A clutch disc is also called a friction disc.

Splines resemble a series of keyways cut into a shaft or hole. Parts splined together rotate as a unit, but the parts are free to move somewhat along the shaft's centerline.

A rivet is a headed pin used to unite two or more pieces by passing the shank through a hole in the pieces and securing them by forming a head on the opposite end of the pin.

Asbestos is a mineral fiber composed of a silicate of calcium and magnesium that occurs in long threadlike fibers. It has a high resistance to heat.

The facings of a disc are often referred to as the lining of the disc.

Clutch Disc

The clutch disc is a steel plate covered with frictional material that fits between the flywheel and the pressure plate. In the center of the disc is the hub, which is splined to fit over the splines of the input shaft. When the clutch is engaged, the disc is firmly squeezed between the flywheel and pressure plate and power from the engine is transmitted by the disc's hub to the transmission's input shaft. The width of the hub prevents the disc from rocking on the shaft while it moves between the flywheel and the pressure plate.

A clutch disc (Figure 3-5) has frictional material riveted or bonded to both sides. Frictional facings are either woven or molded. Molded facings are preferred because they can withstand high pressure plate loading forces without being damaged. Woven facings are used when additional cushioning during clutch engagement is desired. Grooves are cut across the face of the friction facings to allow for smooth clutch action, increased cooling, and a place for the facing dust to go as the clutch disc wears. Like brake lining material, the frictional facing wears as the clutch is engaged. Asbestos wire-woven material was the most common facing for clutch discs. Due to recent awareness of the health hazards resulting from asbestos, new lining materials have been developed and are widely used on newer vehicles. The most commonly used are paper-based and ceramic materials that are strengthened by the addition of cotton and brass particles and wire. These increase the torsional strength of the facings and prolong the life of the clutch disc.

Figure 3-5 Typical clutch disc.

The facings are attached to wave springs, which cause the contact pressure on the facings to rise gradually as the springs flatten out when the clutch is engaged. These springs eliminate chatter when the clutch is engaged and also help to move the disc away from the flywheel when the clutch is disengaged. The wave springs and facings are attached to the steel disc.

CAUTION: Like many types of brake-lining material, most friction discs contain asbestos fibers. Always follow safety precautions when handling asbestos.

There are two types of clutch discs: rigid and flexible. A rigid clutch disc is a solid circular disc fastened directly to a center splined hub. The flexible clutch disc is easily recognized by the torsional dampener springs that circle the center hub. The dampener is a shock-absorbing feature

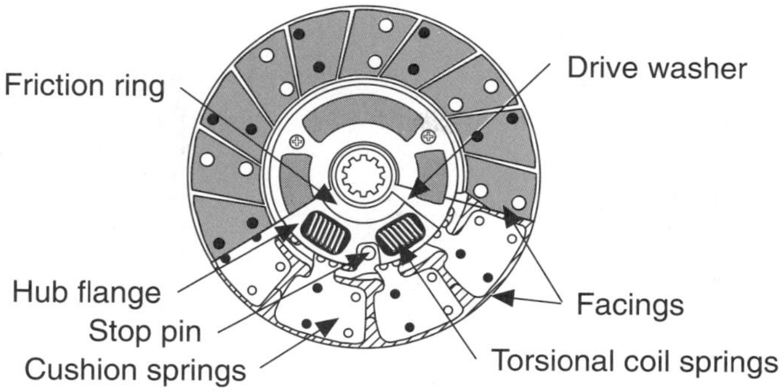

Figure 3-6 Major components of a clutch disc.

built into a flexible clutch disc. The primary purpose of a flexible disc is to absorb power impulses from the engine that would otherwise be transmitted directly to the gears in the transmission. A flexible clutch disc has torsion springs and friction discs between the plate and hub of the clutch. When the clutch is engaged, the springs cushion the sudden loading by flexing and allowing some twist between the hub and plate. When the "surge" is past, the springs release and the disc transmits power normally. The number and tension of these springs is determined by the amount of engine torque and the weight of the vehicle. Stop pins limit this torsional movement to approximately 3/8 inch (Figure 3-6).

The clutch disc's wave springs are also called cushioning springs.

A hub is the center part of a wheel or disc.

Some high performance clutch assemblies use multiple clutch discs. An intermediate plate is used in these assemblies to separate the clutch discs. When the clutch is engaged, the first clutch disc is held between the clutch pressure plate and intermediate plate, and the second clutch disc is held between the intermediate plate and the flywheel. When disengaged, the intermediate plate, flywheel, and pressure plate assembly rotate as a unit, while the clutch discs, which are not in contact with the plates, rotate freely within the assembly and do not transmit power to the transmission.

Pressure Plate Assembly

The pressure plate (Figure 3-7) squeezes the clutch disc onto the flywheel when the clutch is engaged and moves away from the disc when the clutch is disengaged. These actions allow

Shop Manual
Chapter 3, page 56

Some manufacturers refer to a pressure plate as a clutch cover.

Figure 3-7 Typical pressure plate.

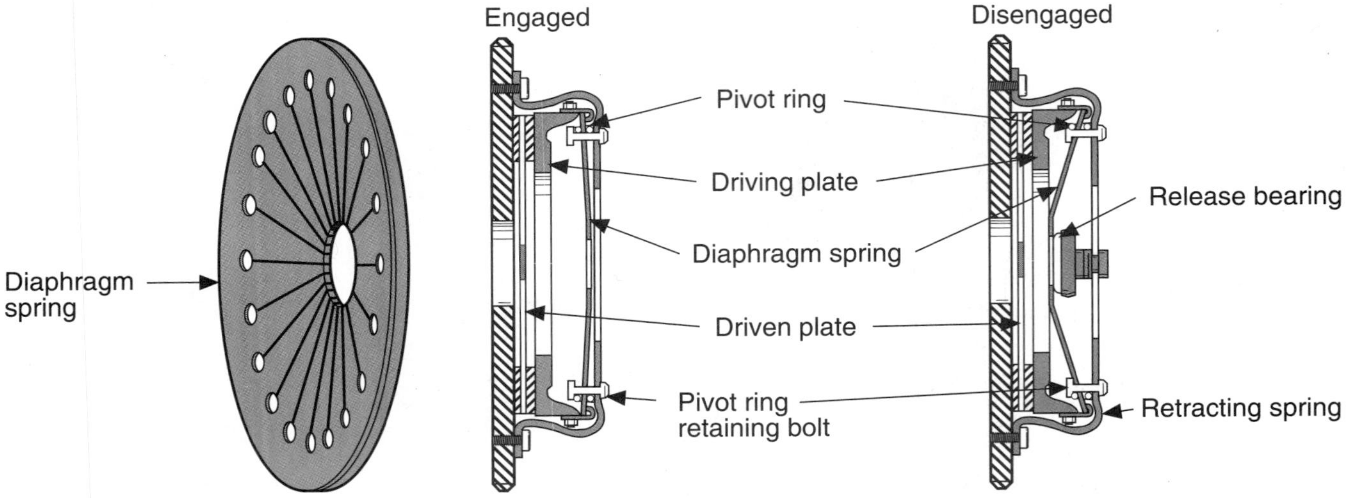

Figure 3-8 A typical diaphragm spring clutch.

A Belleville spring is thin sheet of metal formed into a cone shape.

Some diaphragm springs are fitted with a thrust pad at the ends of the release fingers.

A fulcrum is the support that provides a pivoting point for a lever.

The fulcrum ring is also referred to as the pivot ring.

A coil spring clutch assembly uses coil springs to hold the pressure plate against the friction disc.

A coil spring-type pressure plate is also called a Borg and Beck.

the clutch disc to transmit or not transmit the engine's torque to the transmission. A pressure plate is basically a large spring-loaded clamp that is bolted to and rotates with the flywheel. A pressure plate assembly includes a sheet metal cover, heavy release springs, a metal pressure ring that provides a friction surface for the clutch disc, a thrust ring or fingers for the release bearing, and release levers. The release levers release the holding force of the springs when the clutch is disengaged. The spring used in most pressure plates is a single Belleville or diaphragm-type spring, however a few use multiple coil springs. Some pressure plates are of the semicentrifugal design and use centrifugal weights that increase the clamping force on the thrust springs as engine speed increases.

Diaphragm spring pressure plate (Figure 3-8) assemblies use a cone-shaped diaphragm spring between the pressure plate and the cover to clamp the pressure plate against the clutch disc. This spring is normally secured to the cover by rivets. When pressure is exerted on the center of the spring, the outer diameter of the spring tends to straighten out. As soon as the pressure is released, the spring resumes its normal cone shape. The center portion of the spring is slit into numerous fingers that act as release levers. When the clutch is disengaged, these fingers are depressed by the release bearing. The diaphragm spring pivots over the fulcrum ring and its outer rim moves away from the flywheel. The retracting springs pull the pressure plate away from the clutch disc, thereby disengaging the clutch (Figure 3-9).

When the clutch is engaged, the release bearing is moved away from the diaphragm spring's release fingers. As the spring pivots over the fulcrum ring, its outer rim forces the pressure plate tightly against the clutch disc. At this point the clutch disc is clamped between the flywheel and pressure plate. At the outer rim, the spring is moved outward and the pressure plate is forced against the clutch disc. Diaphragm-type pressure plates are preferred because they are compact, lightweight, require less pedal effort, and have few moving parts to wear.

Coil spring pressure plate assemblies (Figure 3-10) use helical springs that are evenly spaced around the inside of the pressure plate cover. These springs exert pressure to hold the pressure plate tightly against the flywheel. During clutch disengagement, release levers release the holding force of the springs and the clutch disc no longer rotates with the pressure plate and flywheel. Normally these pressure plates are equipped with three release levers and each lever has two pivot points. One pivot point attaches the lever to a pedestal cast into the pressure plate and the other attaches the lever to a release yoke that is bolted to the cover. The levers pivot on the pedestals and release lever yokes to move the pressure plate through its engagement and disengagement operations (Figure 3-11).

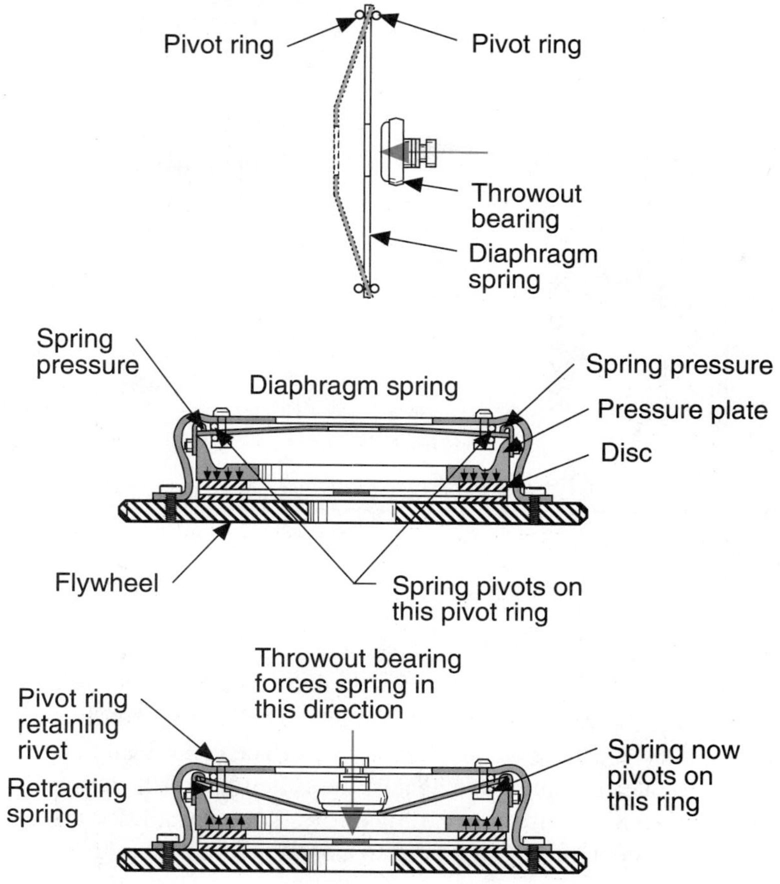

Figure 3-9 Action of a diaphragm pressure plate spring when the clutch is engaged and disengaged.

To disengage the clutch, the release bearing pushes the inner ends of the release levers toward the flywheel. The release levers act as a fulcrum for the levers and the outer ends of the release levers move to pull the pressure plate away from the clutch disc. This action compresses the coil springs and disengages the clutch. When the clutch is engaged, the release bearing moves and allows the springs to exert pressure to hold the pressure plate against the clutch disc. This forces the disc against the flywheel and the engine's power is transmitted to the transmission through the clutch disc.

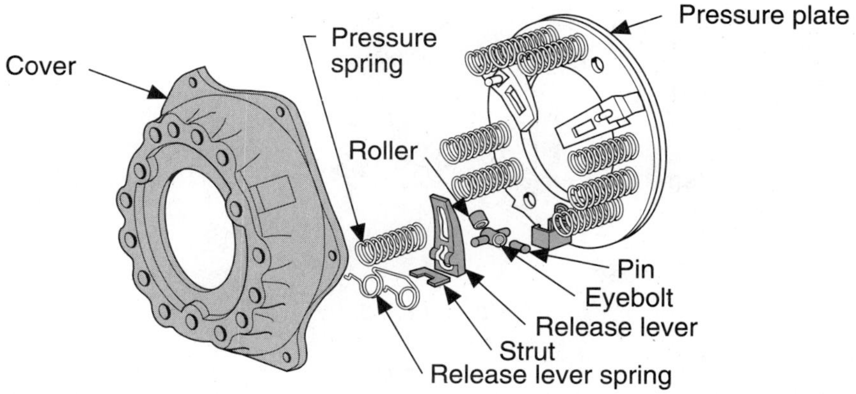

Figure 3-10 Coil spring pressure plate assembly.

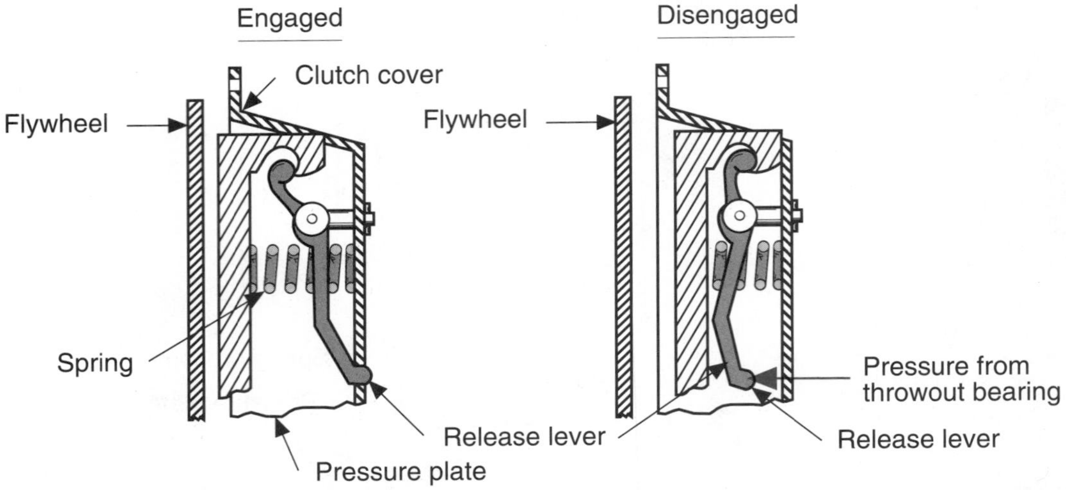

Figure 3-11 Action of a coil-spring pressure plate's release levers.

The holding power of a clutch assembly is determined by its diameter, coefficient of friction, spring pressure, and the number of clutch discs in the assembly.

Centrifugal force is an outward pull from the center of a rotating axis.

A semicentrifugal pressure plate is a design variation of a coil-spring pressure plate. These assemblies increase and decrease the holding force on the clutch disc according to engine speed. A weighted end on the release levers (Figure 3-12) act to increase centrifugal force as the rotational speed of the pressure plate increases (Figure 3-13). The centrifugal force adds to the spring pressure to produce greater holding force against the clutch disc. This design allows for the use of coil springs with less tension, as the centrifugal force compensates for the decrease in spring tension. Therefore, less pedal effort is required to operate the clutch without a loss of clamping pressure on the disc.

The individual parts of a pressure plate assembly are contained in the cover. Some covers are vented to allow heat to escape and air to enter. Other covers are designed to provide a fan action to force air circulation around the clutch assembly. The effectiveness of the clutch is affected by heat, therefore by allowing the assembly to cool it is able to work better.

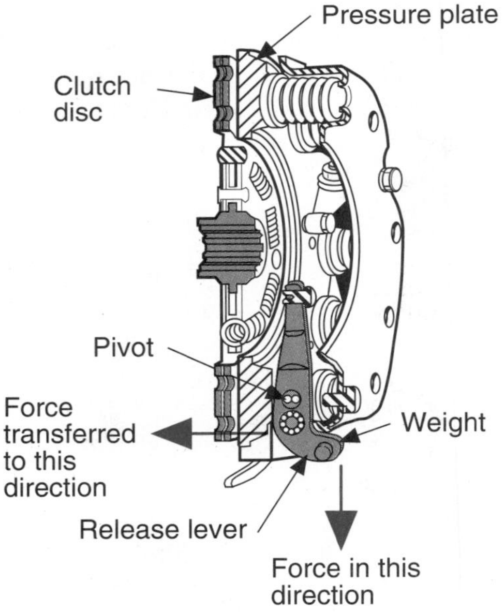

Figure 3-12 Semicentrifugal pressure plate assembly. (Reprinted with the permission of Ford Motor Company)

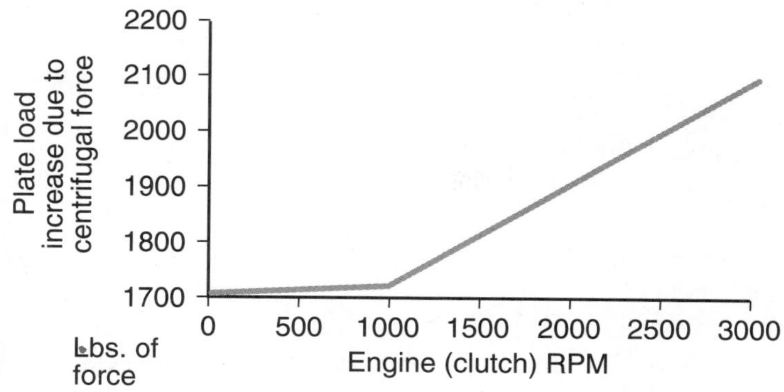

Figure 3-13 Graph showing how a semicentrifugal pressure plate changes plate loading according to engine speed. (Reprinted with the permission of Ford Motor Company)

Release Bearing

The clutch release bearing (Figure 3-14) is a carbon-type or ball-type bearing located in the bell housing and operated by the clutch linkage (Figure 3-15). Release bearings are usually sealed and prelubricated to provide smooth and quiet operation as they move against the pressure plate to disengage the clutch. When the clutch pedal is depressed, the release bearing moves toward

Shop Manual
Chapter 3, page 70

A clutch release bearing is often referred to as a throw-out bearing.

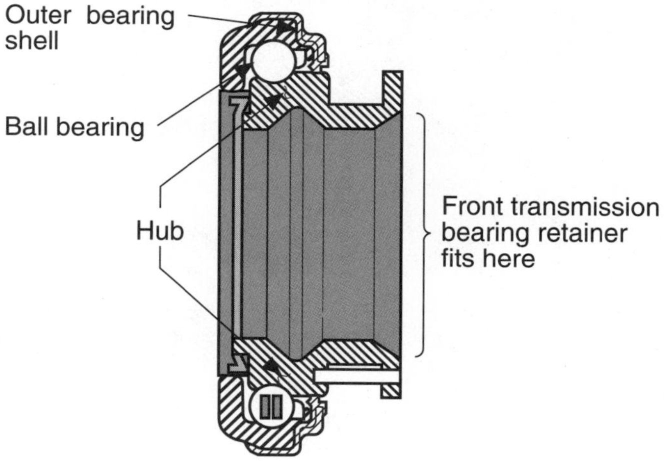

Figure 3-14 A typical clutch release bearing. (Courtesy of Chevrolet)

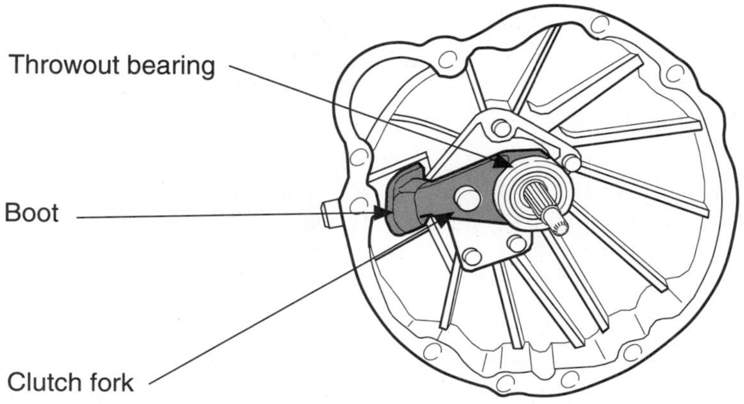

Figure 3-15 Location of clutch fork and release bearing. (Courtesy of Nissan Motors)

the flywheel, depressing the pressure plate's release fingers or thrust pad and moving the pressure plate fingers or levers against pressure plate spring force. This action moves the pressure plate away from the clutch disc, thus interrupting power flow.

The hollow shaft on the front of the front bearing retainer is often referred to as the quill shaft.

Release bearings are mounted on an iron casting called a hub, which slides on a hollow shaft at the front of the transmission housing. This hollow shaft is part of the transmission's front bearing retainer.

The release bearing is mounted on a sleeve that is designed to slide back and forth on the transmission's bearing retainer (Figure 3-16). The sleeve is grooved or has raised flat surfaces and retaining springs that hold the inner ends of the release fork in place on the release bearing assembly. The fork and connecting linkage convert the movement of the clutch pedal to the back and forth movement of the clutch release bearing.

As the release bearing pushes against the pressure plate to release the clutch, the pushing force is against the rear of the crankshaft. A small main bearing, the thrust bearing, in the engine stops the crankshaft from coming out of the front of the engine.

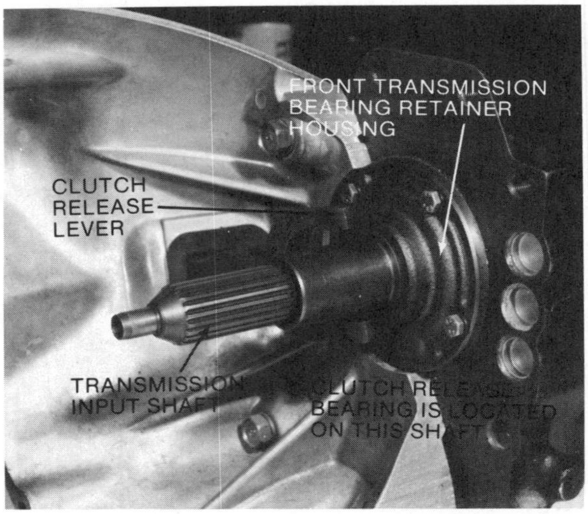

Figure 3-16 The release bearing slides on the hollow shaft nose of the transmission's front bearing retainer.

To disengage a clutch, the release bearing is moved toward the flywheel by the clutch fork. As the bearing contacts the release levers or fingers, it begins to rotate with the pressure plate assembly. As the release bearing continues to move forward, the pressure on the release levers or fingers causes the force of the pressure plate's spring to move away from the clutch disc.

To engage the clutch, the clutch pedal is released and the release bearing moves away from the pressure plate. This action allows the pressure plate's springs to force against the clutch disc, engaging the clutch to the flywheel. Once the clutch is fully engaged, the release bearing is normally stationary and does not rotate with the pressure plate.

Most release bearings must be adjusted so they do not touch the release fingers, beveled springs, or thrust pad when the pedal is released. However, some release bearings are designed to ride lightly against the pressure plate assembly; these are called constant running release bearings (Figure 3-17) and are used on transmissions equipped with self-adjusting or hydraulically operated clutch linkages.

The clutch linkage connects the driver-operated clutch pedal to a bell housing-mounted release fork that acts directly on the release bearing. The release bearing slides back and forth on the transmission's front bearing retainer in response to clutch pedal movement.

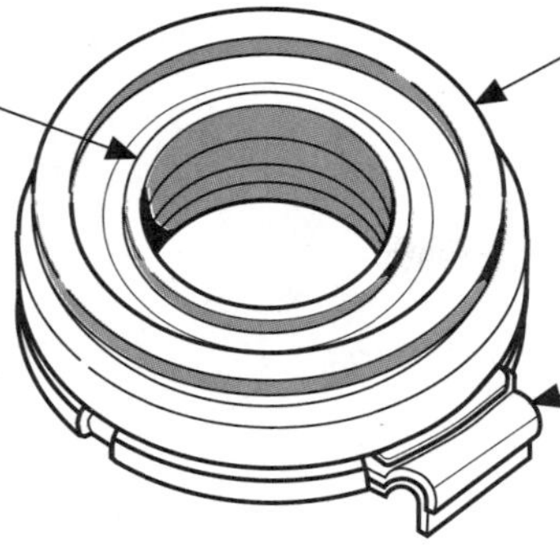

Figure 3-17 Typical constant running release bearing.

Clutch Linkages

Shop Manual
Chapter 3, page 45

Clutches are normally operated by either mechanical or hydraulic linkages. Two types of mechanical linkages are used: the cable type and the shaft and lever type. The shaft and lever clutch link age has many parts and pivot points and transfers the movement of the clutch pedal to the release bearing via shafts, levers, and bell cranks. In older vehicles, the pivot points were equipped with grease fittings. Current systems pivot on low-friction plastic grommets and bushings. As the pivot points wear, the extra play in the linkage makes precise clutch pedal free-play adjustments difficult.

Clutch pedal free-play is the amount the pedal can move without applying pressure on the pressure plate.

A typical shaft and lever clutch control assembly (Figure 3-18) includes a release lever and rod, an equalizer or cross shaft, a pedal to equalizer rod, an assist or overcenter spring, and the pedal assembly. Depressing the pedal moves the equalizer that, in turn, moves the release rod. When the pedal is released, the assist spring returns the linkage to its normal position and removes the pressure on the release rod. This action causes the release bearing to move away from the pressure plate.

A control cable is an assembly with a flexible outer housing anchored at the upper and lower ends. Moving back and forth inside the housing is a braided stainless steel wire cable that transfers pedal movement to the release lever.

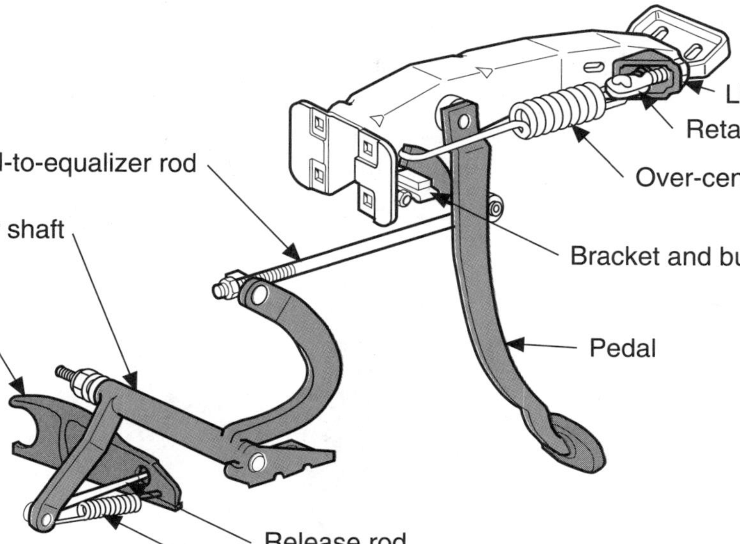

Figure 3-18 Typical shaft and lever clutch control. (Reprinted with the permission of Ford Motor Company)

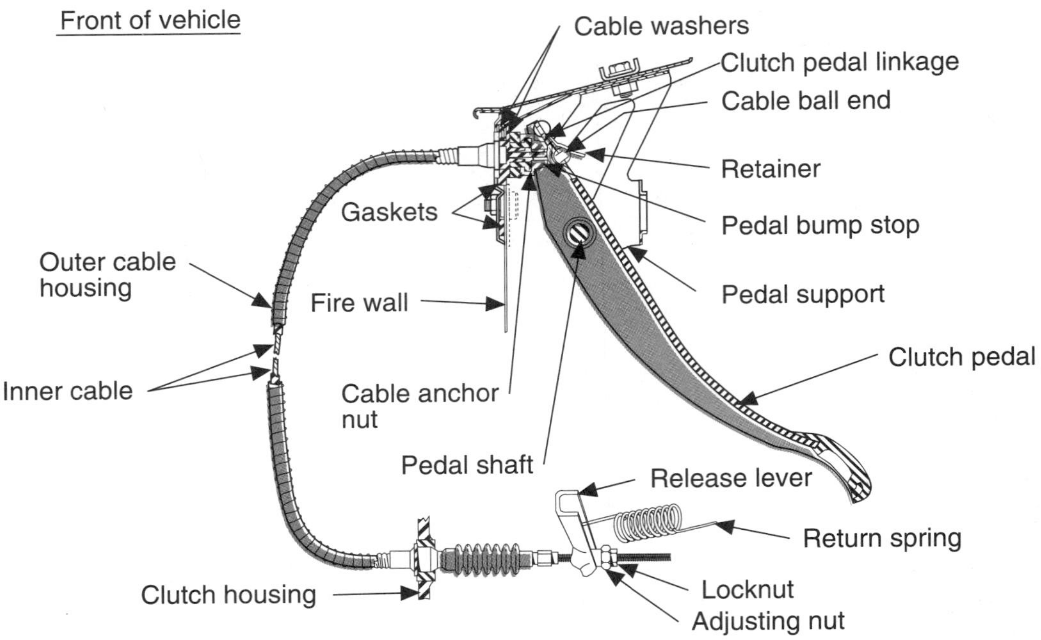

Figure 3-19 Typical clutch pedal and cable assembly. (Reprinted with the permission of Ford Motor Company)

A cable-type clutch linkage is simple and lightweight. Normally the cable connects the pivot of the clutch pedal directly to the release fork (Figure 3-19). This simple set-up is compact, flexible, and eliminates the wearing pivot points of a shaft and lever linkage. However, cables will gradually stretch and can break due to electrolysis.

The clutch fork is a Y-shaped member into which the release bearing is attached.

Typically, one end of the cable is connected to the pedal assembly. At the assembly is a spring that keeps the pedal in the up position. The cable is held under tension by the spring and a cable stop located on the firewall. The other end of the cable is connected to the outer end of the clutch release fork. This end is threaded and fitted with an adjusting nut and locknut that allows for pedal free-play adjustments. When the clutch pedal is depressed, the cable pulls on the clutch fork, which causes the release bearing to move against the pressure plate.

On many late model vehicles, the cable is self-adjusting. At the pedal pivot, the cable is wrapped around and attached to a toothed wheel (Figure 3-20). Slight contact of the release

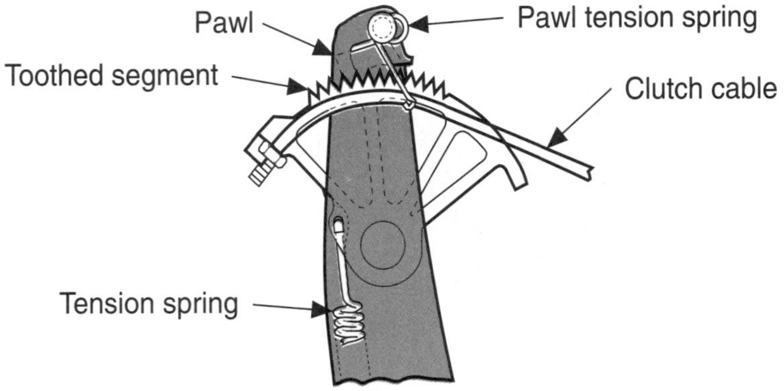

Figure 3-20 Typical automatic clutch cable adjusting mechanism. (Reprinted with the permis sion of Ford Motor Company)

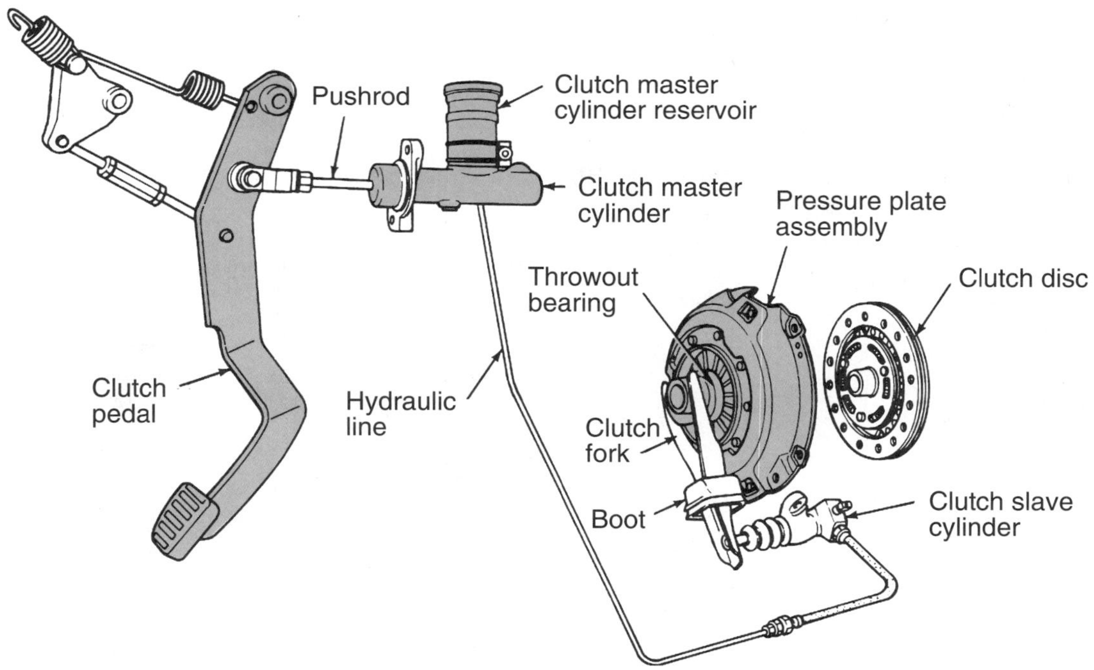

Figure 3-21 Typical hydraulic clutch linkage. (Courtesy of Nissan Motors)

Shop Manual
Chapter 3, page 51

bearing on the pressure plate is maintained by a ratcheting, spring-loaded pawl that is engaged to the toothed wheel. When the clutch pedal is released, the pawl takes any slack out of the cable by engaging the next tooth of the wheel. Self-adjusting clutches use a constant running release bearing and have no built-in free-play.

Hydraulic clutch linkage systems are often called hydraulic clutches.

A hydraulic clutch linkage, like a brake system, consists of a master cylinder, hydraulic tubing, and a slave cylinder (Figure 3-21). The master cylinder is attached to and activated by the clutch pedal through the use of an actuator rod. The slave cylinder is connected to the master cylinder by flexible pressure hose or metal tubing. The slave cylinder is positioned so that it can work directly on the clutch release yoke lever.

The slave cylinder is appropriately named, as it only works in response to the master cylinder.

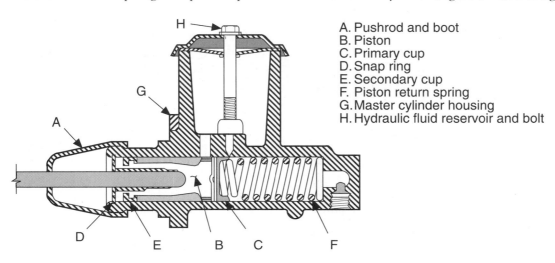

Depressing the clutch pedal pushes the actuator rod into the bore of the master cylinder. This action forces a plunger or piston up the bore of the master cylinder (Figure 3-22). During

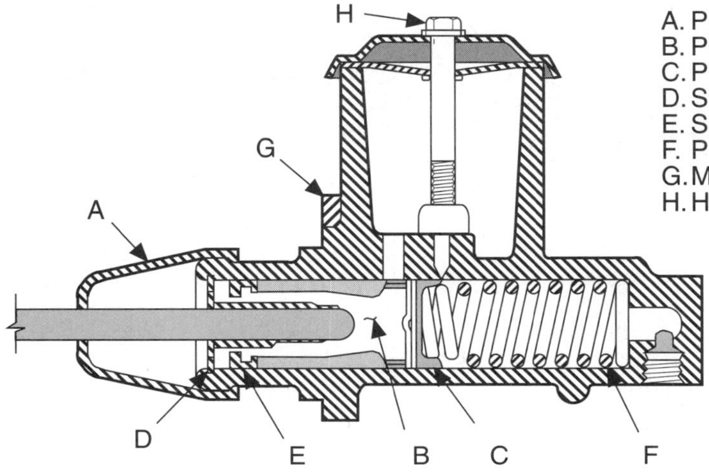

Figure 3-22 Parts of a clutch master cylinder. (Courtesy of General Motors Corporation)

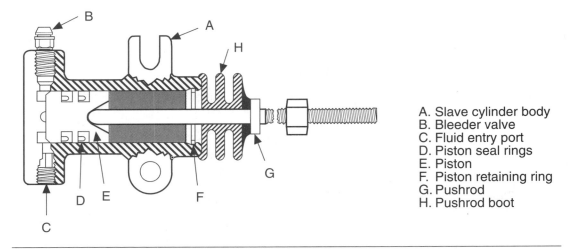

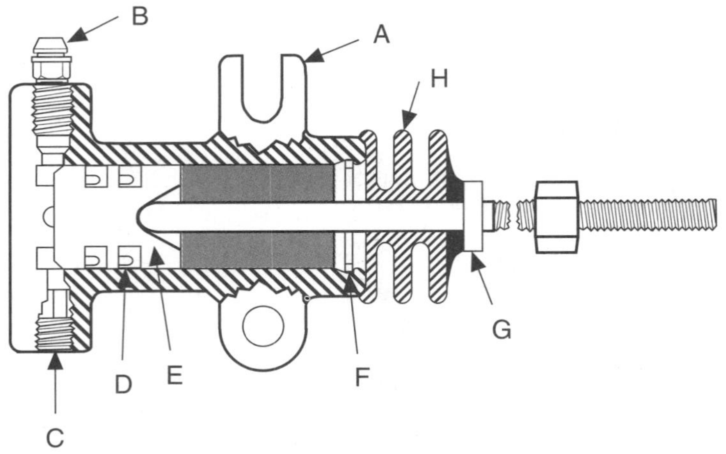

Figure 3-23 Parts of a clutch slave cylinder. (Courtesy of General Motors Corporation)

the initial 1/32 inch of pedal travel, the valve seal at the end of the master cylinder bore closes the port to the fluid reservoir. As the pedal is further depressed, the movement of the plunger forces fluid from the master cylinder through a line to the slave cylinder (Figure 3-23). This fluid is under pressure and causes the piston of the slave cylinder to move. Moving the slave cylinder's piston causes its pushrod to move against the release fork and bearing, thus disengaging the clutch. When the clutch pedal is released, the springs of the pressure plate push the slave cylinder's pushrod back, forcing the hydraulic fluid back into the master cylinder. The final movement of the clutch pedal and the master cylinder's piston opens the valve seal and fluid flows from the reservoir into the master cylinder.

A clutch master cylinder performs the following functions: moves fluid through the hydraulic line to the slave cylinder, compensates for temperature change and minimal fluid loss to maintain the correct fluid volume through the use of a bleed port and compensation port, and compensates for a worn clutch disc and pressure plate by displacing fluid through the reservoir bleed port, thereby eliminating the need for periodic adjustment.

The slave cylinder disengages the clutch by extending the slave cylinder pushrod and makes sure the clutch release bearing is contacting the pressure plate through the use of the slave cylinder preload spring.

Hydraulic clutch controls are normally used on large units that require high pedal pressure to disengage. Hydraulic clutch linkages are also used in vehicles where mechanical linkage would be difficult to route, due to the compactness of the vehicle's design. By using hydraulics, a small force can be multiplied and used effectively to disengage the clutch.

Clutch Operation

The clutch disc is sandwiched between the flywheel and pressure plate assembly to become the driven member of the clutch assembly (Figure 3-24). The flywheel drives the front side while the pressure plate drives the rear side of the clutch disc. In the center of the clutch disc is a splined hub that meshes with the splines on the clutch shaft. The pressure plate is held in contact with the rear friction facing of the clutch disc by spring tension. When the clutch is disengaged, the pressure plate is released by a release bearing, which utilizes a lever action to pull the pressure plate away from the clutch disc. The release yoke, with the release bearing clipped to it, is

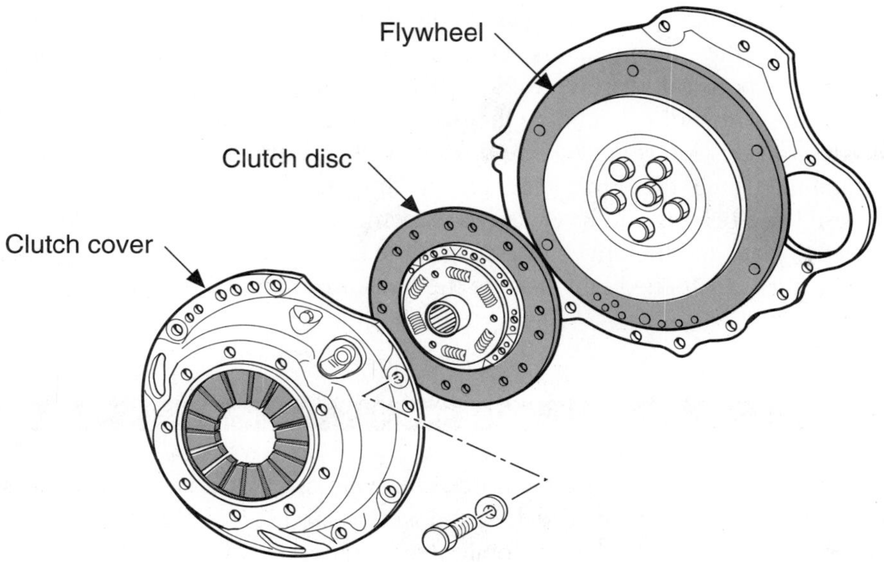
Flywheel
Clutch disc
Clutch cover

Figure 3-24 When engaged, the clutch disc is sandwiched between the pressure plate and flywheel to become the driven member of the clutch assembly. (Courtesy of Nissan Motors)

mounted on a pivot located inside the clutch housing and is operated by the clutch linkage and pedal assembly.

To disengage the clutch, the driver presses the clutch pedal. The linkage forces the release bearing and release yoke forward to move the pressure plate away from the disc (Figure 3-25).

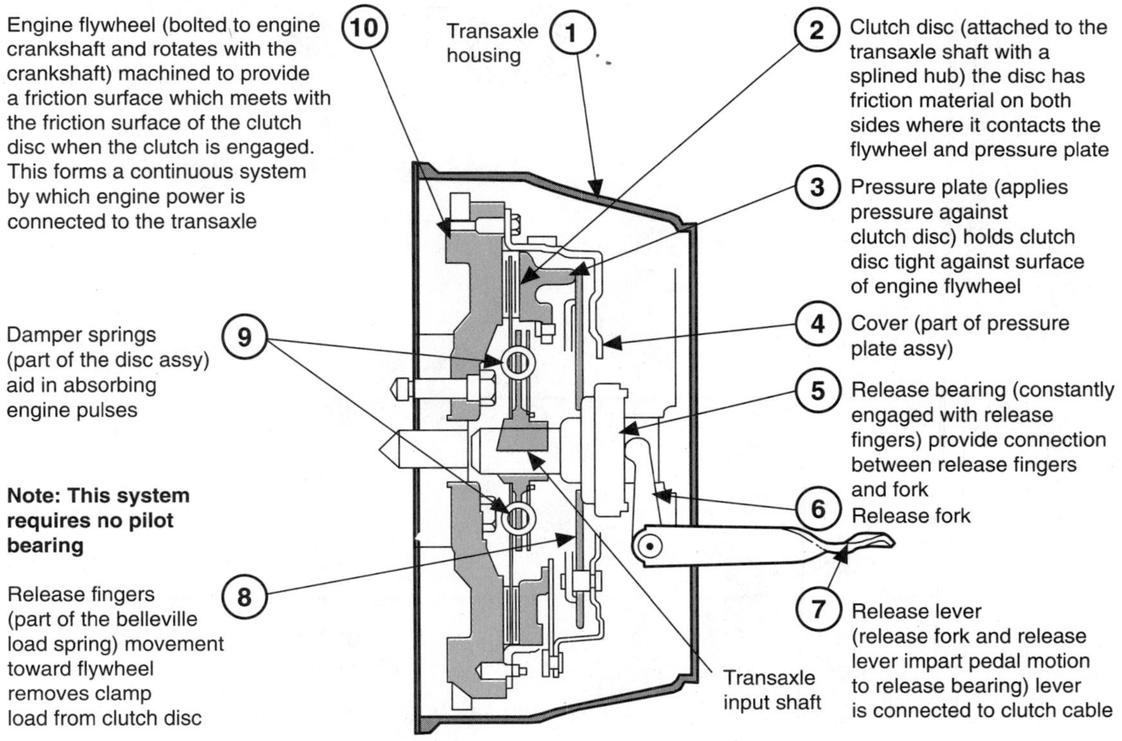
Engine flywheel (bolted to engine crankshaft and rotates with the crankshaft) machined to provide a friction surface which meets with the friction surface of the clutch disc when the clutch is engaged. This forms a continuous system by which engine power is connected to the transaxle
10
Transaxle housing
1
2
Clutch disc (attached to the transaxle shaft with a splined hub) the disc has friction material on both sides where it contacts the flywheel and pressure plate
3
Pressure plate (applies pressure against clutch disc) holds clutch disc tight against surface of engine flywheel
4
Cover (part of pressure plate assy)
9
Damper springs (part of the disc assy) aid in absorbing engine pulses
5
Release bearing (constantly engaged with release fingers) provide connection between release fingers and fork
6
Release fork
Note: This system requires no pilot bearing
8
Release fingers (part of the belleville load spring) movement toward flywheel removes clamp load from clutch disc
7
Release lever (release fork and release lever impart pedal motion to release bearing) lever is connected to clutch cable
Transaxle input shaft

Figure 3-25 Purpose of the major components of a clutch assembly. (Reprinted with the permission of Ford Motor Company)

Because the disc is no longer in contact with the flywheel and pressure plate, the clutch is said to be disengaged. When the clutch pedal is depressed, the flywheel, clutch disc, and pressure plate are disengaged, thus power flow is interrupted. As the clutch pedal is released, the pressure plate moves closer to the clutch disc, clamping the disc between the pressure plate and the flywheel. Therefore, if the transmission is in gear, the drive wheels will turn when the clutch disc turns.

To engage the clutch, the driver eases the clutch pedal up from the floor. The control linkage moves the release bearing and lever rearward, permitting the pressure plate spring tension to force the pressure plate and the driven disc against the flywheel. Engine torque again acts on the disc's friction facings and splined hub to drive the transmission input shaft.

Cone clutches are still used in some limited-slip differentials.

Band-type clutches and multiple disc wet clutches are used in automatic transmissions.

A BIT OF HISTORY

Cone clutches (Figure 3-26) were used almost exclusively on early automobiles. As technology changed, expanding shoe, band-type clutches, and wet or dry disc and plate-type clutches became more prominent. By 1950, nearly all automobiles were equipped with a dry disc clutch system.

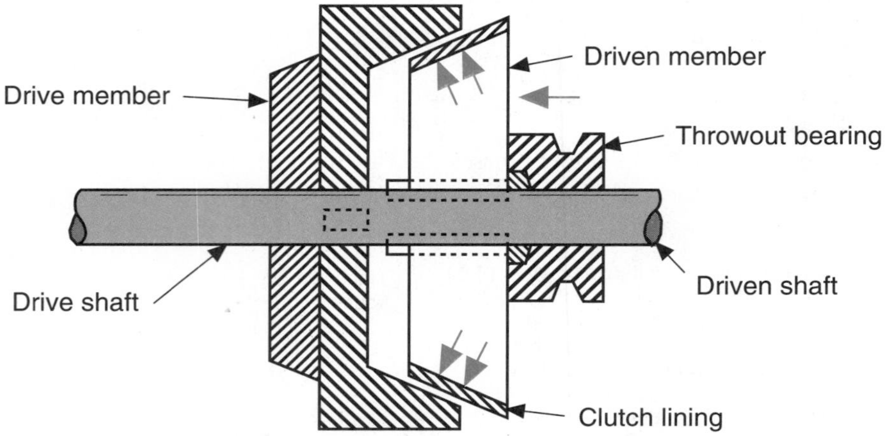

Figure 3-26 A typical cone clutch.

Summary

- The main parts of the clutch assembly are the clutch housing, flywheel, input shaft, disc, pressure plate assembly, release bearing, and linkage.
- The flywheel acts as a balancer and smoothens out, or dampens, engine vibrations caused by firing pulses and adds inertia to the rotating crankshaft of the engine.

Terms to Know

Asbestos

Belleville spring

- ❑ The flywheel also provides a machined surface for the clutch friction disc.
- ❑ Vehicles with automatic transmission are equipped with a drive-plate or flex-plate rather than a heavy flywheel.
- ❑ The clutch disc is splined to the input shaft, which allows the disc to move without rocking on the shaft.
- ❑ The clutch disc is a steel plate with friction material bonded to both sides that fits between the flywheel and the pressure plate.
- ❑ Most friction discs contain asbestos fibers. Always follow safety precautions when handling asbestos.
- ❑ A rigid or solid clutch disc is a solid disc fastened directly to a center splined hub.
- ❑ A flexible clutch disc has torsional dampener springs in its center hub.
- ❑ The primary purpose of a flexible disc is to absorb power impulses from the engine that would otherwise be transmitted directly to the transmission.
- ❑ The pressure plate is a large spring-loaded plate that engages the clutch by pressing the disc against the flywheel surface.
- ❑ The pressure plate moves away from the flywheel when the clutch pedal is depressed, releasing the clamping force and stopping engine torque from reaching the transmission.
- ❑ The clutch release bearing is operated by the clutch linkage.
- ❑ When the clutch pedal is depressed, the bearing moves toward the flywheel, depressing the pressure plate fingers or thrust pad and moving the pressure plate away from the clutch disc.
- ❑ The clutch linkage connects the clutch pedal to a release fork that acts on the release bearing.
- ❑ The clutch is usually located between the engine and the transmission.
- ❑ Clutches are mostly operated by either mechanical or hydraulic linkages.
- ❑ A mechanical clutch linkage transfers the clutch pedal movement to the release bearing via shafts, levers, and bell cranks, or by a cable.
- ❑ A hydraulic clutch linkage consists of a master cylinder, hydraulic tubing, and a slave cylinder.

Terms to Know (continued)

Clutch control cable
Clutch fork
Clutch housing
Clutch linkage
Clutch shaft
Clutch slippage
Coil spring clutch
Cone clutch
Diaphragm spring clutch
Flywheel ring gear
Friction disc
Hub
Hydraulic clutch
Linkage
One-way clutch
Overcenter spring
Overrunning clutch
Pressure plate
Release bearing
Release levers
Rivet
Self-adjusting clutch linkage
Semicentrifugal clutch
Splined hub
Sprag clutch
Throwout bearing
Torsional springs
Wet-disc clutch

Review Questions

Short Answer Essays

1. Define the purpose of a clutch assembly.
2. Describe the major components of a clutch assembly.

3. Describe the operation of a clutch.
4. Compare and contrast the operation of a coil spring pressure plate and a diaphragm spring pressure plate.
5. Define the role of each major component in a clutch.
6. Describe the operation of a mechanical lever-type clutch linkage.
7. Describe the operation of a cable-type clutch linkage.
8. Describe the operation of a hydraulic clutch linkage.
9. Explain why some vehicles are equipped with a semicentrifugal pressure plate.
10. Describe the construction of a flexible clutch disc and state why they are different than a rigid disc.

Fill-in-the-Blanks

1. A clutch housing is also called a ____________________ - ____________________.
2. The teeth around the outside of the flywheel form a ring gear for the ____________________ - ____________________.
3. The clutch disc slides over the splines of the ____________________ shaft.
4. There are two types of dry clutch discs: ____________________ and ____________________.
5. The pressure plate engages and disengages the ____________________ ____________________.
6. A flexible clutch disc has ____________________ springs in its center hub.
7. The pressure plate moves away from the flywheel when the clutch pedal is ____________________.
8. When the clutch is disengaged, the pressure plate is released by a ____________________ ____________________.
9. A hydraulic clutch linkage consists of a ____________________ ____________________, ____________________ ____________________ and a ____________________ ____________________.
10. Tension is maintained in a cable-type clutch linkage by the use of a ____________________ and a toothed ____________________.

ASE Style Review Questions

1. While discussing where a clutch disc is normally mounted:
   Technician A says it is bolted to the flywheel.
   Technician B says it is splined to the transmission input shaft.
   Who is correct?
   A. A only
   B. B only
   C. Both A and B
   D. Neither A nor B

2. While discussing the purpose of the torsional springs in a clutch disc:
   Technician A says they cushion the sudden load on the disc when it is quickly engaged.
   Technician B says they allow the clutch disc and hub to twist slightly.
   Who is correct?
   A. A only
   B. B only
   C. Both A and B
   D. Neither A nor B

3. While discussing the purpose of the flywheel:
   Technician A says it serves as a heat sink for the clutch disc.
   Technician B says it serves as a vibration dampener for the engine.
   Who is correct?
   A. A only
   B. B only
   C. Both A and B
   D. Neither A nor B

4. While discussing the operation of the clutch assembly:
   Technician A says that when the pedal is depressed, the release bearing is pushed into the center of the pressure plate, which releases the plate's pressure on the clutch disc.
   Technician B says that normally the clutch disc is pressed against the flywheel by the pressure plate.
   Who is correct?
   A. A only
   B. B only
   C. Both A and B
   D. Neither A nor B

5. While discussing the major components of a hydraulic clutch linkage:
   Technician A says the master cylinder for the car's brakes is also used for the clutch.
   Technician B says the slave cylinder is connected to the clutch pedal and increases hydraulic pressure as the pedal is depressed.
   Who is correct?
   A. A only
   B. B only
   C. Both A and B
   D. Neither A nor B

6. While discussing the causes for clutch slippage:
   Technician A says a weak pressure plate spring could be the cause.
   Technician B says a warped pressure plate could be the cause.
   Who is correct?
   A. A only
   B. B only
   C. Both A and B
   D. Neither A nor B

7. While discussing cable-type clutch linkages:
   Technician A says they are not commonly used because they are expensive and complicated.
   Technician B says most late model cars use self-adjusting cables for their clutch linkage.
   Who is correct?
   A. A only
   B. B only
   C. Both A and B
   D. Neither A nor B

8. While discussing the different types of clutches:
   Technician A says that flexible clutch discs are most commonly used.
   Technician B says rigid clutch discs are only used when high pedal effort is desirable.
   Who is correct?
   A. A only
   B. B only
   C. Both A and B
   D. Neither A nor B

9. While discussing the purpose of a pilot bearing or bushing:
Technician A says they are only necessary when the transmission's input shaft is long.
Technician B says they serve as a low-friction pivot point for the clutch linkage.
Who is correct?
A. A only
B. B only
C. Both A and B
D. Neither A nor B

10. While discussing the different types of pressure plates:
Technician A says coil spring types are not commonly used because they require great pedal effort when their springs are strong.
Technician B says Belleville-type pressure plates are not commonly used because they require excessive space in the bell housing.
Who is correct?
A. A only
B. B only
C. Both A and B
D. Neither A nor B

CHAPTER 4

Manual Transmissions/Transaxles

Upon completion and review of this chapter, you should be able to:

- ❑ Discuss the purpose and operation of typical manual transmissions.
- ❑ Describe the purpose, design, and operation of a synchronizer assembly.
- ❑ Discuss the flow of power through a manual transmission.
- ❑ Compare and contrast the design and operation of a transmission and a transaxle.
- ❑ Discuss the flow of power through a manual transaxle.
- ❑ Describe how gears are shifted in a manual transmission/transaxle.
- ❑ Identify the accessories controlled by a manual transmission/transaxle.
- ❑ Discuss the importance of gear oil in a manual transmission/transaxle.

Transmissions and transaxles serve basically the same purpose and operate by the same basic principles. Although the assembly of a transaxle is different than a transmission, the fundamentals and basic components are the same. In fact, a transaxle is basically a transmission with other drive line components housed within the assembly. All basics covered in this chapter will refer to both a transaxle and a transmission unless otherwise noted.

A transmission (Figure 4-1) is a system of gears that transfers the engine's power to the drive wheels of the car. The transmission receives torque from the engine through its input shaft when the clutch is engaged. The torque is then transferred through of a set of gears that either multiply it or transfer it directly. The resultant torque turns the transmission's output shaft, which is indirectly connected to the drive wheels. All transmissions have two primary purposes: They select different speed ratios for a variety of conditions and provide a way to reverse the movement of the vehicle.

The purpose of a transmission is to use various sized gears to provide the engine with a mechanical advantage over the vehicle's drive wheels.

A gear transfers power and motion from one shaft to another shaft.

Transmission gears are normally helical-type gears, however some reverse speed gears are spur-type gears.

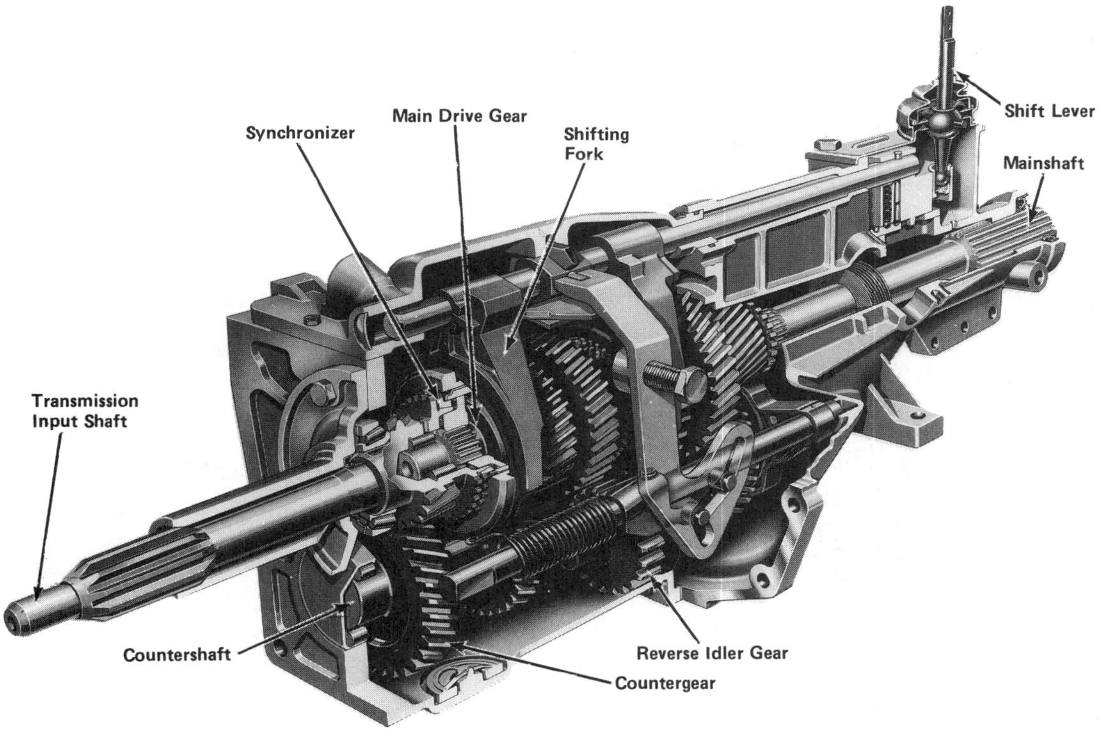

Figure 4-1 Typical manual transmission

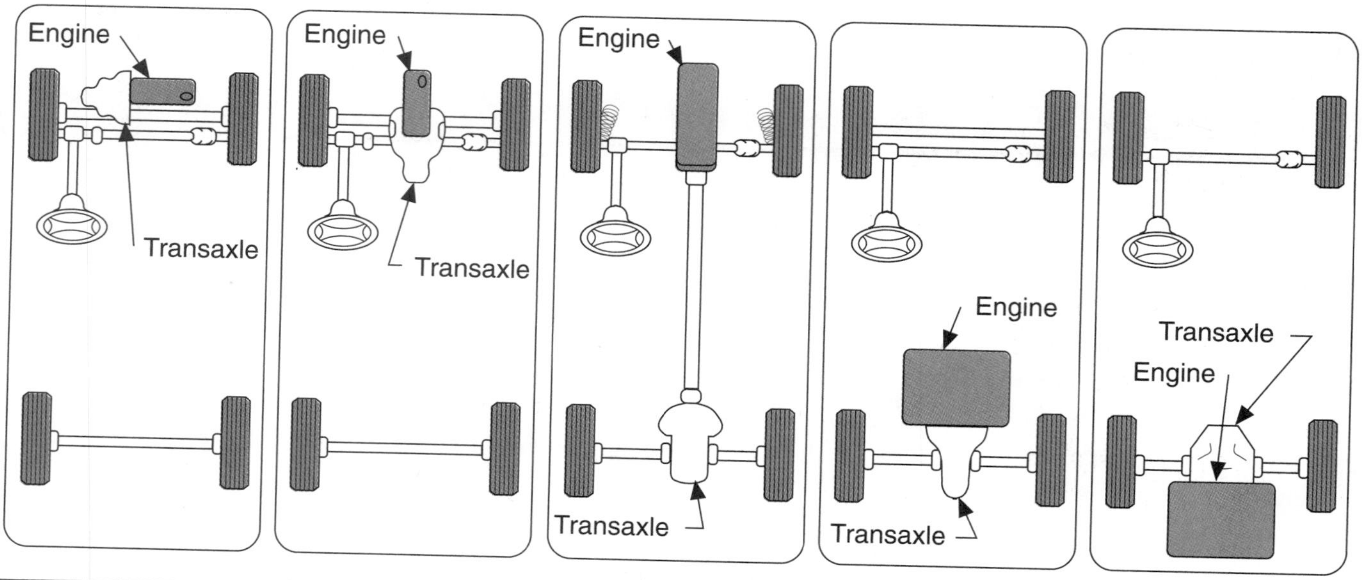

Figure 4-2 Different transaxle locations.

A transaxle is a final drive unit and transmission housed in a single unit.

A manual transaxle is a single unit composed of a manual transmission, differential, and drive axles. Most front-wheel-drive (FWD) cars are equipped with a transaxle. Transaxles are also found on some front engined and rear-wheel-drive (RWD) and four-wheel-drive (4WD) cars and on rear-engined and rear-wheel-drive cars (Figure 4-2).

With a transaxle, the engine is normally mounted transversely across the front of front-wheel-drive (FWD) cars (Figure 4-3). The compactness of a transaxle allows designers to offer increased passenger room, reduce vehicle weight, and reduce vibration and alignment problems normally caused by the long drive shafts of rear-wheel-drive cars.

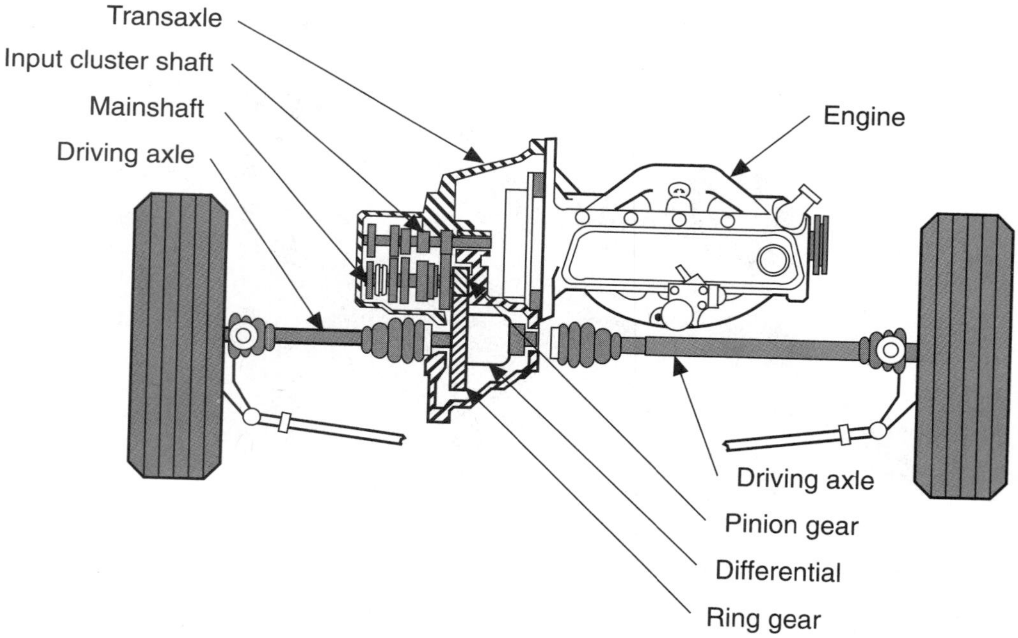

Figure 4-3 Typical drive line with a transaxle.

Types of Manual Transmissions/Transaxles

Three major types of manual transmissions have been used by the automotive industry: the sliding gear, collar shift, and synchromesh designs.

Sliding Gear Transmissions

A sliding gear transmission uses two or more shafts mounted in parallel, with sliding spur gears arranged to mesh with each other to provide a change in speed or direction. The driver moves the shifter, which in turn moves the appropriate gear into mesh with its mate. If either of the gears are rotating, the changing of gears is very difficult and will normally grind into mesh. This type of transmission is currently used only on farm and industrial machines. However, some transmissions use a sliding gear mechanism for the engagement of reverse gear.

Collar Shift Transmissions

A collar shift transmission has parallel shafts fitted with gears in constant mesh. The change of gear ratios is accomplished by locking the free-running gears to their shafts using sliding collars. The sliding collars connect and lock the appropriate gear to its shaft. One side of the gears has short splines into which the internal splines of the sliding collar mesh. Because the collar is splined to the shaft, the connection of the collar to the gear connects the gear to the rotating shaft. Although the splines on the gears and collars have rounded ends for easier shifting, gear clashing will occur whenever the gears are changed or engaged.

Free-running gears are gears that rotate independently on their shaft.

Synchromesh Transmissions

A synchromesh transmission is a constant mesh (Figure 4-4), collar shift transmission equipped with synchronizers, which equalize the speed of the shaft and gear before they are engaged. The action of the synchronizer eliminates gear clashing and allows for smooth changing of gears. Synchromesh transmissions are used on all current models of cars and are commonly found in other machines wherever shifting while moving is required.

Constant mesh implies that the teeth of the gears are always in mesh.

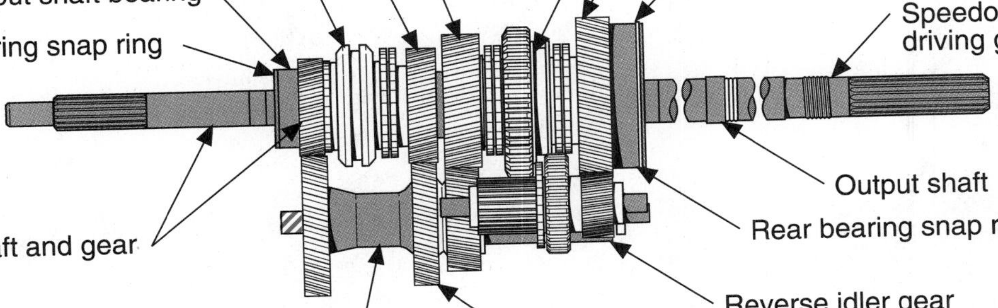

Figure 4-4 Most transmissions are of a constant mesh, fully synchronized design. The gears are all meshed and gear selection is made through the engagement and disengagement of synchronizers. (Reprinted with the permission of Ford Motor Company)

A fully synchronized transmission is a transmission in which all of its forward gears are equipped with a synchronizer assembly

The gear on the input shaft is called the input gear or the clutch gear.

The mainshaft is also referred to as the output shaft on most transmissions.

Some older transmissions and truck transmissions were not equipped with synchronizers on first or reverse gears. These gears could only be easily engaged when the vehicle was stopped.

Shop Manual
Chapter 4, page 99

Block synchronizers are commonly referred to as cone synchronizers.

A synchronizer hub is also called a clutch hub.

Engine torque is applied to the transmission's input shaft when the clutch is engaged. The input shaft enters the transmission case, where it is supported by a large ball bearing and fitted with a gear. The output shaft (mainshaft) is inserted into, but rotates independently of the input shaft. The mainshaft is supported by the input shaft bearing and a bearing at the rear of the transmission case. The various speed gears rotate on the mainshaft. Located below or to the side of the input and mainshaft assembly is a counter shaft that is fitted with several sized gears. All of these gears, except one, are in constant mesh with the gears on the mainshaft. The remaining gear is in constant mesh with the input gear.

Gear changes occur when a gear is selected by the driver and is locked or connected to the mainshaft. This is accomplished by the movement of a collar that connects the gear to the shaft. Smooth and quiet shifting can only be possible when the gears and shaft are rotating at the same speed. This is the primary function of the synchronizers.

Synchronizers

A synchronizer's primary purpose is to bring components that are rotating at different speeds to one synchronized speed. It also serves to lock these parts together. The forward gears of all current automotive transmissions are synchronized. A single synchronizer is placed between two different speed gears, therefore transmissions have two or three synchronizer assemblies. Reverse gear is not normally synchronized because gear rotation is required for synchronizer action and reverse is normally selected when the car is not moving.

Synchronizer Designs

There are four types of synchronizers used in synchromesh transmissions: block, disc and plate, plain, and pin. The most commonly used type on current transmissions is the block type. All synchronizers use friction to synchronize the speed of the gear and shaft before the connection is made.

Block synchronizers consist of a hub, sleeve, blocking ring, and inserts or spring-and-ball detent devices (Figure 4-5). The synchronizer sleeve surrounds the synchronizer assembly and meshes with the external splines of the hub. The hub is internally splined to the transmission's mainshaft. The outside of the sleeve is grooved to accept the shifting fork. Three slots are equally spaced around the outside of the hub and are fitted with the synchronizer's inserts or spring-and-ball detent assemblies.

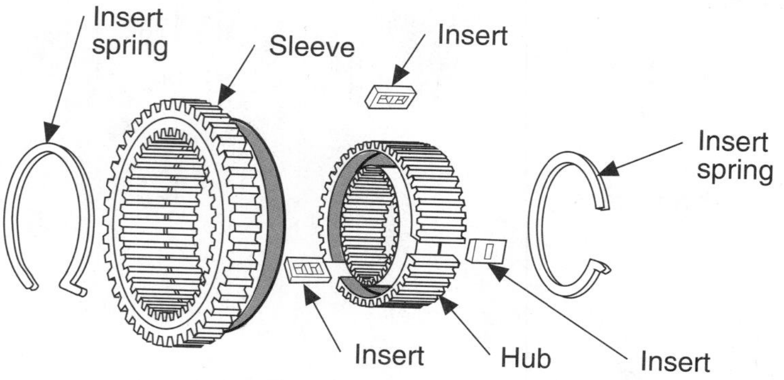

Figure 4-5 Typical block synchronizer assembly. (Reprinted with the permission of Ford Motor Company)

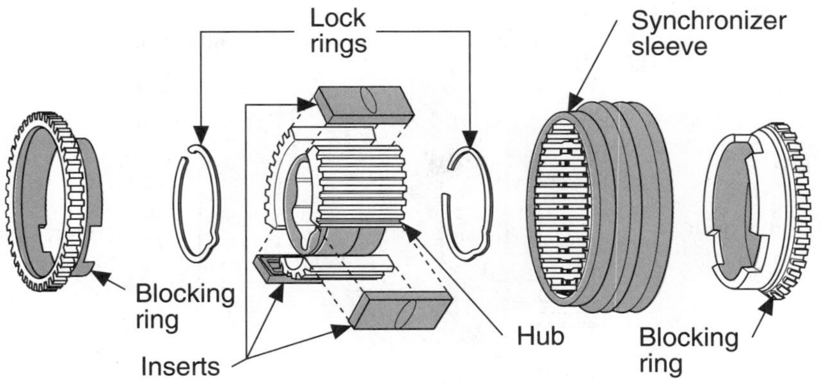

Figure 4-6 The notches in the blocking ring correspond with the spacing of the inserts.

These inserts are able to freely slide back and forth in the slots. The inserts are designed with a ridge on their outer surface and insert springs hold this ridge in contact with an internal groove in the synchronizer sleeve. When the transmission is in the "neutral" position, the inserts keep the sleeve lightly locked into position on the hub. If the synchronizer assembly uses spring-and-ball detents, the balls are held in this groove by their spring. The sleeve is machined to allow it to slide smoothly on the hub.

Bronze, brass, fiber faced blocking rings are positioned at the front and rear of each synchronizer assembly. Each blocking ring has three notches equally spaced to correspond with the three insert keys (Figure 4-6). Around the outside of the blocking ring is a set of beveled dog teeth. These teeth are used for alignment during the shift sequence. The inside of the blocking ring is shaped like a cone, the surface of which has many sharp grooves. These inner surfaces of the blocking rings match the conical shape of the shoulders of the driven gear. These cone-shaped surfaces serve as the frictional surfaces for the synchronizer. The shoulder of the gear also has a ring of beveled dog teeth designed to align with the dog teeth on the blocking ring (Figure 4-7).

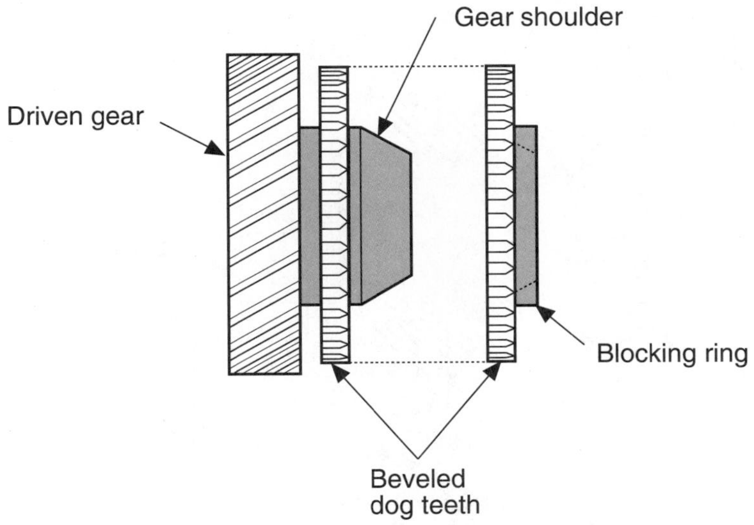

Figure 4-7 Beveled dog teeth on the speed gear and the blocking ring.

Synchronizer inserts are often referred to as keys.

Because of the high heat produced by the friction, blocking rings are made of brass or bronze. The use of these metals minimizes the wear on the hardened steel gear's cone. A common trend in the automotive industry today is to reduce frictional losses as much as possible. The materials used to manufacture blocking rings has not been overlooked. Powdered metal and organic frictional materials are currently being used and/or being tested by a few manufacturers.

The term *dog teeth* is used to describe a set of gear teeth that has gaps between the teeth and are not close together.

The term *synchro* is a commonly used slang word for synchronizer.

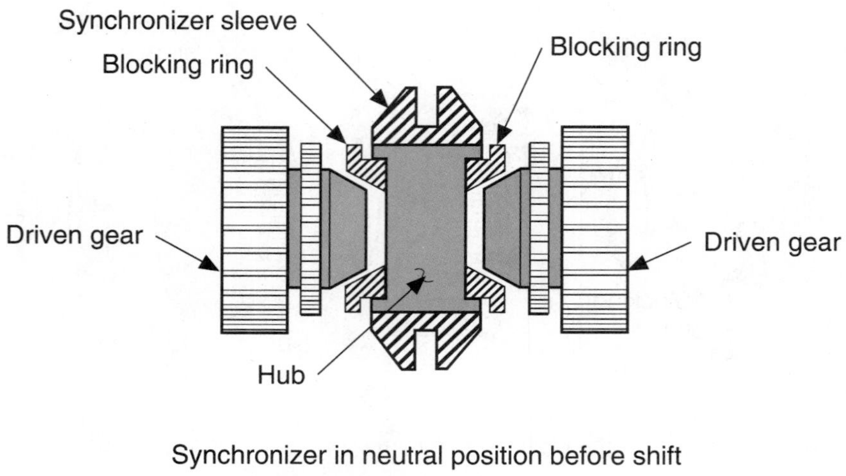

Figure 4-8 A block synchronizer in its neutral position.

Synchronizer Operation

When the transmission is in neutral (Figure 4-8), the synchronizers are in their neutral position and are not rotating with the mainshaft. The mainshaft's gears are meshed with the counter gears and are rotating with the counter shaft. However, they turn freely, at various speeds, on the mainshaft and do not cause the shaft to rotate because they are not connected to it.

When a gear is selected, the shifting fork forces the sleeve toward the selected gear. As the sleeve moves, so do the inserts because they are locked in the sleeve's internal groove. The movement of the inserts pushes the blocking ring into contact with the shoulder of the driven gear. When this contact is made, the grooves on the blocking ring's cone cut through the film of lubrication on the gear's shoulder. If the film of lubrication is not cut by the grooves, synchronization cannot take place. Destroying the film allows for metal-to-metal contact and begins the speed synchronization of the two parts. (Figure 4-9). The resultant friction between the two brings the gear's cone to the blocking ring cone's speed.

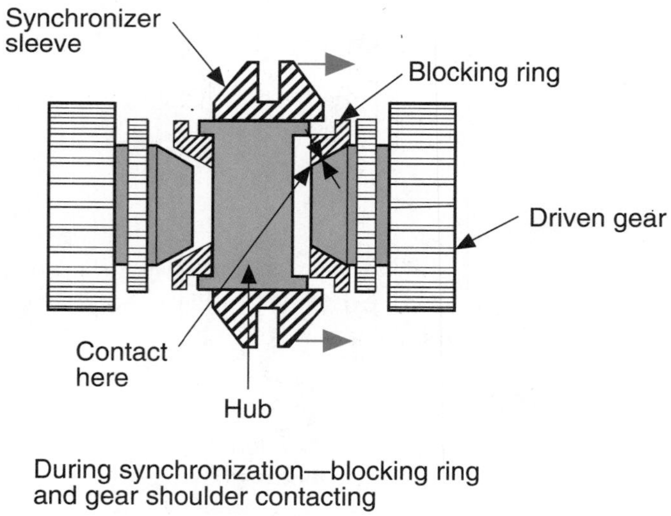

Figure 4-9 Initial contact of the blocking ring to the gear's cone.

As the components reach the same speed, the synchronizer sleeve can now slide over the external dog teeth on the blocking ring and then over the dog teeth on the speed gear's shoulder (Figure 4-10). This completes the engagement of the synchronizer and the gear is now locked to the mainshaft. The blocking ring will only allow the sleeve to mesh with the gear when its teeth are lined up with the locking teeth on the gear.

Power now flows from input gear to the counter gear, back up to the speed gear locked by the synchronizer. Power then flows from the gear through the locking teeth to the sleeve, then to the hub, and finally to the mainshaft.

To disengage a gear, the shifter is moved to the neutral position, which causes the synchronizer sleeve to move away from the previous gear, thereby disconnecting it from the shaft.

In summary, synchronization occurs in three stages. In the first stage, the sleeve is moved toward the gear by the shift lever and engages the hub assembly. In the second stage, the movement of the sleeve causes the inserts to press the blocking ring onto the cone of the gear. In the third stage, the synchronizer ring completes its friction fit over the gear cone and the gear is brought up to the same speed as the synchronizer assembly. The sleeve slides onto the gear's teeth and locks the gear and its synchronizer assembly to the mainshaft.

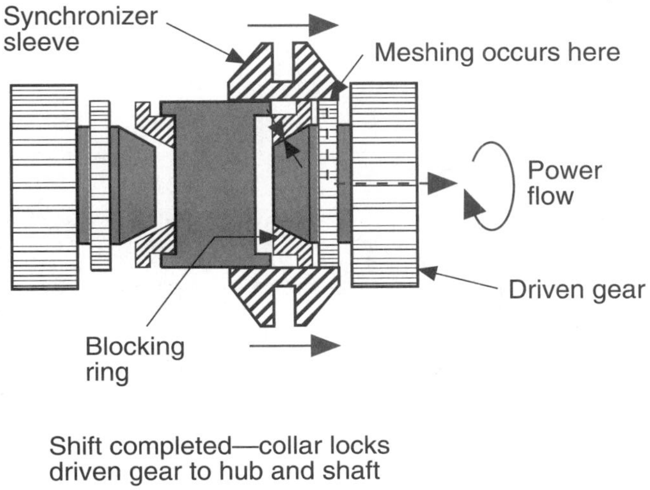

Figure 4-10 Shift collar movement locks the gear to the synchronizer hub.

Transmission Designs

All automotive transmissions/transaxles are equipped with a varied number of forward speed gears and one reverse speed. Transmissions can be divided into groupings according to the number of forward gears (speeds) it has. For many years, the three-speed manual transmission was the most commonly used; four-speed transmissions were found only in heavy-duty or utility vehicles and high-performance cars. The growing concern for improved fuel mileage led to smaller engines with four-speed transmissions. The additional gear allowed the smaller engines to perform better by matching the engine's torque curve with vehicle speeds.

Currently, five-speed transmissions are common. Some of these units are actually four-speeds with an additional overdrive gear. This fifth gear reduces engine rpm at highway speeds and therefore increases its fuel economy. Other five- and six-speed transmissions incorporate fifth and sixth gears in their main assembly.

Three-Speed Transmissions

In a typical three-speed transmission, one synchronizer assembly shifts first gear and reverse and another synchronizer shifts second and third gears (Figure 4-11). Actually, there is no third gear. Instead of a separate gear, the input shaft and the mainshaft are coupled together by the synchronizer. This gives a 1:1 gear ratio. The benefit of directly coupling the shafts is greater efficiency. A direct coupling transmits about 99% of its received torque, and a gear coupling is about 96% efficient.

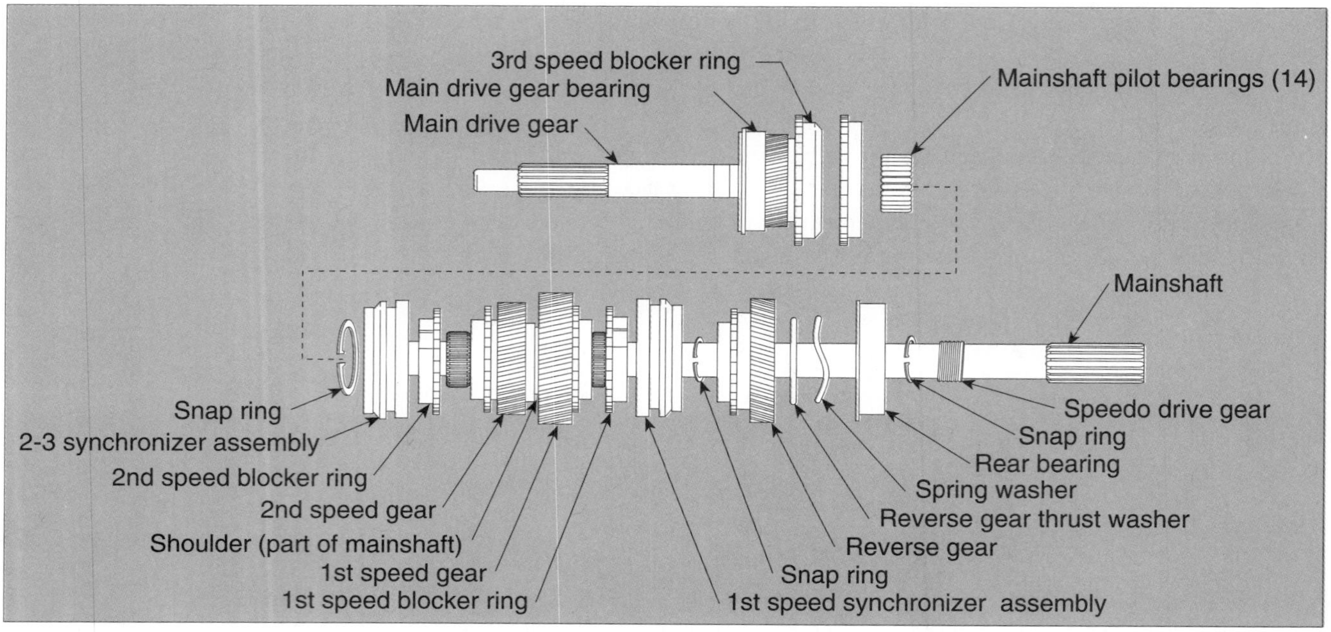

Figure 4-11 Typical three-speed transmission. (Courtesy of Oldsmobile Div.—GMC)

The locking teeth on a gear are also called the gear's dog or clutch teeth.

When a three-speed transmission is shifted out of neutral into first gear, the first/reverse synchronizer sleeve slides onto the first gear's locking teeth. When it is shifted into second, the first/reverse sleeve slides into its neutral position, and the second/third sleeve slides onto the second gear's locking teeth. When third is selected, the second/third sleeve moves off second gear, slides through neutral, and engages the locking teeth on the input shaft. To engage reverse, the second/third synchronizer is pulled into its neutral position. To engage reverse gear, the clutch must be disengaged to stop the turning of the input and counter gears. To stop the mainshaft from turning, the vehicle must be brought to a complete stop. Then the first/reverse sleeve can be pushed onto reverse gear's locking teeth. All shafts must be stopped because there is no synchronizing action on the reverse gear.

Four-Speed Transmissions

Fourth gear is commonly called "top" or "high" gear.

Fourth gear is a 1:1 ratio like the third gear of a three-speed transmission. The additional gear is used to provide an extra speed ratio in the first and second gear range of the three-speed gearbox (Figure 4-12). Synchronizer assemblies are placed between first and second gears and between third and fourth gears. A third hub and sleeve assembly is added to shift the reverse gear. Some designs use a sliding reverse gear that is moved by the shift fork until it meshes with the reverse idler gear.

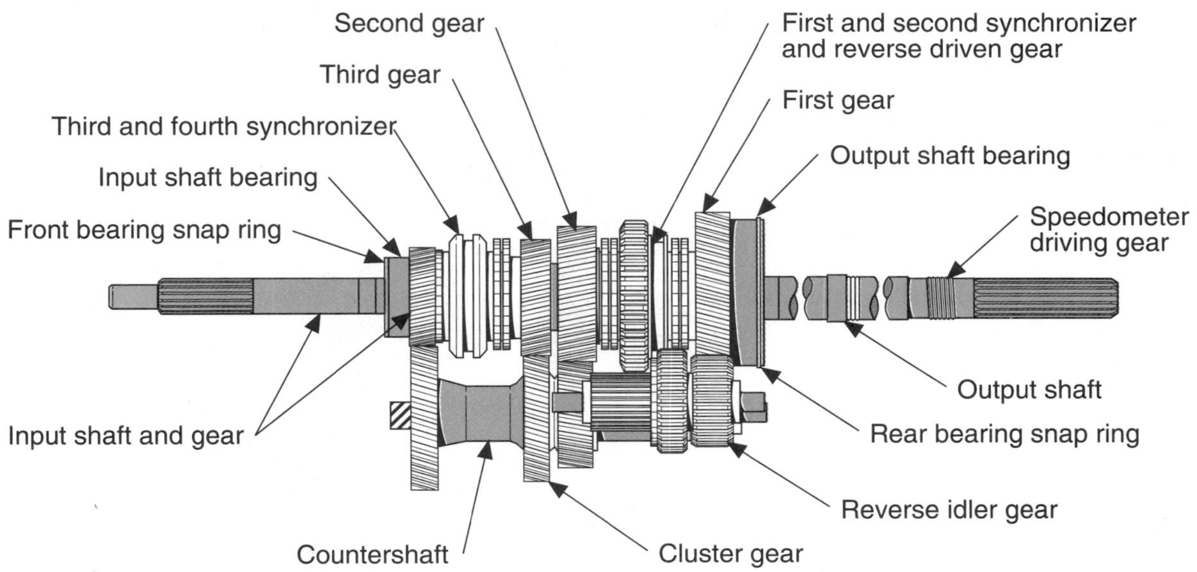

Figure 4-12 Typical four-speed transmission. (Reprinted with the permission of Ford Motor Company)

Five-Speed Transmissions

Shop Manual
Chapter 4, page 99

A five-speed transmission is usually a four-speed plus an overdrive gear. The fifth gear is added to the rear of the mainshaft, near the reverse gear (Figure 4-13). The hub-and-sleeve shifting assembly used only for reverse in the four-speed transmission becomes the fifth/reverse synchronizer. The 1:1 ratio of fourth gear is usually retained, and fifth gear is made into an overdrive. Typical fifth-gear overdrive ratios range from 0.70 to 0.90:1. Such ratios greatly reduce engine rpm at freeway speeds and increase fuel mileage and engine life.

Not all transmissions are designed exactly as described here. Some four-speed transmissions have overdrive fourth gears; thus, they actually have a fourth gear. Some five-speeds have two overdrive gears, and therefore have five gears, not four.

Overdrive Units

Overdrive gears can be an integral part of a transmission, as in a four- or five-speed transmission, or they can be add-on units. Overdrive units were popular before the widespread use of four- and five-speed transmissions. They gave three-speed manual transmissions a freeway cruising gear for higher top speeds and better fuel mileage. Currently a few high performance cars are equipped with five- or six-speed transmissions with an overdrive unit. These units allow the driver to select a mode of operation, between fuel efficiency and power. While operating in the fuel efficient mode, the transmission operates normally but the torque on the output shaft is reduced by the overdrive unit as it flows to the drive wheels. The unit effectively changes the gear ratio of all gears in the transmission. When the power mode is selected, the overdrive unit is disengaged and the torque on the output shaft is not altered as it flows to the drive wheels.

A BIT OF HISTORY

The first American car to use a four-speed gear box was the Locomobile in 1903.

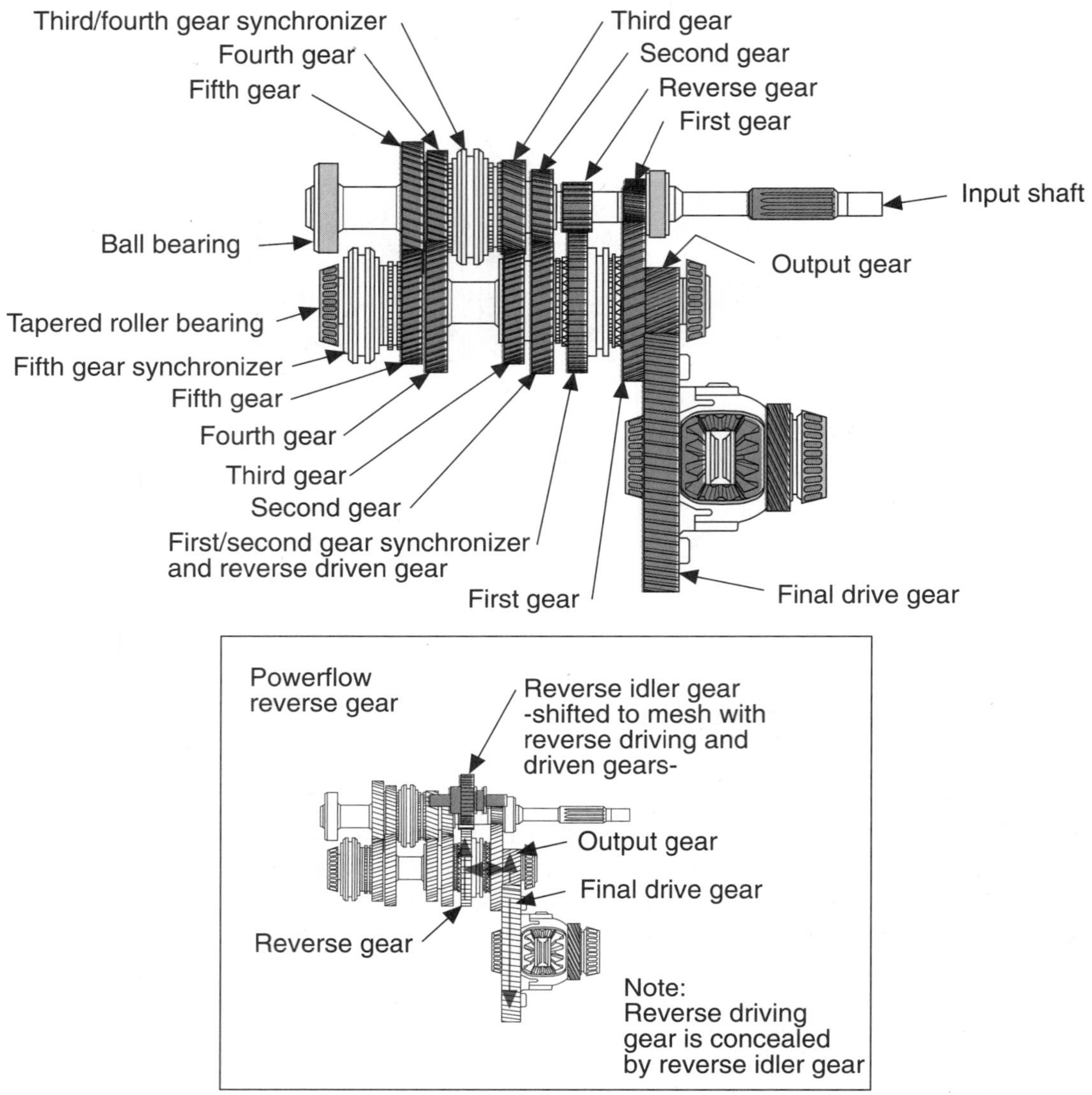

Figure 4-13 Typical five-speed transaxle. (Reprinted with the permission of Ford Motor Company)

Classroom Manual
Chapter 4, page 65

Shop Manual
Chapter 4, page 97

Basic Operation of Manual Transmissions

All manual transmissions function in much the same way and have similar parts. (Although transaxles operate in basically the same way as transmissions, there are enough differences to warrant a separate discussion. The operation of transaxles is explained after this section on transmissions.) Before you can completely understand how a transmission works, you must be familiar with the names, purposes, and descriptions of the major components of all transmissions.

The transmission case is normally a cast aluminum case that houses most of the gears in the transmission and has machined surfaces for a cover, rear extension housing, and mounting to the clutch's bell housing (Figure 4-14).

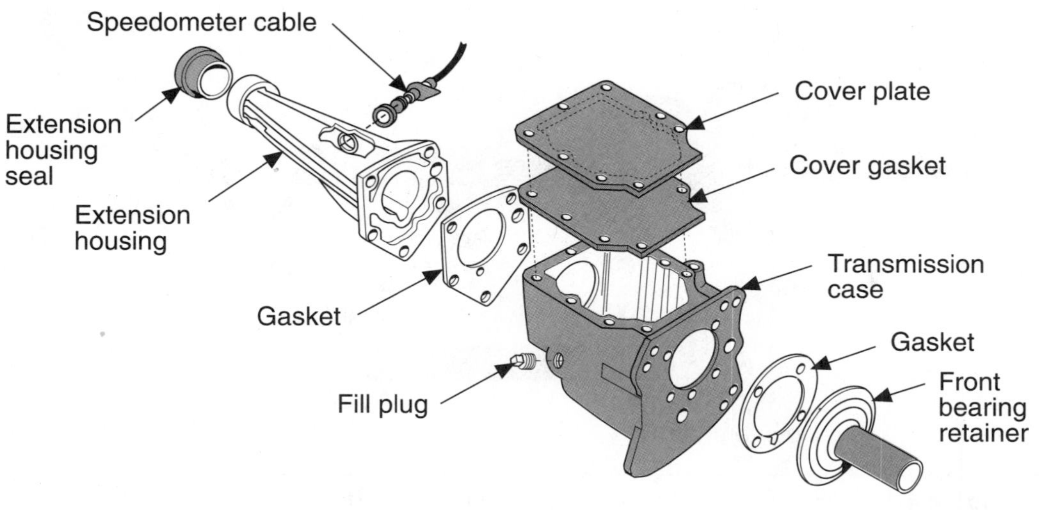

Figure 4-14 Typical transmission case with attachments. (Reprinted with the permission of Ford Motor Company)

The front bearing retainer is a cast-iron piece bolted to the front of the transmission case. The retainer serves several functions: It houses an oil seal that prevents oil from leaking out the input shaft and into the clutch disc area, it holds the input shaft bearing rigid, and because the input shaft is pressed into the bearing, it prevents all undesired movements of the input shaft (Figure 4-15).

The counter gear assembly is normally located in the lower portion of the transmission case and is constantly in mesh with the input gear. The counter shaft has several different sized gears on it, that rotate as one solid assembly. Normally the counter gear assembly rotates on the counter shaft using several rows of roller or needle bearings (Figure 4-16).

The counter gear assembly is often referred to as the cluster gear assembly.

The speed gears are located on the mainshaft (Figure 4-17). These gears are not fastened to the mainshaft and rotate freely on the mainshaft journals. The gears are constantly in mesh with the counter gears and are turned by the counter gears. A worm gear is machined into or pressed onto the rear of the mainshaft to drive a speedometer pinion gear. The outer end of the mainshaft has splines for the slip-joint yoke of the drive shaft.

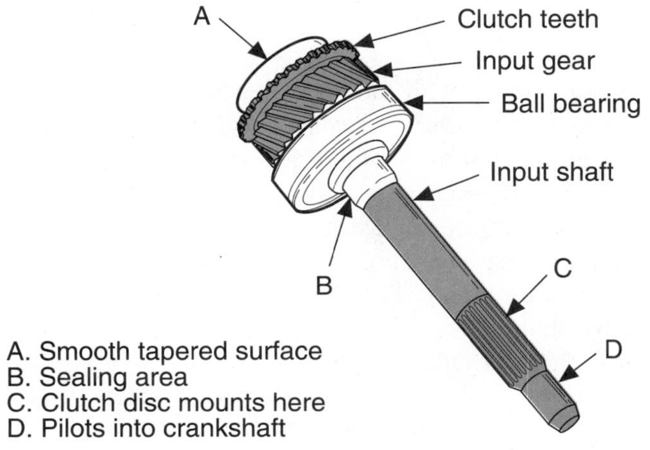

Figure 4-15 Typical input shaft assembly. (Courtesy of Chrysler Corporation)

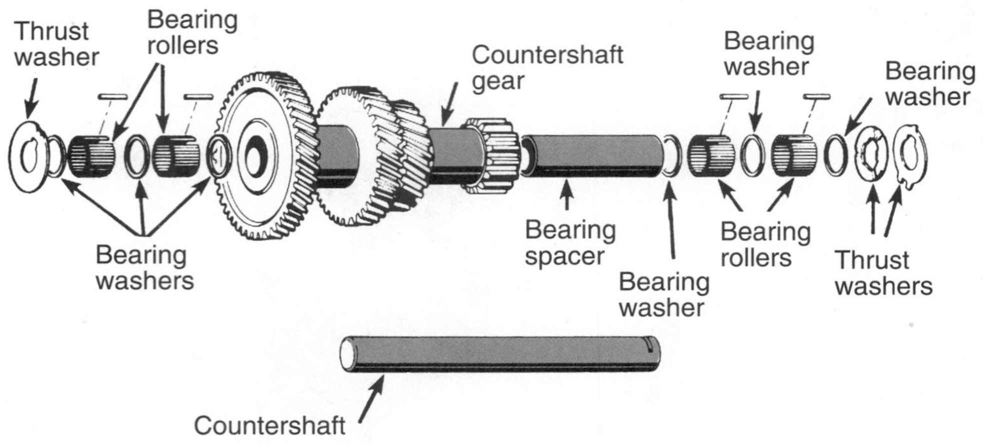

Figure 4-16 Typical counter shaft assembly. (Courtesy of Chrysler Corporation)

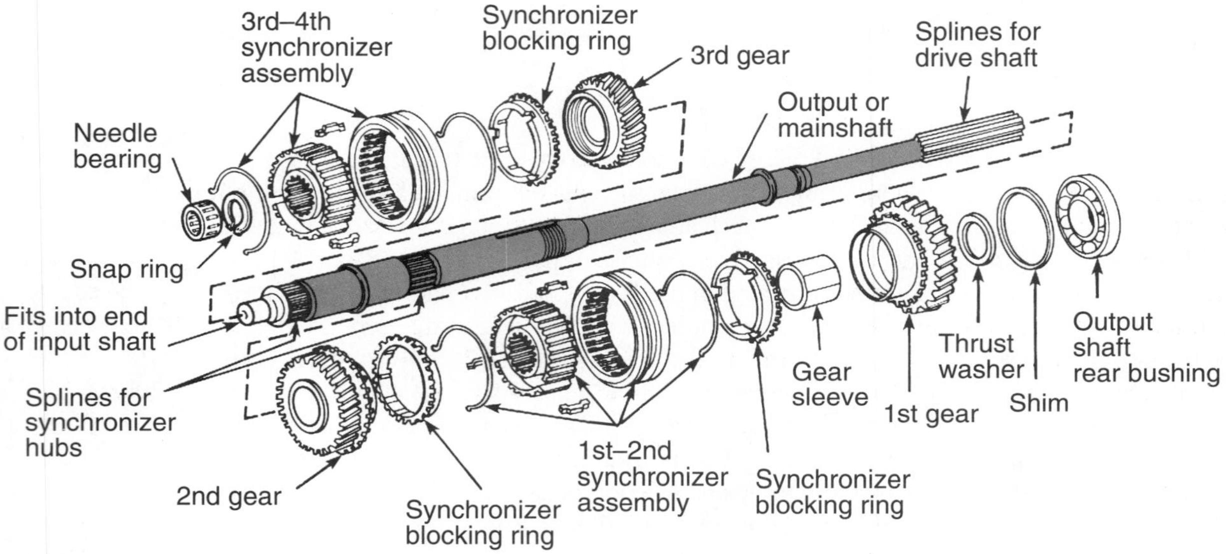

Figure 4-17 Typical main shaft assembly. (Courtesy of Mazda)

The name reverse idler gear defines its purpose. This gear has no effect on the speed or torque of the reverse gear, it simply causes the direction of rotation to change.

A reverse gear is not meshed with the counter gear like the forward gears; rather, the reverse idler gear is (Figure 4-18). Normally, reverse gear is engaged by sliding it into mesh with the reverse idler gear. The addition of this third gear causes the reverse gear to rotate in the opposite direction as the forward gears.

Shift forks move the synchronizer sleeves to engage and disengage gears. Most shift forks have two fingers that ride in the groove on the outside of the sleeve. The forks are bored to fit over shift rails. Tapered pins are commonly used to fasten the shift forks to the rails (Figure 4-19). Each of the shift rails have shift lugs that the shift lever fits into. These lugs are also fastened to the rails by tapered pins. The shift rails slide back and forth in bores of the transmission case. Each shift rail has three notches, which a spring-loaded ball or bullet rides in to give a detent feel to the shift lever and locate the proper position of the shift fork during gear changes. The shift rails also have notches cut in their sides for interlock plates or pins to fit in (Figure 4-20). Interlock plates prevent the engagement of two gears at the same time. The lower portion of the shifter assembly fits into the shift rail lugs and moves the shift rails for gear selection.

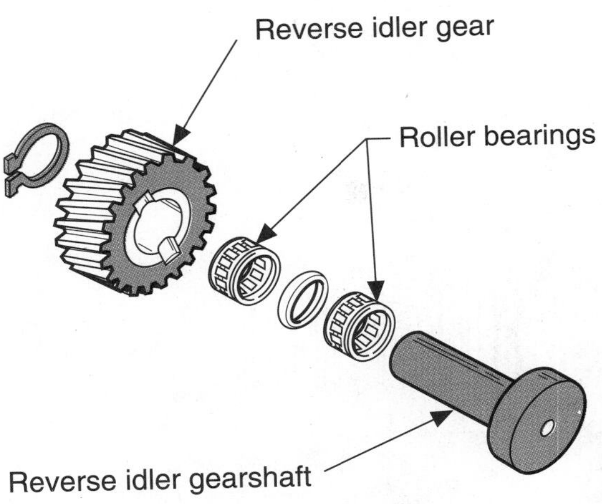

Figure 4-18 Reverse idler gear assembly. (Courtesy of Chrysler Corporation)

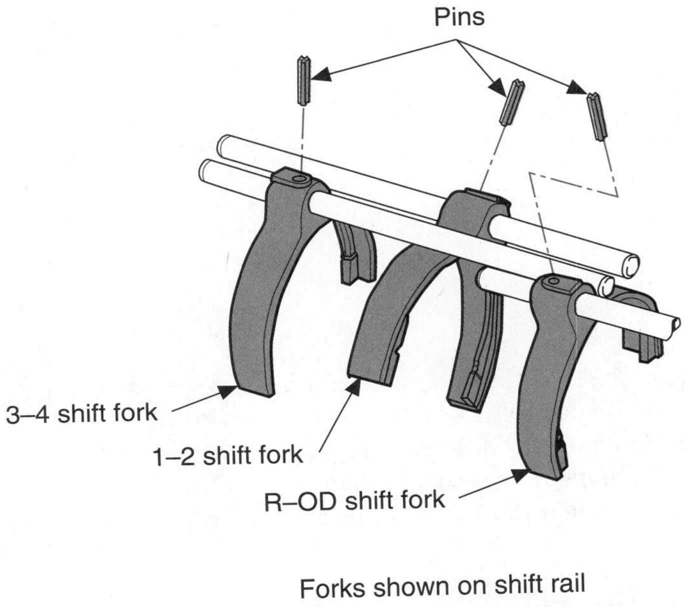

Figure 4-19 Typical shift fork and rail arrangement. (Courtesy of Chrysler Corporation)

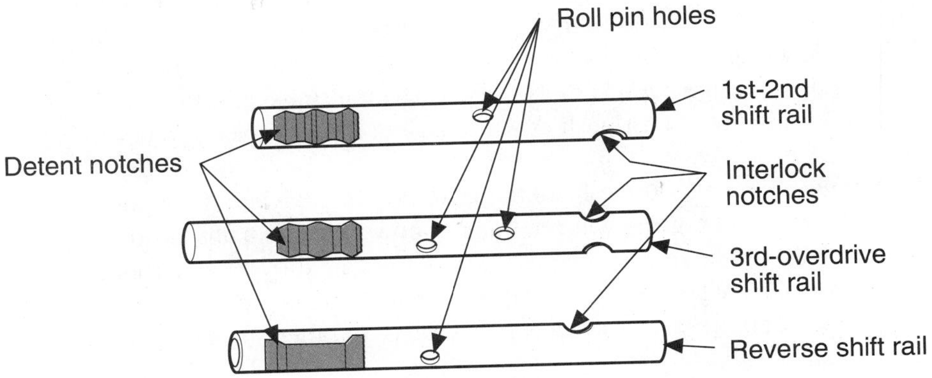

Figure 4-20 Notches in typical shift rails. (Reprinted with the permision of Ford Motor Company)

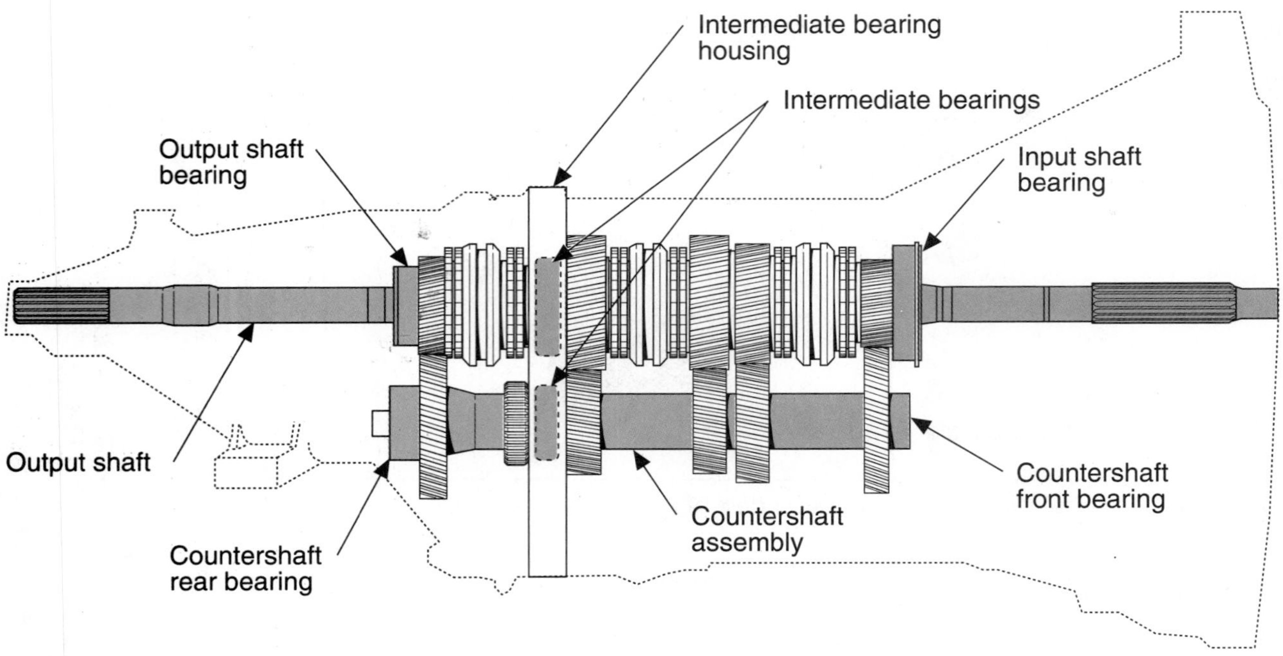

Figure 4-21 Location of various bearings in a typical transmission. (Reprinted with the permision of Ford Motor Company)

The arrangement of the shafts in a transmission require bearings to support them. The ends of the shafts are fitted with large roller or ball bearings pressed onto the shaft. Some transmissions with long shafts use an intermediate bearing, which is fitted into the intermediate bearing-housing, to give added strength to the shaft (Figure 4-21). Small roller or needle bearings are often used on the counter shaft, reverse idler gear shaft, and at the connection of the output shaft to the input shaft.

Operation of a Five-Speed Transmission

In a typical five-speed transmission, all five forward helical gear assemblies are in constant mesh. They are activated by the first/second speed synchronizer, the third/fourth speed synchronizer, and the fifth speed synchronizer. Each synchronizer is activated by its own shift fork. All three shifter forks slide along the transmission's single shifter rail. Only the reverse idler gear slides along another rail to engage the reverse spur gear. The transmission's floor-mounted shift lever is spring loaded, so the driver must push down or pull up on the lever in order to engage reverse gear.

With the gears in constant mesh, all the gears will rotate when the input shaft is supplying input power. However, the gears will not transfer power to the mainshaft until one of the synchronizers engages with a gear. If the gears are not engaged by a synchronizer, they are free-wheeling on the mainshaft. The individual output shaft gears are mechanically locked to the output shaft only when the synchronizers are activated. At other times they rotate independently of the output shaft.

The counter gear assembly is a one-piece machined unit, containing first, second, third, and fourth counter gears. The fifth speed counter gear is often a separate assembly that is splined to the counter shaft.

Power Flow in Neutral

In the neutral position (Figure 4-22), the input shaft drives the counter shaft gears but no power is transferred out of the transmission. Because they are in mesh, all of the gears on the mainshaft rotate, but power is not transferred to the output shaft because the synchronizers are not engaged with any of the gears.

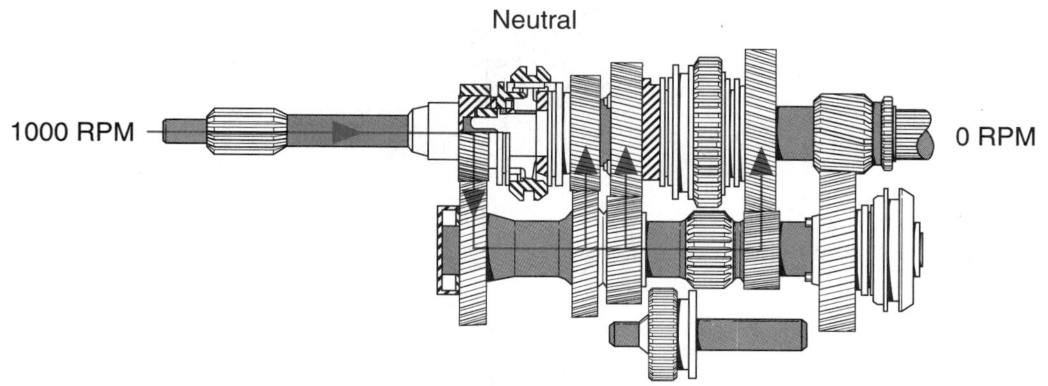

Figure 4-22 Power flow through a five-speed transmission when it is in neutral.

Power Flow in First Gear

In the first gear (Figure 4-23), power enters the transmission through the input shaft and rotates the counter shaft gear. The first/second synchronizer sleeve is engaged with the dog teeth on the first-speed gear, locking the gear to the mainshaft. The power coming in the input shaft transfers through the counter gear and up into the first gear. The gear rotates the synchronizer sleeve, which rotates the hub and mainshaft for output power. All other gears mounted on the mainshaft rotate freely.

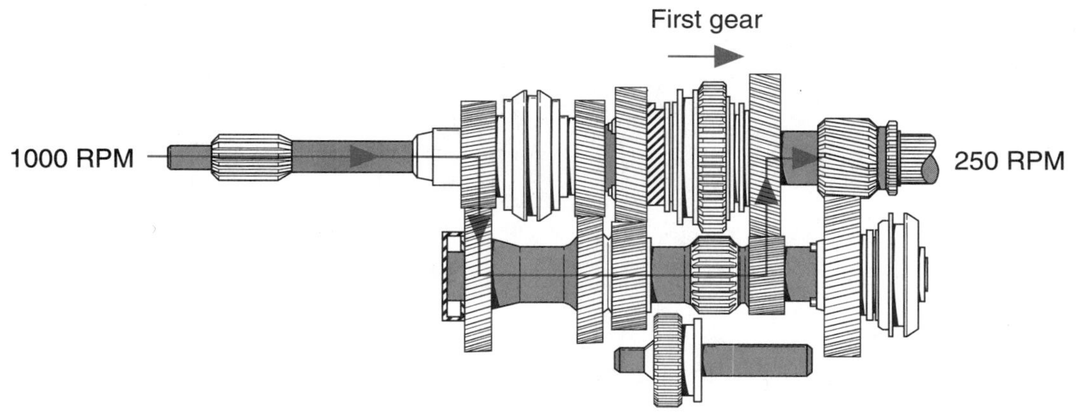

Figure 4-23 Power flow in first gear.

Power Flow in Second Gear

In the second-gear position (Figure 4-24), the input shaft again drives the counter shaft gear. The first/second synchronizer sleeve is moved to engage with the dog teeth of the second gear, lock ing it to the output shaft. Power comes in through the input shaft, down to the counter gear and up to the second gear. The dog teeth of the second gear rotate the synchronizer sleeve, which rotates the hub and mainshaft for output power.

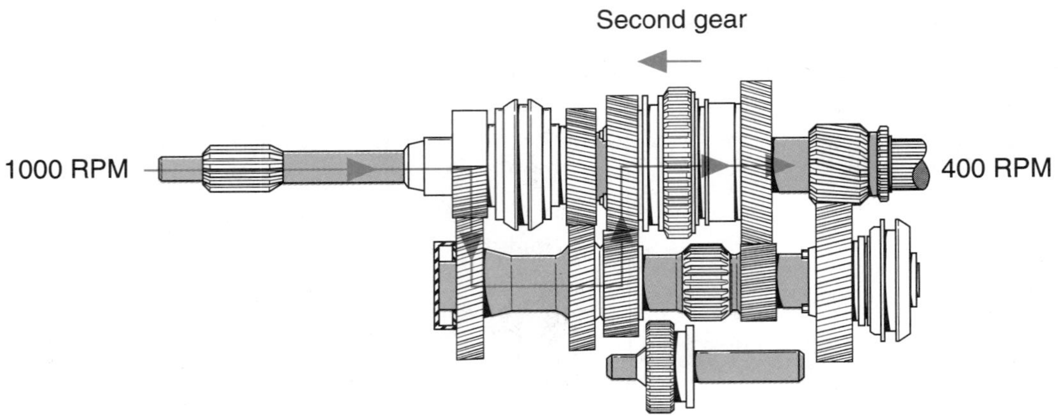

Figure 4-24 Power flow in second gear.

Power Flow in Third Gear

The third-gear position causes the counter shaft gear, which is driven by the input shaft, to be mechanically locked to the third gear on the output shaft (Figure 4-25). The third/fourth synchronizer sleeve is moved to engage with the dog teeth of the third gear. The power comes from the input shaft to the counter gear and then to the third gear. The dog teeth on the third gear rotate the synchronizer sleeve, which rotates the hub and mainshaft for output power.

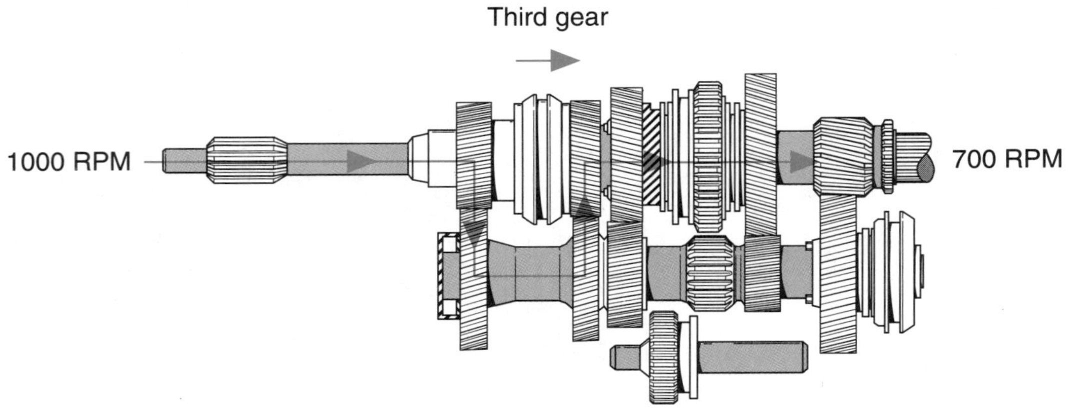

Figure 4-25 Power flow in third gear.

Power Flow in Fourth Gear

The fourth-gear position mechanically locks the output shaft to the input shaft (Figure 4-26). The third/fourth synchronizer sleeve is moved to engage with the dog teeth of the input gear. The power flows in from the input shaft, through the synchronizer sleeve and hub, and then through the mainshaft for output power. This directly links the two shafts, and the output shaft rotates at the same speed as the input shaft to provide for direct drive.

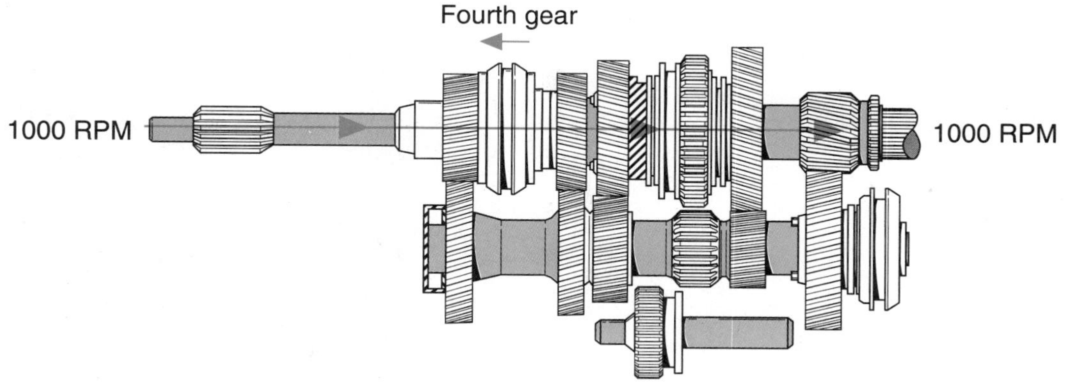

Figure 4-26 Power flow in fourth gear.

Power Flow in Fifth Gear (Overdrive)

The fifth-gear position causes the counter shaft gear, which is driven by the input shaft, to rotate the fifth gear (Figure 4-27). The fifth-gear synchronizer sleeve is moved to engage with the dog teeth on the fifth gear. The power on the input gear transfers to the counter gear and then to the fifth gear. Power transfers through the synchronizer sleeve and hub to the mainshaft for output power. As a result, the output shaft rotates at a higher speed than the input shaft.

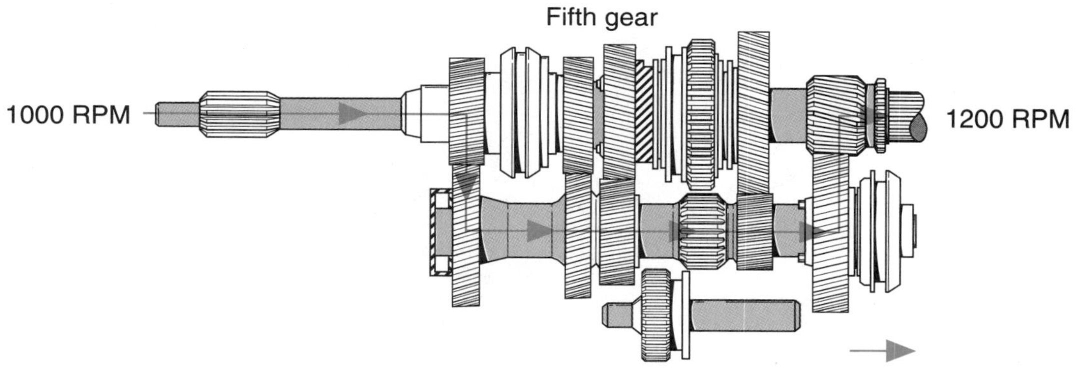

Figure 4-27 Power flow in fifth gear.

Power Flow in Reverse Gear

In the reverse-gear position, if the transmission has a synchronized reverse gear, the reverse synchronizer sleeve moves to engage with the reverse gear. Power comes in through the input shaft, into the counter gear, through the reverse idler gear, and into the reverse gear on the mainshaft. The reverse gear rotates the synchronizer sleeve, which rotates the hub and mainshaft in a reverse direction.

If reverse is a nonsynchronized gear, a reverse-gear shift relay lever slides the reverse idler gear into contact with the counter shaft reverse gear and the reverse gear on the output shaft (Figure 4-28). The reverse idler gear causes the reverse output shaft gear to rotate counterclockwise. The counter shaft rotates counterclockwise and causes the output shaft to rotate clockwise to produce forward motion of the vehicle.

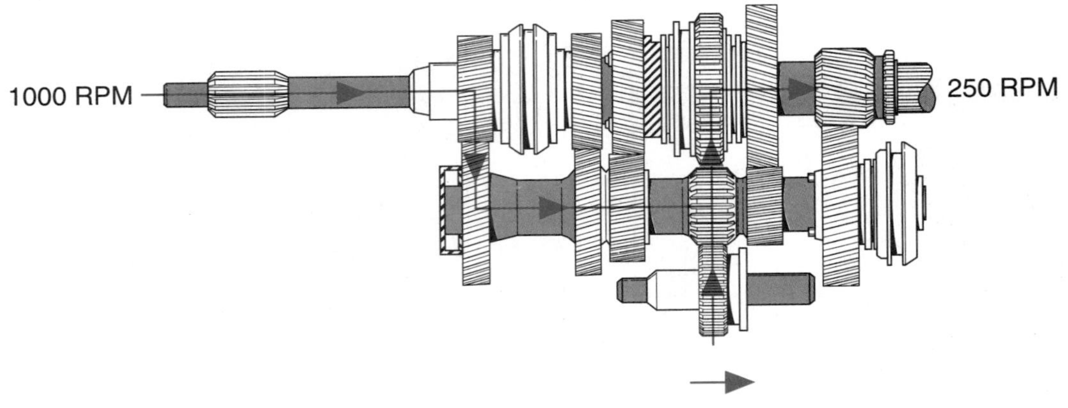

Figure 4-28 Power flow in reverse gear.

Gearshift Linkages

Shop Manual
Chapter 4, page 92

Shift rails are machined with interlock and detent notches. The interlock notches prevent the selection of more than one gear during operation. The detent notches hold the transmission in the selected gear.

There are two designs of gearshift linkages: Some are internal and others are external to the transmission. Internal linkages are located at the side or top of the transmission housing (Figure 4-29). The control end of the shifter is mounted inside the transmission, as are all of the shift controls. Movement of the shifter moves a shift rail and shift fork toward the desired gear and moves the synchronizer sleeve to lock the speed gear to the shaft.

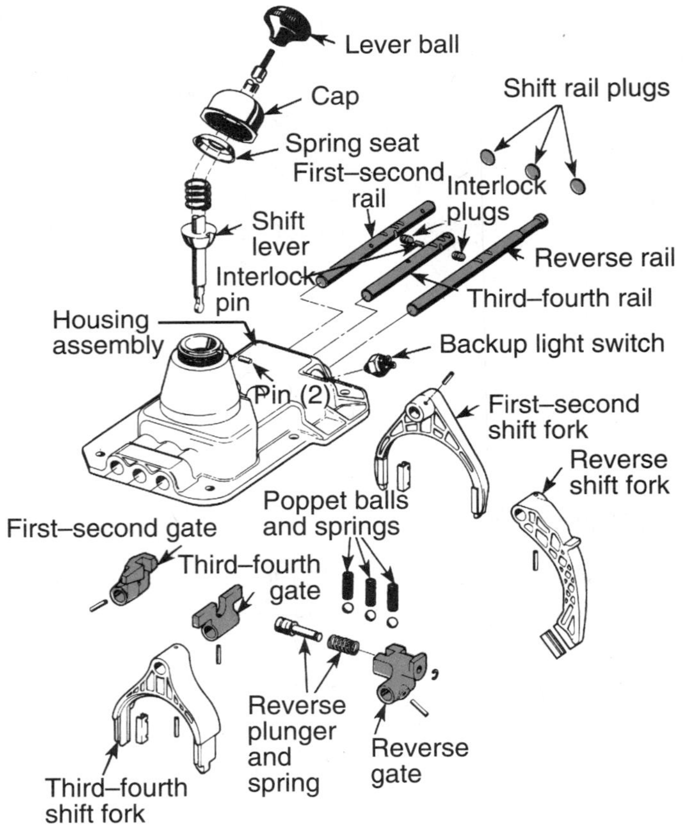

Figure 4-29 Typical internal shift linkage design. (Courtesy of Chrysler Corporation)

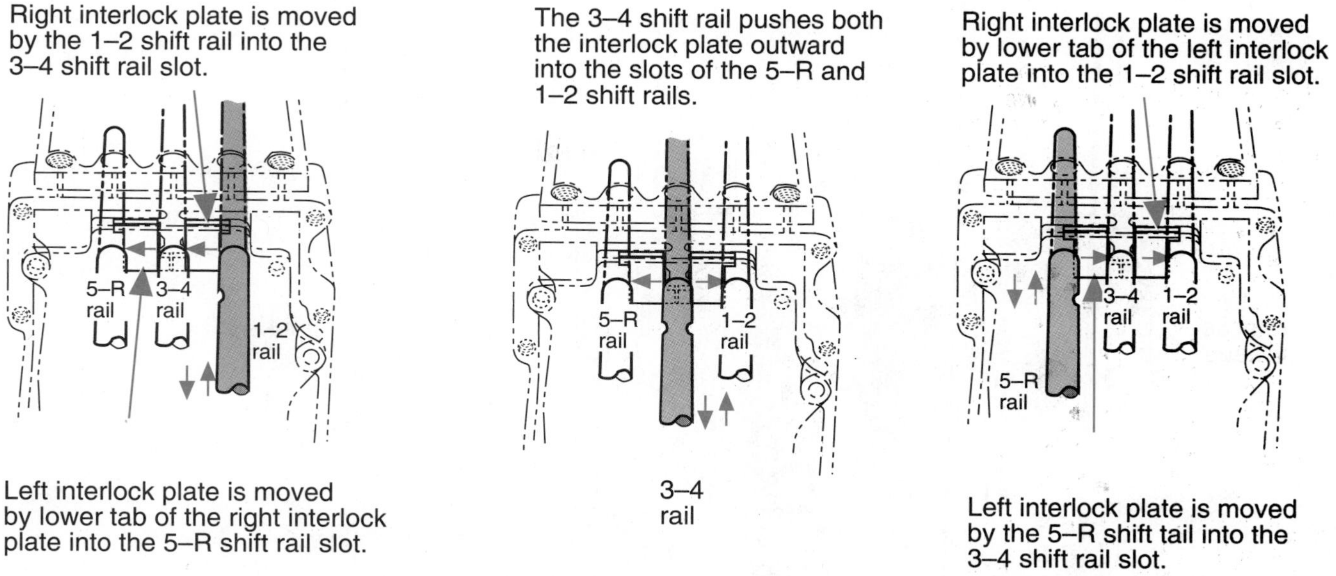
Right interlock plate is moved by the 1–2 shift rail into the 3–4 shift rail slot.
5–R rail
3–4 rail
1–2 rail
Left interlock plate is moved by lower tab of the right interlock plate into the 5–R shift rail slot.
The 3–4 shift rail pushes both the interlock plate outward into the slots of the 5–R and 1–2 shift rails.
5–R rail
1–2 rail
3–4 rail
Right interlock plate is moved by lower tab of the left interlock plate into the 1–2 shift rail slot.
3–4 rail
1–2 rail
5–R rail
Left interlock plate is moved by the 5–R shift tail into the 3–4 shift rail slot.

Figure 4-30 Operation of interlock plates. (Courtesy of Chrysler Corporation)

As the rail moves, a detent ball moves out of its detent notch and drops into the notch for the selected gear as the rail moves. At the same time, an interlock pin moves out of its interlock notch and into the other shift rails (Figure 4-30).

External linkages function in much the same way except rods, external to the transmission, act on levers connected to the internal shift rails of the transmission.

Basic Transaxle Operation

The transmission section of a transaxle is practically identical to RWD transmissions. It provides for torque multiplication, allows for gear shifting, and is synchronized. They also use many of the design and operating principles found in transmissions. However, a transaxle also contains the differential gear sets and the connections for the drive axles (Figure 4-31).

Shop Manual
Chapter 4, page 99

Transaxles normally use fully synchronized, constant mesh helical gears for all forward speeds and spur gears for reverse. To keep a transaxle compact, many designs use pressed-fit synchronizer hubs and narrower gears. Transaxles differ from RWD transmissions in that the cluster gear assembly is eliminated. The input shaft's gears drive the output shaft gears directly. The output shaft rotates according to each synchro-activated gear's operating ratio. The shafts usually ride on roller, tapered roller, or ball-type bearings (Figure 4-32). Shaft endplay is normally controlled by using thrust washers or by the placement of shims or spacers between the end plate and case.

A BIT OF HISTORY

Not many years ago, a new convenience for the driver was introduced: a column mounted gear shifter. For many years the shifter was mounted to the floor and took up valuable space in the interior of cars. By mounting the shifter on the column, that floor space was reclaimed. This trend was reversed in the 1960s and today column shifted vehicles are a rarity.

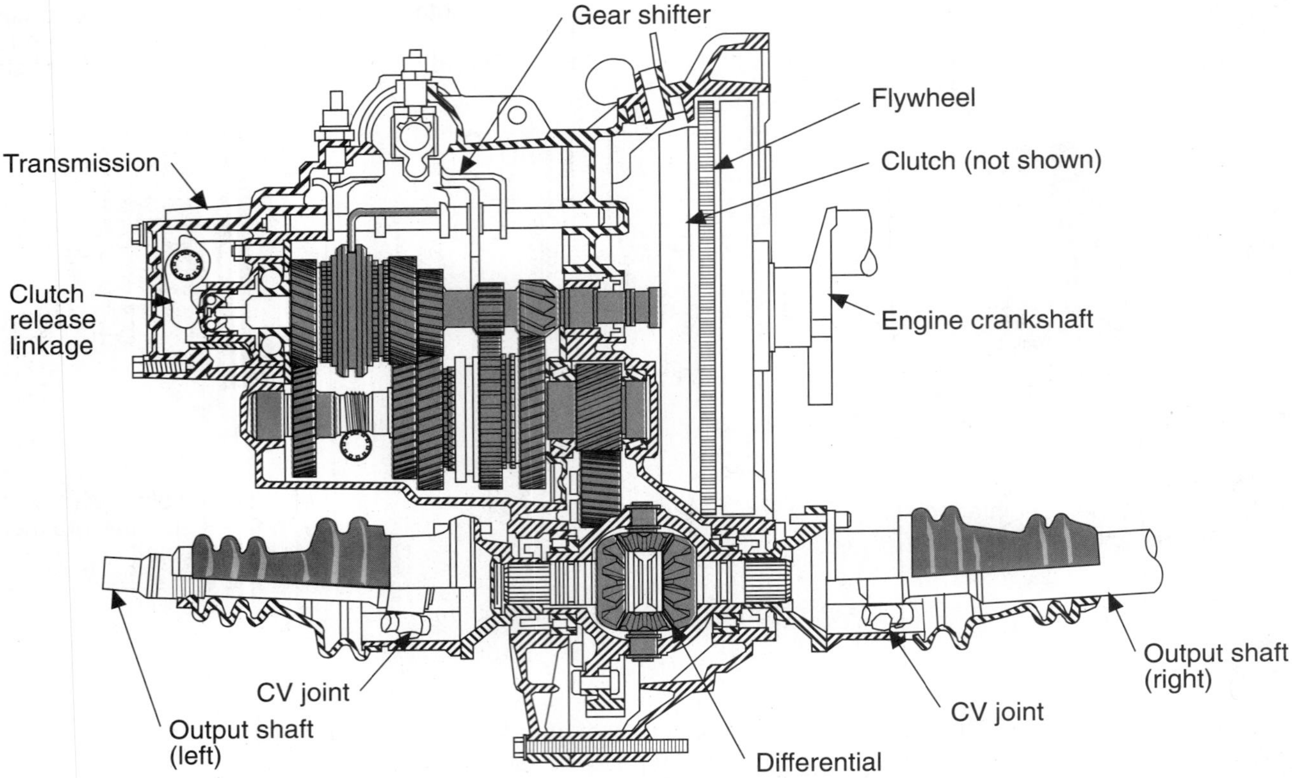

Figure 4-31 Typical transaxle assembly. (Courtesy of Chrysler Corporation)

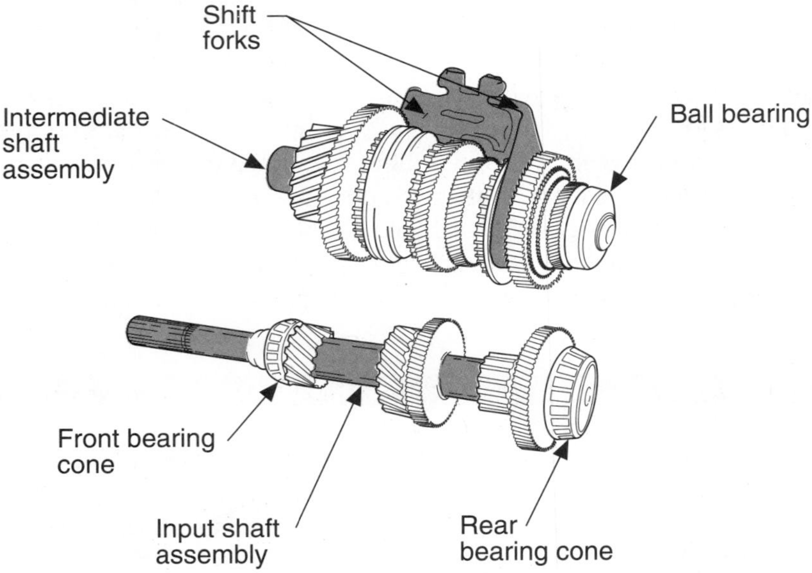

Figure 4-32 Typical transaxle input and output shafts. (Courtesy of Chrysler Corporation)

Item	Description
1.	Mainshaft
2.	Input cluster gear shaft
3.	4th speed gears
4.	3rd speed gears
5.	2nd speed gears
6.	Reverse gears
7.	Reverse idler gears
8.	1st speed gears
9.	Driveshaft
10.	5th speed gear
11.	5th gear driveshaft pinion gear
12.	Mainshaft pinion gear
13.	Differential oil seals
14.	CV shafts
15.	Differential pinion gears
16.	Differential side gears
17.	Final drive ring gear
18.	1st/2nd synchronizer
19.	3rd/4th syncronizer
20.	5th syncronizer

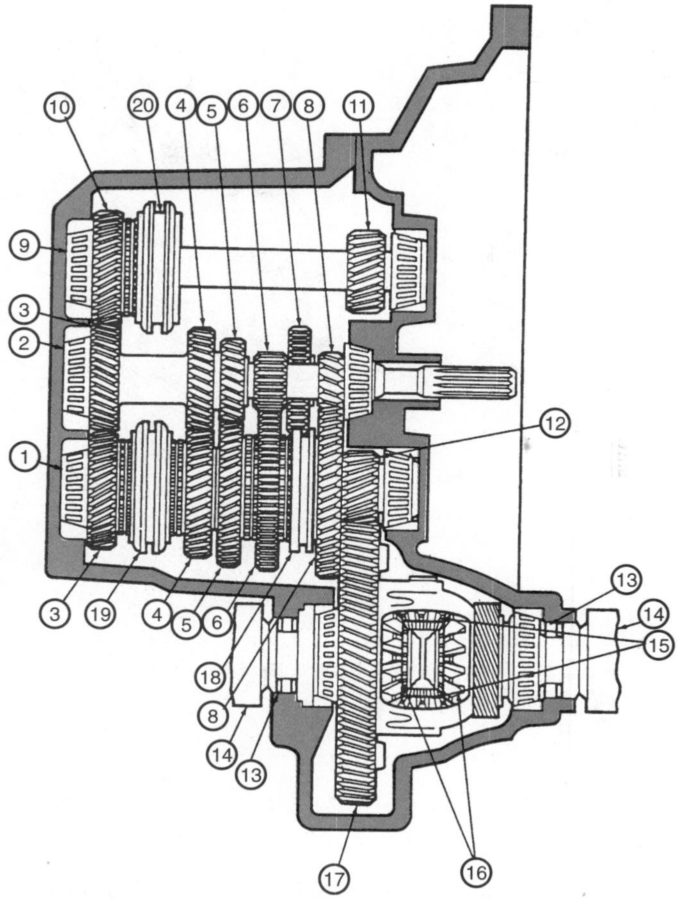

Figure 4-33 Typical five-speed transaxle with three gear shafts. (Reprinted with the permission of Ford Motor Company)

Normally a transaxle has two separate shafts: an input shaft and an output shaft. The engine's torque is applied to the input shaft and the revised torque (due to the transaxles gearing) rotates the output shaft. Normally the input shaft is located above and parallel to the output shaft. The main gears freewheel around the output shaft unless they are locked to the shaft by synchronizers. The main speed gears are in constant mesh with their mating gears on the input shaft and rotate whenever the input shaft rotates.

Some manufacturers refer to the input shaft as the mainshaft and the output shaft as the driven pinion, intermediate, or the mainshaft.

Some transaxles are equipped with an additional shaft designed to offset the power flow on the output shaft (Figure 4-33). Power is transferred from the output shaft to the third shaft using helical gears and by placing the third shaft in parallel with the output shaft and input shaft. The third shaft is added only when an extremely compact transaxle installation is required. Other transaxles with a third shaft use an offset input shaft that receives the engine's power and transmits it to a mainshaft, which serves as an input shaft.

In a transaxle, the transmission and differential are both located in a lightweight housing bolted to the engine. A pinion gear is machined onto the end of the transaxle's output shaft. This pinion gear is in constant mesh with the differential ring gear. When the output shaft rotates, the pinion gear causes the differential ring gear to rotate. The resultant torque rotates the other differential gears, which in turn rotate the vehicle's drive axles and wheels.

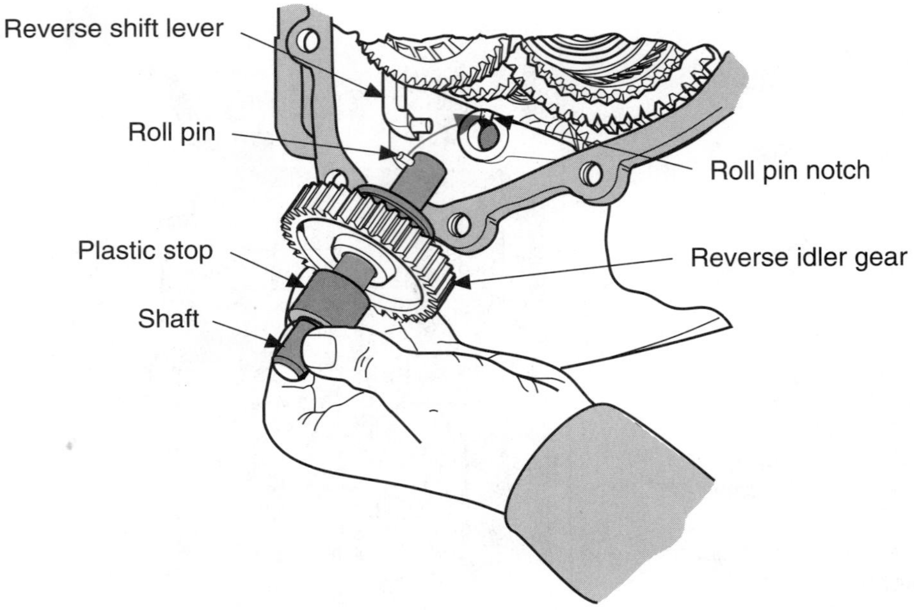

Figure 4-34 A slider-type reverse gear used in many transaxles. (Courtesy of Chrysler Corporation)

Reverse is usually engaged by moving a sliding gear arrangement rather than a sliding synchronizer collar (Figure 4-34), although some five-speed units have synchronized gears in all forward gears and in reverse. The sliding gears have splines through their center bore and rotate at the speed of the shaft. There is no speed-matching action, so nonsynchronized gears tend to clash when shifted into mesh.

The driver changes gears by moving the shift lever to move the internal shift mechanisms of the transaxle. Movement of the shift lever is transmitted to the main shift control shaft and forks to select and engage the forward gears. An internal shift mechanism assembly transfers shift lever movements through the main shift control shaft assembly and the shifting forks (Figure 4-35). A reverse relay lever assembly is commonly used to engage the reverse idler gear with the reverse sliding gear and the input shaft.

The external shift mechanism normally consists of a floor-mounted shift lever that pivots through the shifter boot and is held in place by a stabilizer assembly and bushing (Figure 4-36). Shift lever motion is transmitted to the internal shift mechanism by a shift rod that is connected to and operates the transaxle input shift shaft.

Some transaxles use a two-cable assembly to shift the gears (Figure 4-37). One cable is the transmission selector cable and the other is a shifter cable. The selector cable activates the desired shift fork and the shifter cable causes the engagement of the desired gear.

Power Flow in Neutral

When a transaxle is in its "neutral" position, no power is applied to the differential. Because the synchronizer collars are centered between their gear positions, the meshed drive gears are not locked to the output shaft. Therefore, the gears spin freely on the shaft and the output shaft does not rotate.

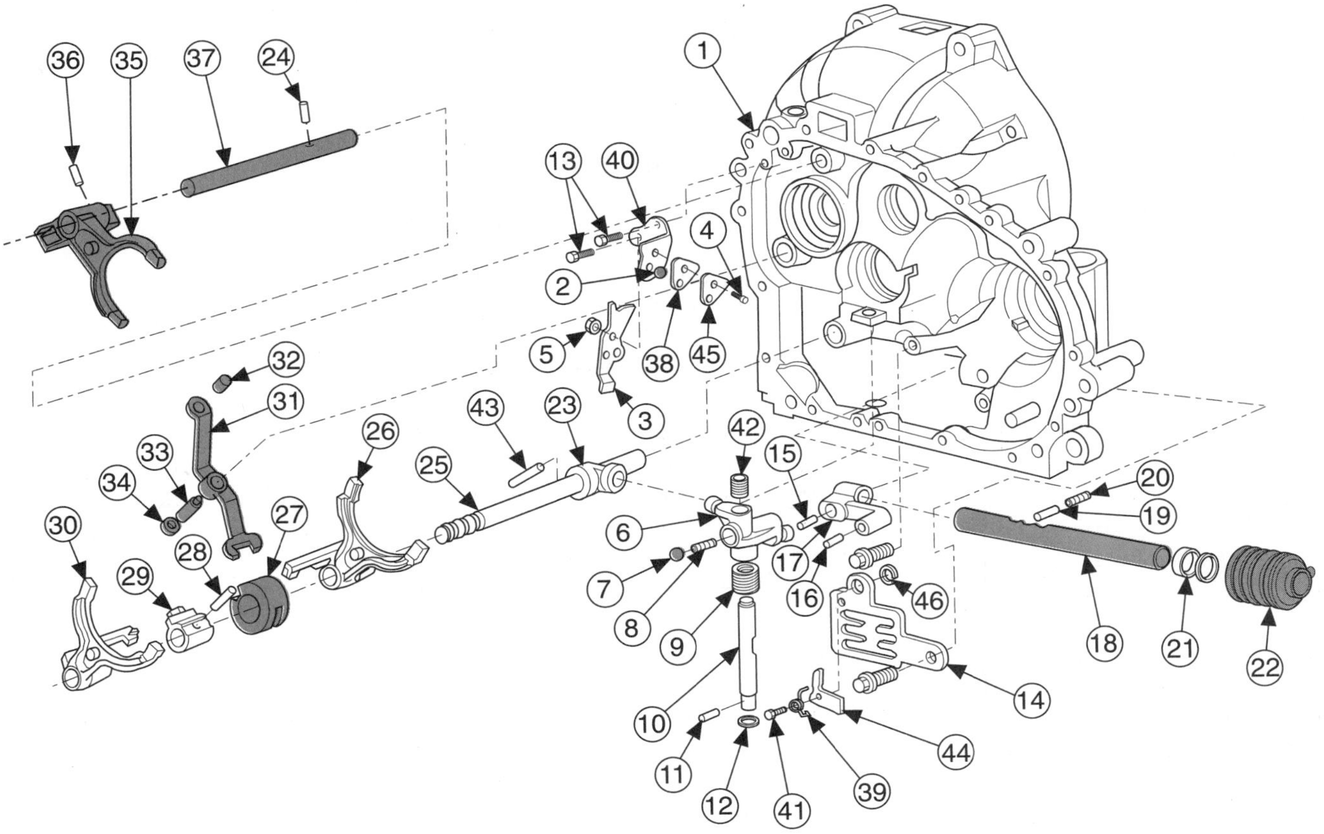

Item	Description
1.	Case–Clutch housing
2.	Ball
3.	Lever–Reverse relay
4.	Pin– Reverse relay lever pivot
5.	Ring–External retaining
6.	Lever–Shift
7.	Ball
8.	Spring–5th/reverse inhibitor
9.	Spring–3rd/4th shift bias
10.	Shaft–Shift lever
11.	Pin–Shift lever
12.	Seal–Shift lever shaft
13.	Bolts–Shift gate attaching
14.	Plate–Shift gate
15.	Roll pin–Selector arm
16.	Pin–Shift gate selector
17.	Arm–Shift gate selector
18.	Shaft–Input shift
19.	Plunger–Shift shaft detent
20.	Spring–Shift shaft detent
21.	Seal assembly–Shift shaft oil
22.	Boot–Shift shaft
23.	Block–Trans input fork control shaft

Item	Description
24.	Pin–Shift gate selector
25.	Shaft–Main shift fork control
26.	Fork–1st/2nd
27.	Sleeve–Fork interlock
28.	Pin–Spring
29.	Arm–Fork selector
30.	Fork–3rd/4th
31.	Lever–5th Shift relay
32.	Pin–Reverse shift relay lever
33.	Pin–5th Relay lever pivot
34.	Ring–External retaining
35.	Fork–5th
36.	Spring pin–5th Retaining
37.	Shaft–5th Fork control
38.	Spring–Trans reverse shift relay lever
39.	Spring–Shift gate pawl
40.	Bracket–Reverse shift relay lever support
41.	Pin–Reverse lockout pawl pivot
42.	Spring–5th/Reverse kickdown
43.	Pin–Reverse relay lever actuating
44.	Pawl–Shift gate plate
45.	Spring–Trans reverse shift relay lever return
46.	C-Clip

Figure 4-35 Internal shift mechanism for a typical transaxle. (Reprinted with the permission of Ford Motor Company)

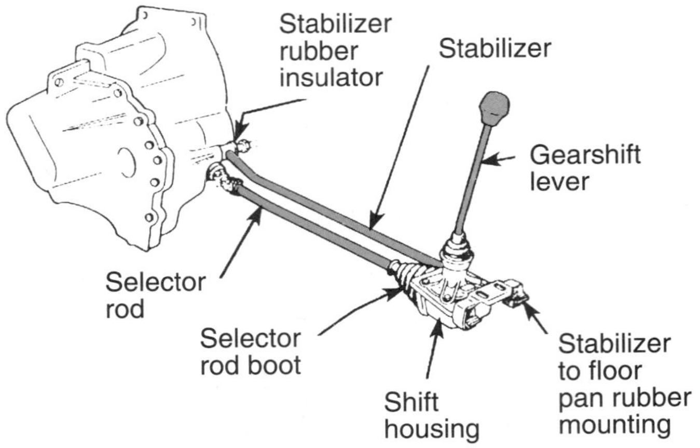

Figure 4-36 External gear linkage. (Reprinted with the permission of Ford Motor Company)

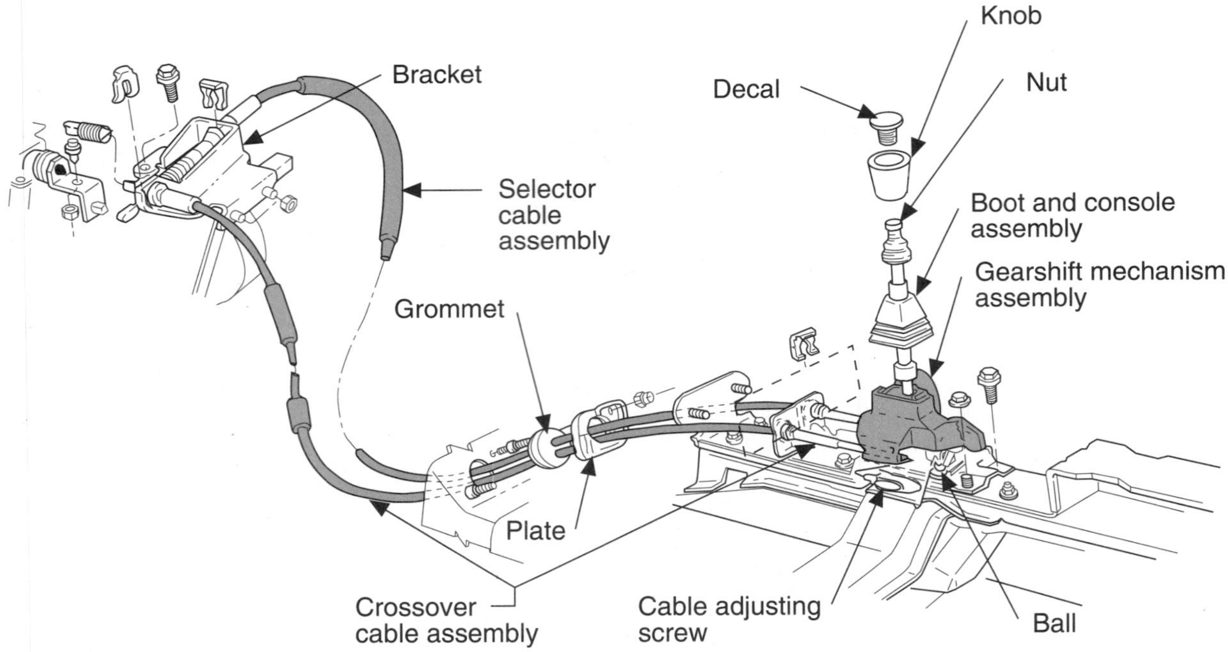

Figure 4-37 Typical two-cable gear linkage. (Courtesy of Chrysler Corporation)

Power Flow through Forward Gears

When first gear is selected (Figure 4-38), the first and second gear synchronizer engages with first gear. Because the synchronizer hub is splined to the output shaft, first gear on the input shaft drives its mating gear (first gear) on the output shaft. This causes the output shaft to rotate at the ratio of first gear, and to drive the differential ring gear at that same ratio.

As the other forward gears are selected, the appropriate shift fork moves to engage the synchronizer with the gear. Because the synchronizer's hub is splined to the output shaft, the desired gear on the input shaft drives its mating gear on the output shaft. This causes the output shaft to rotate at the ratio of the selected gear and drive the differential ring gear at that same ratio (Figure 4-39).

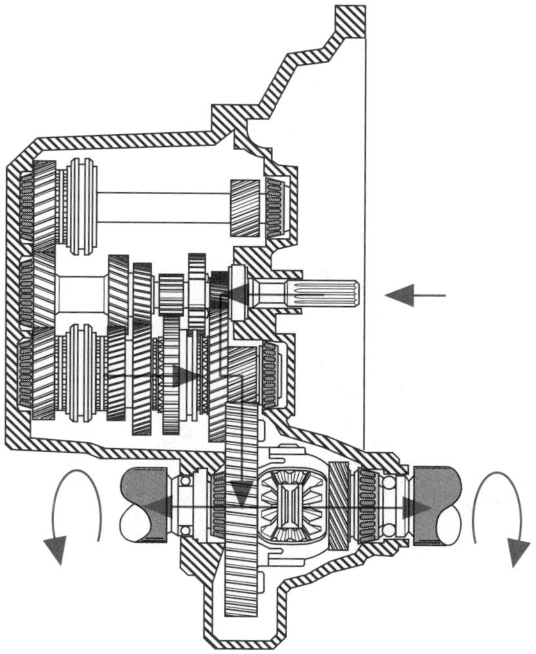

Figure 4-38 Power flow in first gear. (Reprinted with the permission of Ford Motor Company)

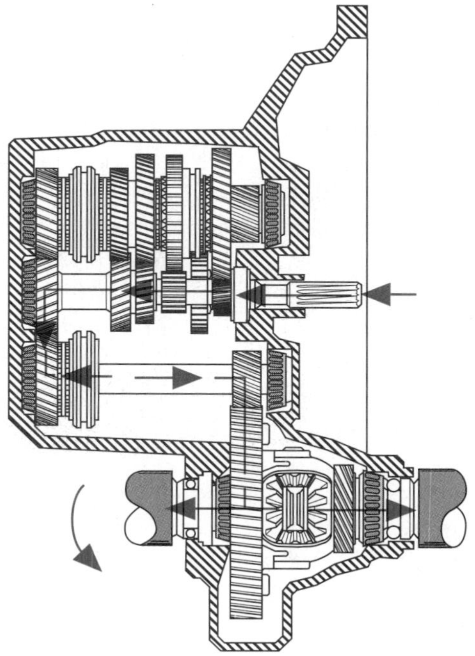

Figure 4-39 Power flow in fifth gear. (Reprinted with the permission of Ford Motor Company)

Power Flow in Reverse

When reverse gear is selected on transaxles that use a sliding reverse gear (Figure 4-40), the shifting fork forces the gear into mesh with the input and output shafts. The addition of this third gear reverses the normal rotation of first gear and allows the car to change direction.

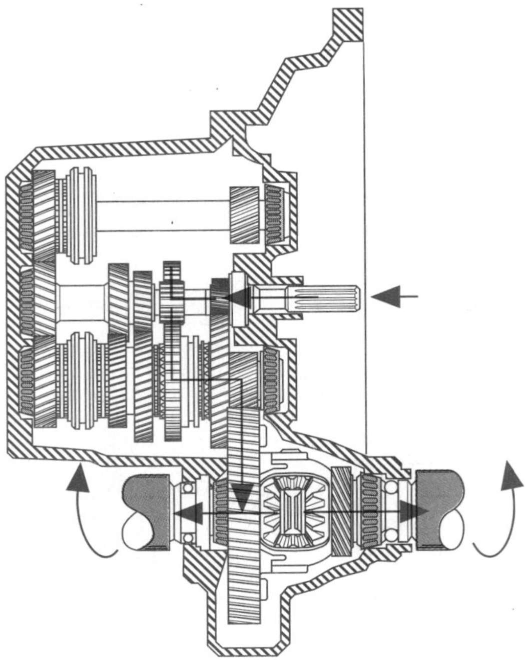

Figure 4-40 Power flow in reverse gear. (Reprinted with the permission of Ford Motor Company)

Classroom Manual
Chapter 7, page 137

Some transaxles need the 90-degree-power-flow change in the differential. These units are used in rear-engine with rear-wheel-drive applications or in longitudinally positioned engines with front-wheel-drive. Alfa Romeo, Volkswagen, and Porsche are good examples of these types of transaxles.

Differential Action

The final drive ring gear is driven by the transaxle's output shaft. The ring gear then transfers the power to the differential case. One major difference between the differential in a RWD car and the differential in a transaxle is power flow. In a rear-wheel-drive differential, the power flow changes 90 degrees between the drive pinion gear and the ring gear. This change in direction is not needed with most front-wheel-drive cars. The transverse engine position places the crankshaft so that it already is rotating in the correct direction. Therefore, the purpose of the differential is only to provide torque multiplication and divide the torque between the drive axle shafts so that they can rotate at different speeds (Figure 4-41).

The final drive or differential case is driven by the output shaft of the transmission. The case holds the ring gear with its mating pinion gear. The differential side gears are connected to inboard constant velocity (CV) joints by splines and are secured in the case with circlips. The drive axles extend out from each side of the differential to rotate the car's wheels (Figure 4-42). The axles are made up of three pieces connected together to allow the wheels to turn for steering and to move up and down with the suspension. A short stub shaft extends from the differential to the inner CV-joint. An axle shaft connects the inner CV-joint and the outer CV-joint. A short spindle shaft that fits into the hub of the wheels extends from the outer CV-joint. To keep dirt and moisture out of the CV-joints, a neoprene boot is installed over each CV-joint assembly.

Because the right and left drive axles are different lengths, FWD cars normally experience torque steer. Manufacturers take different approaches to solve this problem. One solution is to make one axle hollow or to increase its diameter. This changes the ability of the axle to twist and allows it to twist only as far as the other axle. Another solution is to make the longer side in two pieces. The outer part, which is the same length as the axle on the other side of the car, connects the drive wheel to a bearing. The bearing supports the inner part that comes out of the transaxle. With this design, the effective axle lengths are equal and torque steer does not occur.

A BIT OF HISTORY

The first mass-produced FWD passenger car was introduced on an Alvis Model 12/75 produced in England in 1928. The first series-produced FWD cars marketed in the United States were the Ruxton and the Cord in 1929.

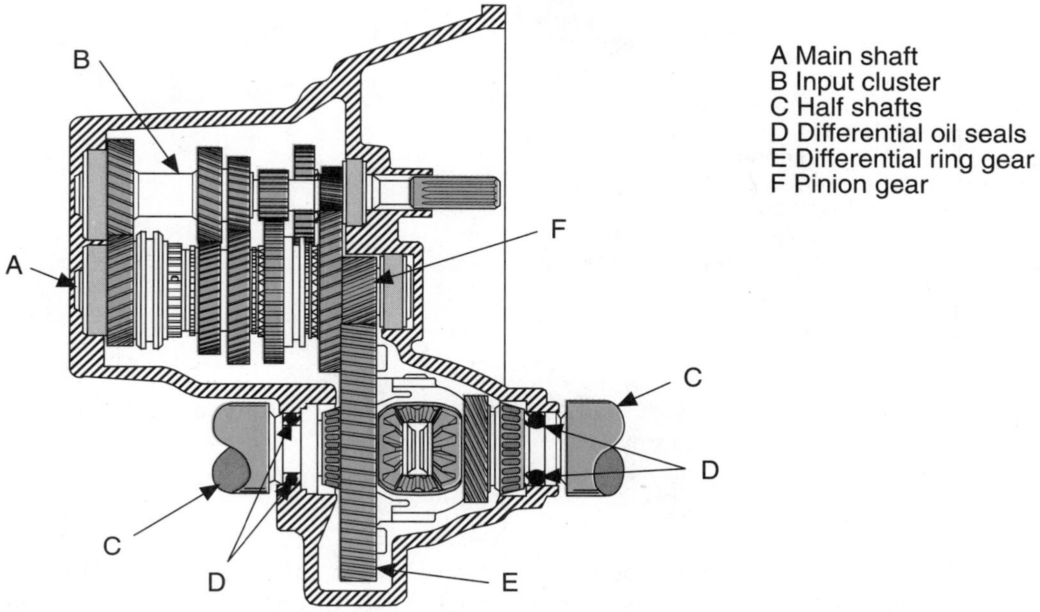

Figure 4-41 Transaxle final drive assembly. (Reprinted with the permission of Ford Motor Company)

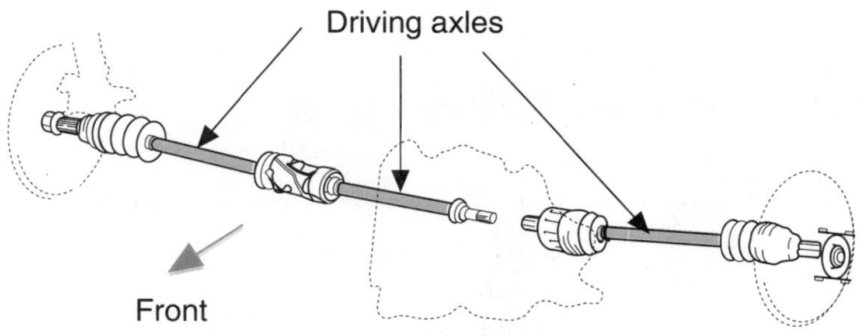

Figure 4-42 FWD drive axles. (Courtesy of Mazda)

Classroom Manual
Chapter 7, page 155

Constant velocity joints are much like the universal joints used in the drive shaft of RWD cars. However, the speed of the driven side of the U-joint may vary with relation to the driving side, depending on the angle of the shaft. A CV-joint maintains an equal speed on both sides of the joint, which helps reduce vibration and wear.

Electrical Connections to a Transmission/Transaxle

Torque steer is a self-induced steering condition in which the axles twist unevenly under engine torque.

Although a transmission is not electrically operated or controlled, a few accessories of the car are controlled by the transmission. Among these accessories are the back-up lights. The switch that activates these light is normally located on the transmission but can be mounted to the shift link-

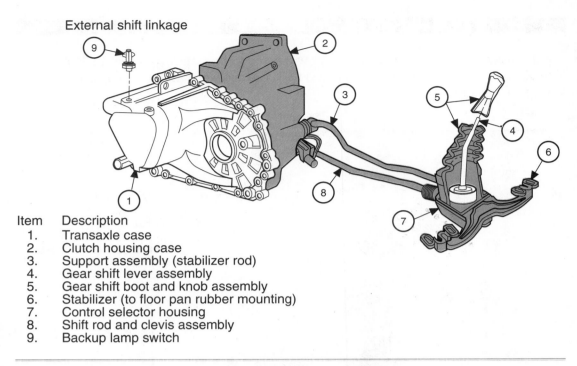

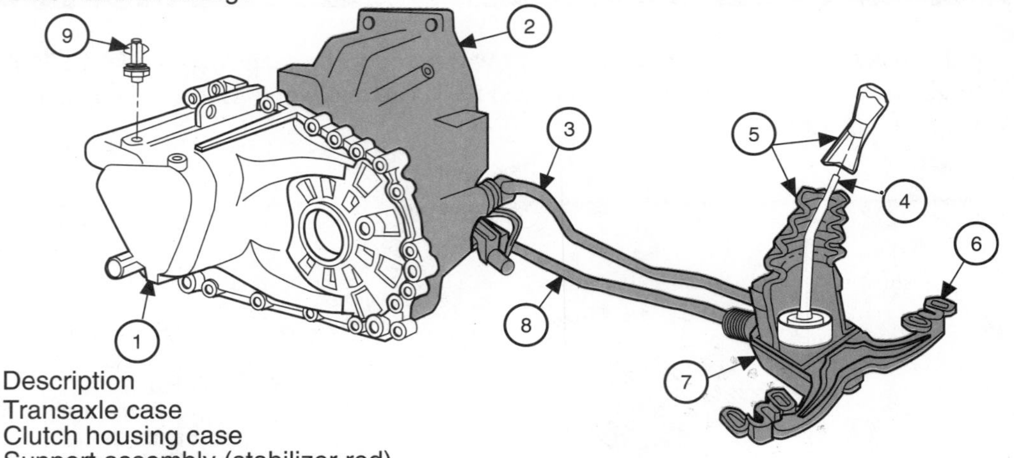

Item	Description
1.	Transaxle case
2.	Clutch housing case
3.	Support assembly (stabilizer rod)
4.	Gear shift lever assembly
5.	Gear shift boot and knob assembly
6.	Stabilizer (to floor pan rubber mounting)
7.	Control selector housing
8.	Shift rod and clevis assembly
9.	Backup lamp switch

Figure 4-43 Location of transaxle mounted back-up light switch. (Reprinted with the permission of Ford Motor Company)

Emission control systems are designed to reduce the pollutants emitted out of the engine's exhaust.

Ignition vacuum advance units change the amount of ignition advance in response to changes in engine load.

age away from the transmission. If mounted in the transmission, the shifting fork will close the switch and complete the circuit whenever the transmission is shifted into reverse gear (Figure 4-43). If the switch is mounted on the linkage, the switch is closed directly by the linkage.

Some early model cars are equipped with a transmission-controlled spark (TCS) switch. This emission control switch closes the circuit only when the transmission is shifted into high gear. Closing the switch allows ignition vacuum advance to operate, thereby preventing vacuum advance during all other gears.

Transmission and Transaxle Maintenance

Transmissions and transaxles require little maintenance if they are operated properly. Like all gear boxes, transmissions and transaxles require clean lubricant to function properly and to be durable. The moving metal parts must not touch each other and should be continuously separated by a thin film of lubricant to prevent excessive wear. Bearings rely on lubricant to allow for smooth and low-friction rotation of the shafts they support. Lubricant is continuously being wiped away as meshed gears rotate. This action causes friction and high heat, and without a constant supply of lubricant, the gears would rapidly wear. Gear oil, by reducing the friction in the transmission/transaxle, also limits the amount of power loss due to friction. Gear oil also protects the gears and related parts from rust and corrosion, keeps the parts cool, and keeps the internal parts clean.

The type and amount of lubricant varies with the different designs (Figure 4-44). Gear oil must have adequate load-carrying capacity to prevent the puncturing of the oil film on the gears. Chemical additives are mixed with gear oil to improve its load-carrying capacity. The typical lubricant for transmissions is a straight mineral oil with an extreme pressure (EP) additive. The gear oils recommended for most cars have a classification of SAE 75W, 75W-80, 80W-90, 85W-90, 90, or 140. Not all transmissions require the use of a heavy gear oil, others use engine oil or automatic transmission fluid (ATF).

EP additives increase the load-carrying capacity of an oil.

The higher the numerical classification of an oil, the thicker the oil is.

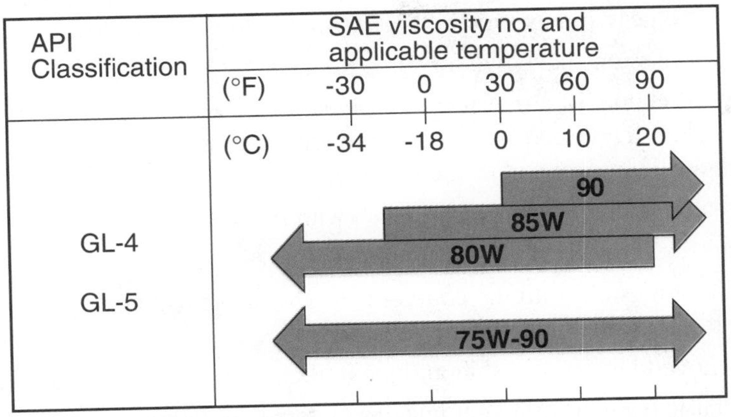

Figure 4-44 Typical transmission/transaxle gear oil classification and viscosity range data. (Courtesy of Subaru of America)

Some transaxles are connected to the engine's oil sump and use the same 30-weight oil as the engine. Other designs have the engine and the transaxle sealed between each other, with each using its own lubricant. A transaxle can require 30W motor oil, 90W gear oil, or ATF. Thinner lubricants, such as 30-weight oil and ATF, are more commonly recommended because these lubricants absorb less horsepower and offer better fuel economy.

CUSTOMER CARE: Today's manual transmissions and transaxles require the use of single- and multiviscosity gear oils, engine oils, or automatic transmission fluids. Always refer to the appropriate service manual to determine the proper lubricant for the vehicle being serviced. Using the correct gear oil will not only prevent premature wear but will also allow the transmission/ transaxle to shift smoothly and operate quietly.

Summary

- ❑ The purpose of a transmission is to use various sized gears to provide the engine with a mechanical advantage over the vehicle's drive wheels.
- ❑ A transmission is a system of gears that transfers the engine's power to the drive wheels of the car.
- ❑ A manual transaxle is a single unit composed of a transmission, differential, and drive axles.
- ❑ Three major types of manual transmissions have been used by the automotive industry: the sliding gear, collar shift, and synchromesh designs.
- ❑ A synchromesh transmission is a constant mesh, collar shift transmission equipped with synchronizers.
- ❑ A synchronizer's primary purpose is to bring components that are rotating at different speeds to one synchronized speed.
- ❑ Block synchronizers consist of a hub, sleeve, blocking ring, and inserts.

Terms to Know

Block synchronizers
Blocking ring
Cluster gear
Clutch gear
Clutch hub
Collar shift
Constant mesh
Constant velocity joints

Terms to Know (continued)

Counter gear
Detent
Dog teeth
EP additives
Free-running gears
High gear
Input gear
Interlock plates/pins
Mainshaft
Output shaft
Pin-type synchronizer
Reverse idler gear
Shift forks
Shift rail
Sliding gear
Synchromesh
Synchronizers
Torque steer

- ❑ Synchronization occurs in three stages. In the first stage, the sleeve is moved toward the gear by the shift lever and engages the hub assembly. In the second stage, the movement of the sleeve causes the inserts to press the blocking ring onto the cone of the gear. In the third stage, the synchronizer ring completes its friction fit over the gear cone and the gear is brought up to the same speed as the synchronizer assembly. The sleeve slides onto the gear's teeth and locks the gear and its synchronizer assembly to the mainshaft.
- ❑ Most current transmissions have five forward speeds.
- ❑ The front bearing retainer of a transmission serves to house an oil seal for the input shaft, holds the input shaft rigid, serves as a centering and holding fixture for the clutch release bearing, and limits the movement of the input shaft.
- ❑ The counter gear assembly is in constant mesh with the speed gears.
- ❑ The speed gears are located on the mainshaft.
- ❑ Reverse gear is not meshed with the counter gear and is normally engaged by sliding the reverse gear into the reverse idler gear.
- ❑ Shift forks are moved by the gear shift and move the synchronizer sleeves to engage a gear.
- ❑ There are two basic types of gear shift linkages: internal and external.
- ❑ One of the primary differences between a transmission and a transaxle is the absence of a counter shaft and gear set in a transaxle.
- ❑ One major difference between the differential of a RWD car and the differential of a FWD car is power flow. In a RWD differential, power flow changes 90 degrees between the drive pinion gear and the ring gear. This change of direction is not needed in a transaxle.
- ❑ A car's back-up lights are activated by a switch that is closed by the movement of the gear shift or shift forks.
- ❑ Transmissions and transaxles require clean lubricant to function properly and to be durable.

Review Questions

Short Answer Essays

1. Define the purpose of a transmission.
2. What is the primary purpose of a synchronizer?
3. Describe the three stages of synchronization.
4. List the functions of a transmission's front bearing retainer.
5. How is reverse speed obtained by most transmissions?
6. Define the primary differences between a transmission and a transaxle.
7. Describe and explain the major difference between the differential of a RWD car and the differential of a FWD car.
8. Explain why transmissions and transaxles require clean lubricant to function properly.
9. Describe the power flow through a transaxle when it is in first gear.
10. What is the purpose of fifth gear in most transaxles and transmissions?

Fill-in-the-Blanks

1. A transmission is a system of ________________ that transfers the ________________ to the ________________ of the car.

2. A manual transaxle is a single unit composed of a ________________, ________________, and ________________.

3. Block synchronizers consist of a ________________, ________________, ________________ ________________, and ________________.

4. The counter gear assembly is in constant mesh with the ________________ ________________.

5. ________________ ________________ are moved by the gear shift and move the synchronizer sleeves to engage a gear.

6. The speed gears are located on the ________________.

7. Three major types of manual transmissions have been used by the automotive industry: the ________________, ________________, and ________________ designs.

8. A synchromesh transmission is a ________________ ________________, ________________ ________________ transmission equipped with synchronizers.

9. The types of fluid normally recommended for transaxles are 30W ________________ oil and ________________ ________________ fluid.

10. Smooth and quiet shifting can only be possible when the ________________ and the ________________ are rotating at the same speed.

ASE Style Review Questions

1. While discussing the operation of a synchronizer:
Technician A says the synchronizer inserts force the blocking ring against the conical face of the gear, which slows the speed of the gear and allows for engagement.
Technician B says that by matching the speed of the blocking ring and the gear, the synchronizer sleeve is able to engage with the gear dog teeth.
Who is correct?
A. A only
B. B only
C. Both A and B
D. Neither A nor B

2. While discussing the construction of a transaxle:
Technician A says transaxles normally have two main gear shafts.
Technician B says most transaxles have a cluster gear set, an input shaft, and an output shaft.
Who is correct?
A. A only
B. B only
C. Both A and B
D. Neither A nor B

3. While discussing the causes of torque steer:
Technician A says it is caused by the position of the engine in the frame, which places more weight on one wheel than the other.
Technician B says it is caused by using right and left drive axles of different lengths.
Who is correct?
A. A only
B. B only
C. Both A and B
D. Neither A nor B

4. While reviewing the procedure for maintaining a transaxle:
Technician A says most manufacturers recommend the use of 90-weight oil in the transaxle.
Technician B says some manufacturers recommend the use of automatic transmission fluid in the transaxle.
Who is correct?
A. A only
B. B only
C. Both A and B
D. Neither A nor B

5. While discussing the various types of transmissions used in automobiles:
Technician A says one type is the sliding gear transmission.
Technician B says another type uses sliding collars.
Who is correct?
A. A only
B. B only
C. Both A and B
D. Neither A nor B

6. Who is correct?
Technician A says most transmissions use two main shafts, the input and output shafts.
Technician B says most transmissions use three shafts, the third shaft is located below the input and output shafts.
A. A only
B. B only
C. Both A and B
D. Neither A nor B

7. While discussing current trends in manual transmissions:
Technician A says five-speed transmissions are commonly used today.
Technician B says transmissions with four speeds and an additional overdrive gear are commonly used.
Who is correct?
A. A only
B. B only
C. Both A and B
D. Neither A nor B

8. Who is correct?
Technician A says that in automotive transmissions, the commonly used type of synchronizer is the block type.
Technician B says that in automotive transmissions, the commonly used type of synchronizer is the plain type.
A. A only
B. B only
C. Both A and B
D. Neither A nor B

9. While discussing overdrive gears:
Technician A says the typical ratio is 0.70 to 0.90:1.
Technician B says the typical ratio is 1.0:1.
Who is correct?
A. A only
B. B only
C. Both A and B
D. Neither A nor B

10. While discussing the power flow through a five-speed transmission while it is in first gear:
Technician A says power enters in on the input shaft, which rotates the counter shaft that is engaged with first gear.
Technician B says the first gear synchronizer engages with the clutching teeth of first gear and locks the gear to the main shaft, allowing power to flow from the input gear through the counter shaft and to first gear and the main shaft.
Who is correct?
A. A only
B. B only
C. Both A and B
D. Neither A nor B

CHAPTER 5

Front Drive Axles

Upon completion and review of this chapter, you should be able to:

- ❑ Explain the purposes of a FWD car's drive axles and joints.
- ❑ Describe the different methods used by manufacturers to offset torque steer.
- ❑ Name and describe the different types of CV-joints currently being used.
- ❑ Name and describe the different designs of CV-joints currently being used.
- ❑ Explain how a ball-type CV-joint functions.
- ❑ Explain how a tripod-type CV-joint functions.

Between 1975 and 1979, only 4%–5% of the total vehicle population had front wheel drive. Some aftermarket analysts predict that by 1995, FWD may account for approximately 90%–95% of the cars sold, and by the year 2000, up to 99% of the vehicle population.

The drive axle assembly transmits torque from the engine and transmission to drive the vehicle's wheels. FWD drive axles transfer engine torque from the transaxle differential to the front wheels. With the engine mounted transversely in the car, the differential does not need to turn the power flow 90 degrees to the drive wheels. However, on a few FWD cars, the differential unit is a separate unit (like RWD cars) and the power flow must be turned. Both arrangements use the same type of drive axle (Figure 5-1).

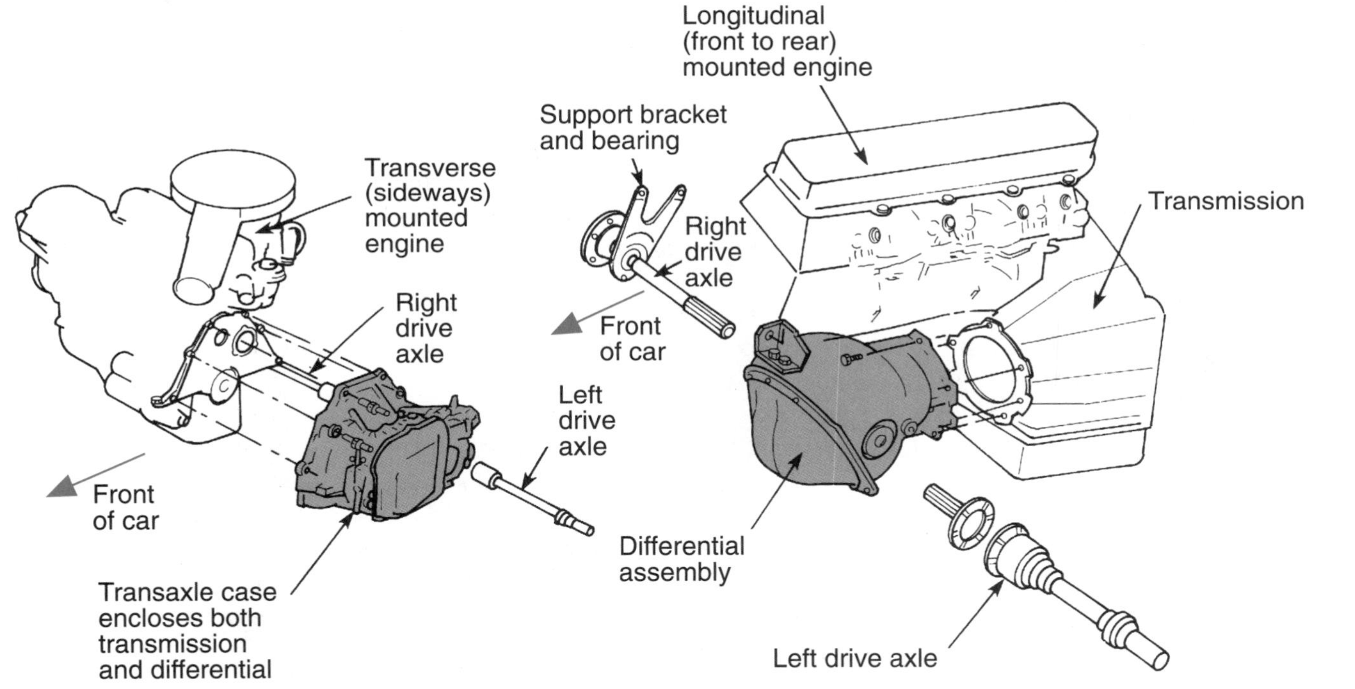

Figure 5-1 The two basic FWD engine and transaxle configurations. (Courtesy of Chevrolet)

Drive Axle Construction

The basic FWD driveline consists of two drive shafts. The shafts extend out from each side of the differential to supply power to the drive wheels. Typically, solid bars of steel are used as the drive axles on the front of four-wheel-drive and front-wheel-drive cars and trucks. In all FWD and some 4WD systems, the transaxle is bolted to the chassis and the axles pivot on CV-joints (Figure 5-2). The outer parts of the axles are supported by the steering knuckles that house the

FWD drive axles are also called axle shafts, drive shafts, and half-shafts.

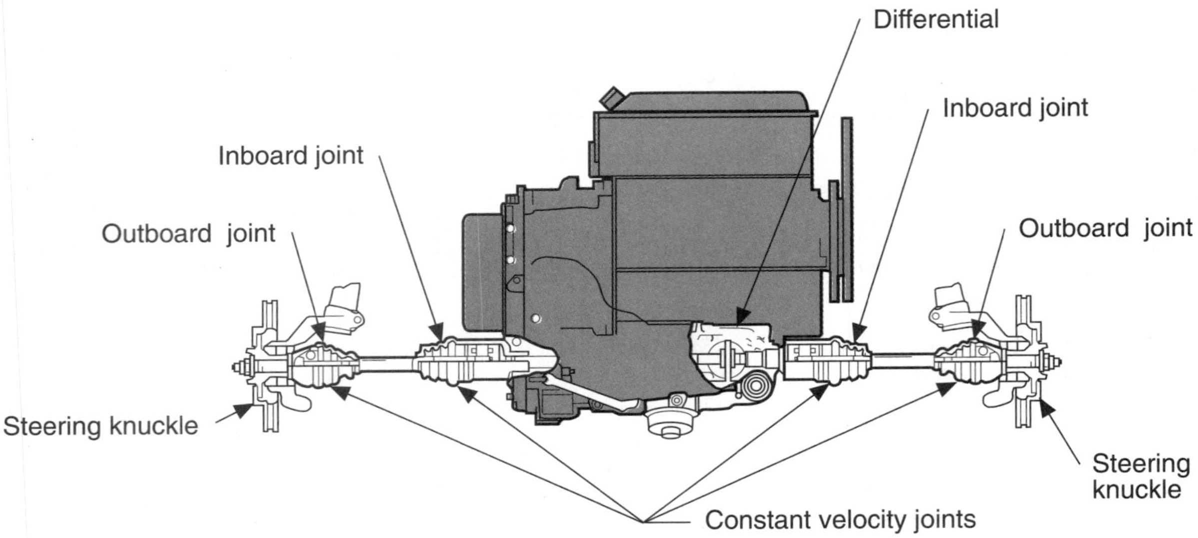

Figure 5-2 Typical FWD drive axle arrangement. (Reprinted with the permission of Ford Motor Company)

The complete drive axle, including the inner and outer CV-joints is typically called a half-shaft.

CV-joints are constant velocity joints.

An axle shaft is the shaft that connects the inner and outer CV-joints.

The spindle shaft is commonly called the stub axle.

The inboard joint is often called the plunge joint and the outboard joint is called the fixed joint.

The twisting of an axle shaft is called the torquing of the shaft.

axle bearings. Steering knuckles serve as suspension components and as the attachment points for the steering gear, brakes, and other suspension parts.

The drive axles used with FWD systems are actually made up of three pieces attached together in such a way as to allow the wheels to turn and move with the suspension. The axle shaft is connected to the differential by the inboard CV-joint. The axle shaft then extends to the outer CV-joint. A short spindle shaft runs from the outer CV-joint to mate with the wheel assembly (Figure 5-3). The steering knuckle connects to the spindle shaft from the outer CV-joint.

A FWD car has two short drive shafts, one on each side of the engine, which drive the wheels and adjust for steering and suspension changes. They operate at angles as high as 40 degrees for turning and 20 degrees for suspension travel (Figure 5-4). Each shaft has two CV-joints, an inboard joint that connects to the differential, and an outboard joint that connects to the wheel. As a front wheel is turned during steering, the outboard CV-joint moves with it around a fixed center. Up and down movements of the suspension system force the inboard joint to slide in and out.

CV-joints are used on the front axles of many 4WD vehicles as well. They have also been used on RWD buses and cars that have the engine mounted in the rear, such as Porsches. Mid engine cars with rear wheel drive, such as Pontiac Fieros and Toyota MR2s, also are equipped with CV-joints. Some RWD cars with independent rear suspension (IRS), such as BMWs, Nissan 300ZXs, and Ford Thunderbirds, use CV-joints on their drive axles (Figure 5-5).

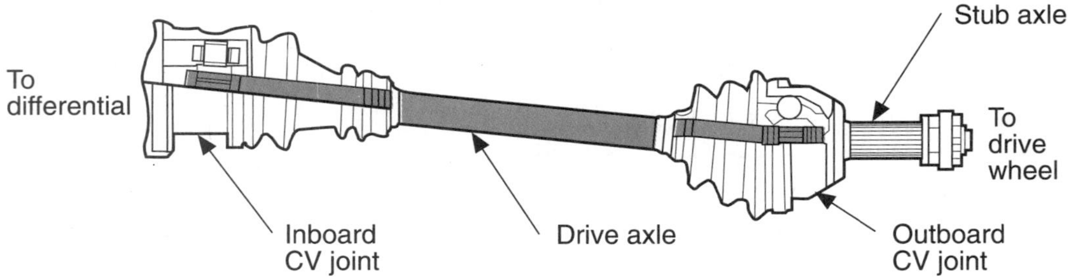

Figure 5-3 Typical FWD drive axle assembly. (Courtesy of Perfect Circle/Dana)

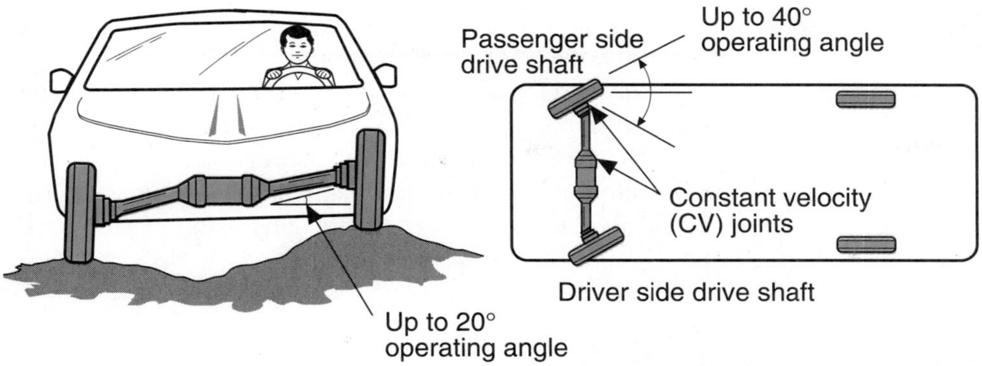

Figure 5-4 FWD drive shaft angles.

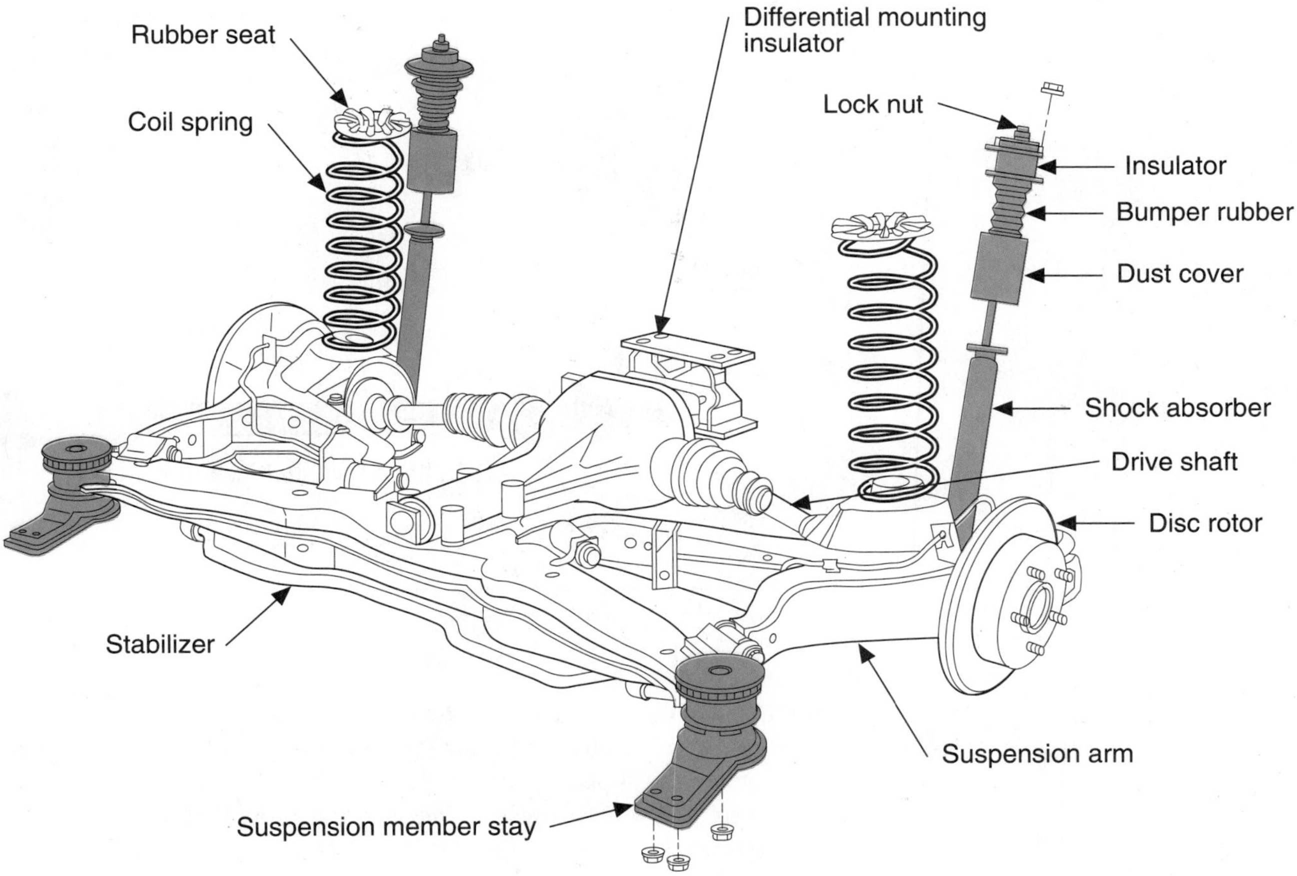

Figure 5-5 Typical IRS RWD power train with CV-joints. (Courtesy of Nissan Motors)

A BIT OF HISTORY

The first patent application in the United States for a front-wheel-drive automobile was made by George Selden in 1879. His gasoline buggy had a three-cylinder engine mounted over the front axle.

Torque steer is a term used to describe a condition in which the car tends to steer or pull in one direction as engine power is applied to the drive wheels.

Drive Axles

With the engine mounted transversely, the transaxle sits to one side of the engine compartment. Because of this offset, one of the drive axles must be longer than the other (Figure 5-6). When torque from the transaxle begins to turn the drive axles, the axles twist before they turn. The longer axle shaft will twist more before it begins to turn than will the shorter shaft. This additional twisting of the longer axle makes it begin to turn slightly later than the shorter axle. This condition is called torque steer.

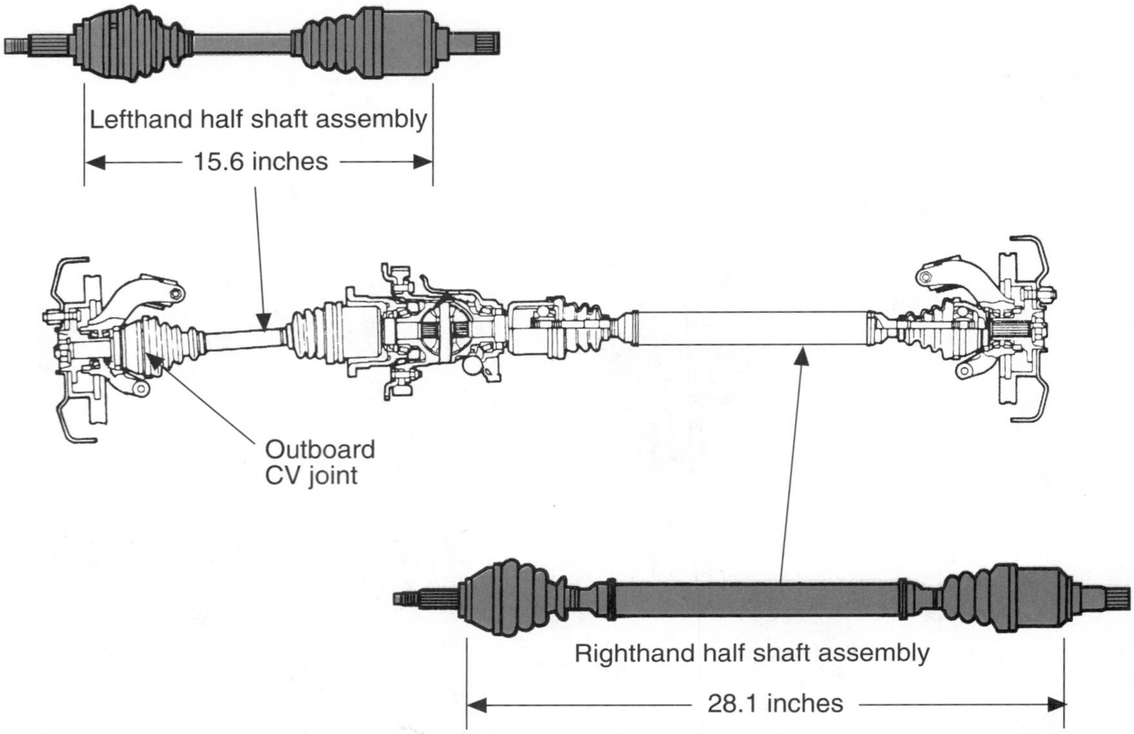

Figure 5-6 FWD drive axles of different lengths.

CUSTOMER CARE: Car owners who have switched from large RWD cars to a FWD car will be very sensitive to torque steer because they have never experienced anything like this before. You should understand the causes of torque steer well enough to be able to explain them to these car owners. They need to know that this condition is normal, so they don't worry about it or feel the car has a defect.

Equal Length Shafts

Equal length shafts are used in some vehicles to help reduce torque steer. This is accomplished by making the longer side into two pieces (Figure 5-7). One piece comes out of the transaxle and is supported by a bearing. The other piece is made to the same length as the shorter side axle, therefore torque steering is not a problem.

The inner piece of the axle shaft is normally called the intermediate shaft or linkshaft.

In these applications, the inner piece is used as a link from the transaxle to the half-shaft. This inner piece is normally linked to the differential by a yoke and ordinary universal joint or is directly splined to the differential. At the outer end is a support bracket and bearing assembly, which is normally fastened to the engine block. The half-shaft is connected to this bearing assembly with a CV-joint.

A universal joint is a mechanism that allows a shaft to rotate at a slight angle.

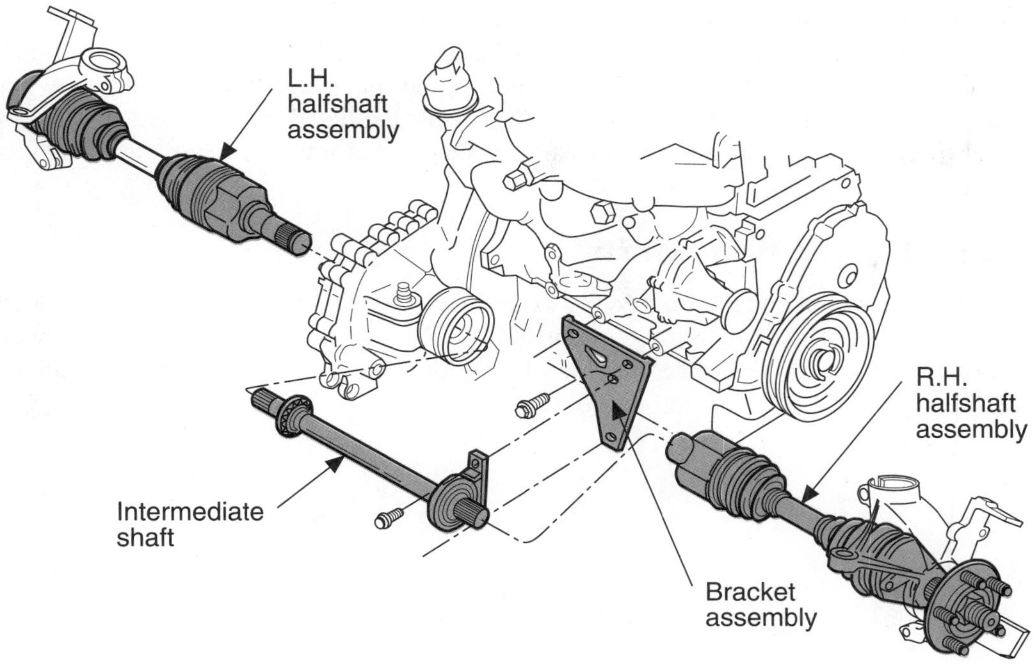

Figure 5-7 Equal length drive axles. (Reprinted with the permission of Ford Motor Company)

Vibration Dampers

A small damper weight is sometimes attached to the long half-shaft to dampen harmonic vibrations in the driveline and to stabilize the shaft as it spins, not to balance the shaft (Figure 5-8). This torsional vibration absorber is splined to the axle shaft and is often held in place by a snap ring.

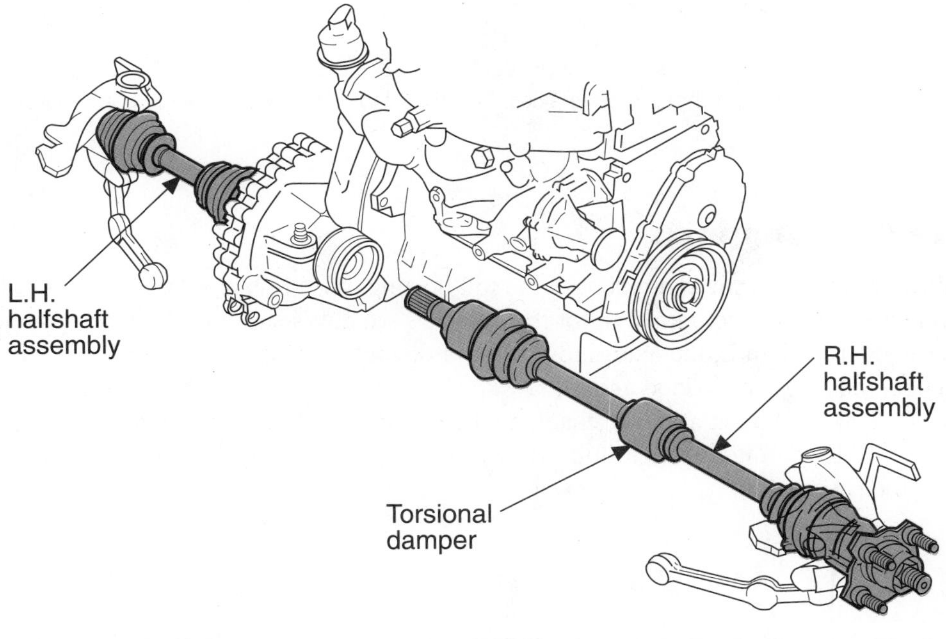

Figure 5-8 The long half-shaft is sometimes fitted with a torsional damper. (Reprinted with the permission of Ford Motor Company)

Unlike the drive shaft of RWD cars, FWD axle shaft balance is not very important because of the relatively slow rotational speeds. In fact, FWD drive shafts operate at about one-third the rotational speed of a RWD drive shaft. This is because the shafts drive the wheels directly, with the final gear reduction already having taken place inside the transaxle's differential. The lower rotational speed has the advantage of eliminating vibrations that sometimes result from high rotational speeds.

Unequal Length Half-Shafts

If the half-shafts are not equal in length, the longer one is usually made thicker than the shorter one or one axle may be solid and the other tubular (Figure 5-9). These combinations would allow both axles to twist the same amount while under engine power. If they twist unequal amounts, the car will experience torque steer.

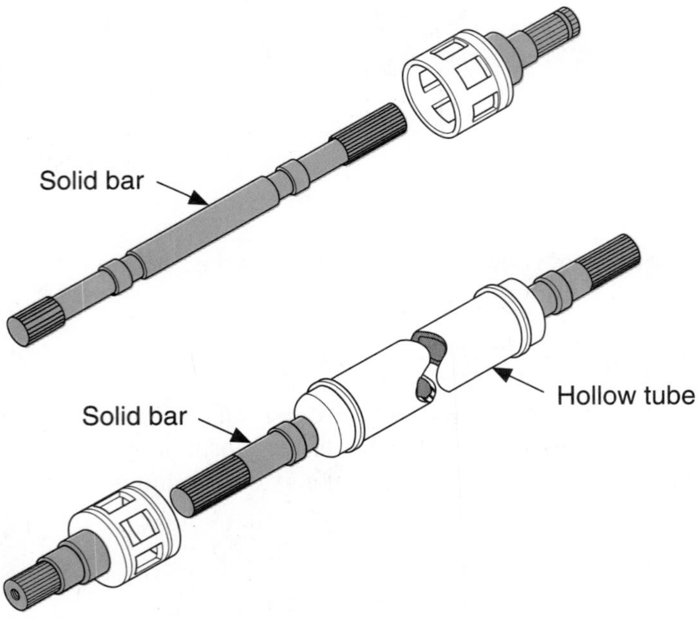

Figure 5-9 Solid and hollow drive axles. (Reprinted with the pemission of Ford Motor Company)

Drive Axle Supports

The drive axles used with transaxles are the full-floating type. This type of axle does not support the weight of the vehicle; rather, all of the vehicle's weight is supported by the suspension. As the car goes over bumps, the axles move up and down, which changes their length. Because of this action, the axles are splined and fitted with CV-joints at both ends of the axles. The inner CV-joints let the axles slide in and out, changing their length, as they move up and down. The outer joints allow the steering system to turn the wheels, as well as allow for the up and down movement of the suspension.

CV-Joints

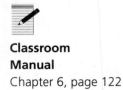

Classroom Manual
Chapter 6, page 122

One of the most important components of a FWD drive axle is the constant velocity joint (Figure 5-10). These joints are used to transfer a uniform torque and a constant speed, while operating

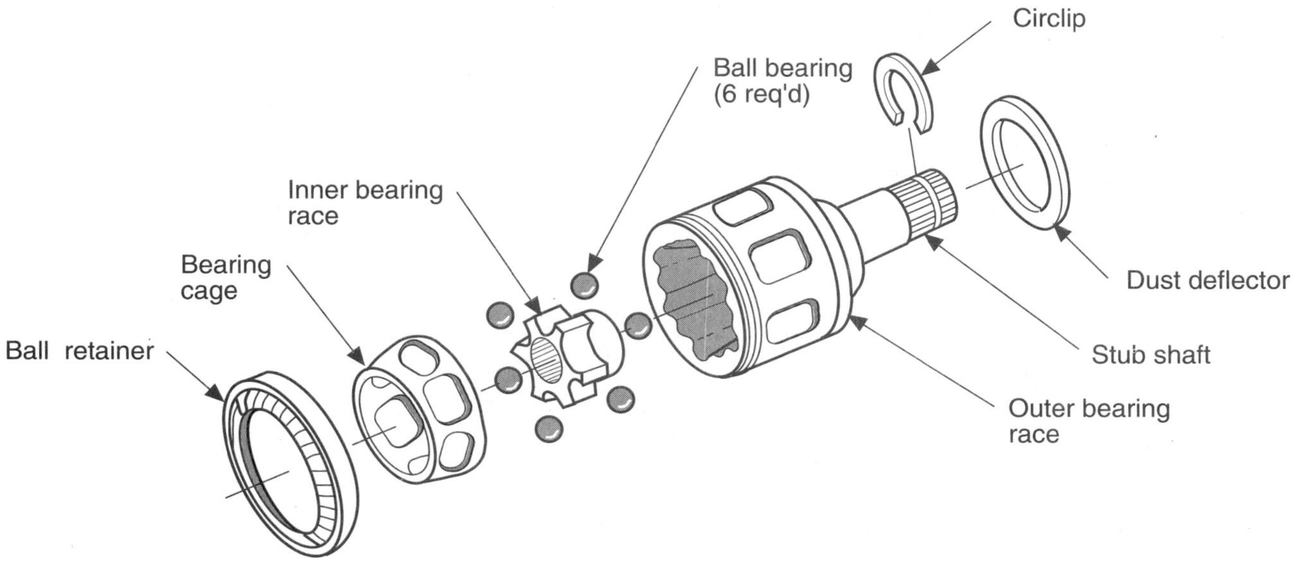

Figure 5-10 An exploded view of a CV-joint. (Reprinted with the permission of Ford Motor Company)

through a wide range of angles. On FWD and 4WD cars, operating angles of as much as 40 degrees are common. The drive axles must transmit power from the engine to the front wheels that must also drive, steer, and cope with the severe angles caused by the up and down movement of the vehicle's suspension. CV-joints are compact joints that allow the drive axles to rotate at a constant velocity, regardless of the operating angle.

CV-joints do the same job as the universal joints of front-engined RWD cars. The drive-shaft of these cars is fitted with universal joints at each end of the shaft. The joints allow the drive shaft to move with the suspension as it transfers power to the drive wheels. As the shaft rotates on an angle, the first U-joint sets up a vibration in the drive shaft and then a second U-joint, at the other end of the shaft, cancels the vibration before it reaches the axle. The ability of the joints to cancel the inherent vibrations lessens as the angle of the shaft increases. U-joints only work well if the shaft angle is 3 to 6 degrees. Two sets of U-joints are often used on drive shafts with greater operating angles. When the U-joints respond to the changes in the angle of the drive shaft, the speed of the shaft changes during each revolution. This change in speed causes the drive shaft to vibrate or pulse as it rotates. Constant velocity joints turn at the same speed during all operating angles and therefore can smoothly deliver power to the wheels.

Universal joints are commonly referred to as U-joints.

A BIT OF HISTORY

In the 1920s an engineer from Ford, Alfred Rzeppa, developed a compact constant velocity joint using ball bearings between two bearing races. Through the years, this joint was further developed and improved. The improved design, similar to what is being used today, appeared on British and German FWD cars in 1959.

Shop Manual
Chapter 5, page 148

A screw boot is a type of boot that is held together by screws and does not require the disassembly or removal of the half-shaft to install.

A split boot is a style of CV boot that may be installed without disassembly or removal of the half-shaft.

The two commonly used boot clamps are the Band-it and Oetiker clamps.

CV Boots

Bellows-type neoprene boots are installed over each CV-joint to retain lubricant and to keep out moisture and dirt (Figure 5-11). These boots must be maintained in good condition. If they need replacement, either the axle must be removed and the new boot or screw boot installed, or a split boot can be laced onto the axle. Each end of the boot is sealed tightly against the shaft or housing by a retaining clamp or strap (Figure 5-12). These straps may be metal or plastic and are available in many sizes and designs (Figure 5-13).

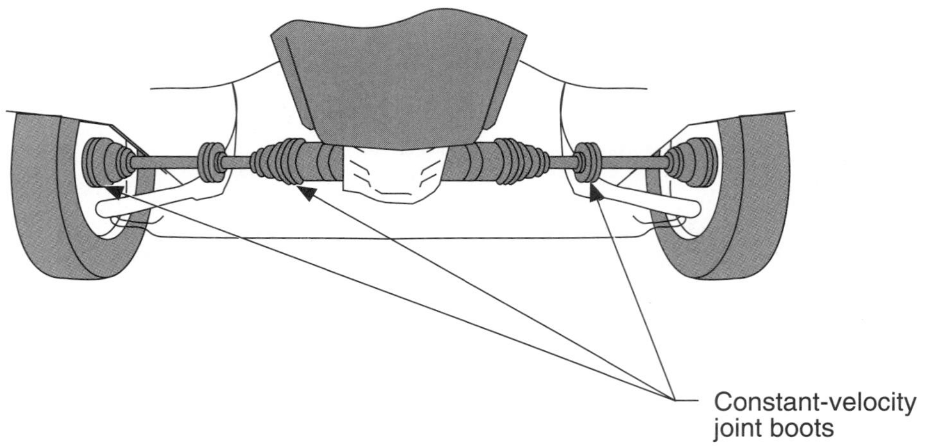

Figure 5-11 Location of CV-joint boots.

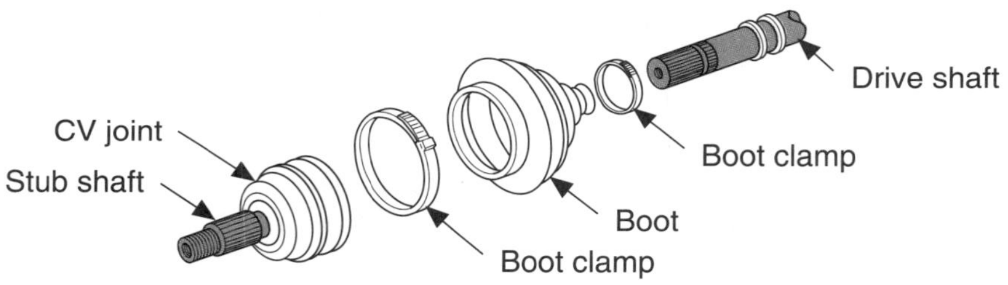

Figure 5-12 Clamps seal the boot against the CV-joint and drive axle.

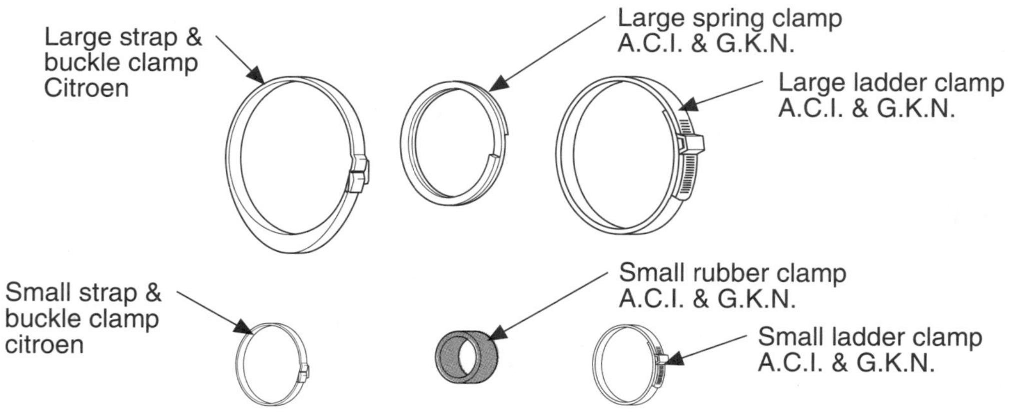

Figure 5-13 Various sizes and designs of boot clamps. (Courtesy of Chrysler Corporation).

In some applications, the inboard CV-joints operate very close to the engine's exhaust system. Special rubber boots of silicone or thermoplastic materials are required to withstand the temperatures. In these extremes, a special high temperature lubricant will be specified by the manufacturer.

If the CV-joint boot keeps the joint properly sealed, the joint can last more than 100,000 miles.

Types of CV-joints

To satisfy the needs of different applications, CV-joints come in a variety of styles. The different types of joints can be referred to by position (inboard or outboard), by function (fixed or plunging), or by design (ball type or tripod).

Inboard and Outboard Joints

In FWD drivelines, two CV-joints are used on each half-shaft. The joint nearer the transaxle is the inner or inboard joint, and the one nearer the wheel is the outer or outboard joint. In a RWD vehicle with independent suspension (IRS), the joint nearer the differential can also be referred to as the inboard joint. The one closer to the wheel is the outboard joint.

There are two basic types of outboard CV-joints, the Rzeppa fixed CV-joint and the fixed tripod joint. Three basic types of inboard CV-joints are used: double-offset CV-joint (DOJ), plunging tripod CV-joint, and the cross-groove plunge joint. The applications of these vary with car make and model (Figure 5-14).

The Rzeppa joint is also called a ball-type CV-joint.

CV-joints are held onto the axle shafts by three different methods: nonpositive, positive, and single retention. Most inner joints use positive retention, whereas outboard joints may be held by any one of the three methods. Nonpositive retention is accomplished by the slight inter-

Nonpositive retention is used on the outer CV-joints of all models of Ford and Chrysler cars, as well as some German-made cars.

Positive retention is used on most inner CV-joints and on the outer joints of some European cars.

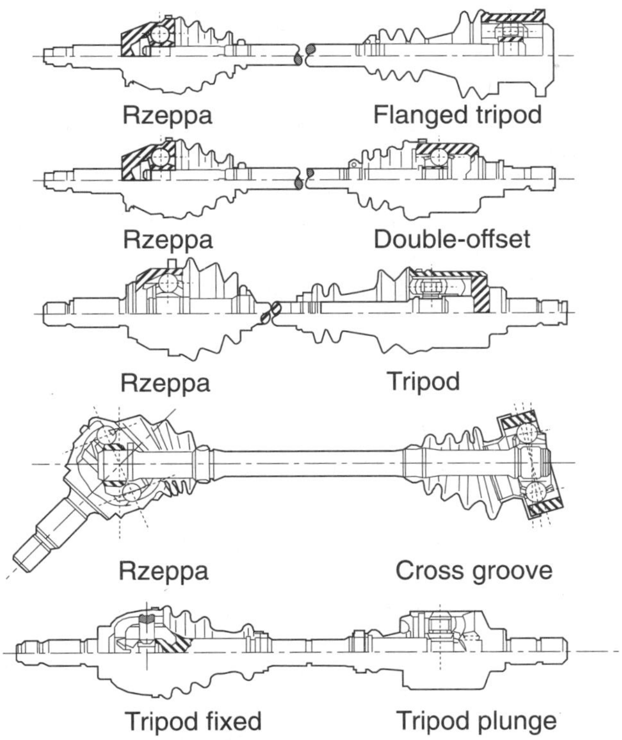

Figure 5-14 Typical CV-joint combinations on FWD drive axles.

The outer CV-joints of most Asian import cars use a modified single retention method for securing the CV-joints to the shaft.

ference fit of the joint onto the shaft. All positively retained joints use a snap ring to secure the joint to the shaft. Single retention is accomplished by a very tight press-fit connection of the joint onto the shaft. Often, joints retained in this way cannot be removed without destroying the joint or the shaft.

Fixed and Plunging Joints

CV-joints can also be categorized by function. They are either fixed (meaning they do not plunge in and out to compensate for changes in length) or plunging (one that is capable of in and out movement) joints.

Inboard CV-joints are often called the plunging joint because they allow the drive shaft to move in and out of the transaxle.

In response to the suspension of the car, the drive axles' effective length changes as the distance from the inboard to the outboard CV-joint changes. The inboard CV-joint must allow the drive shaft to freely move in and out of the transaxle as the front wheels go up and down (Figure 5-15).

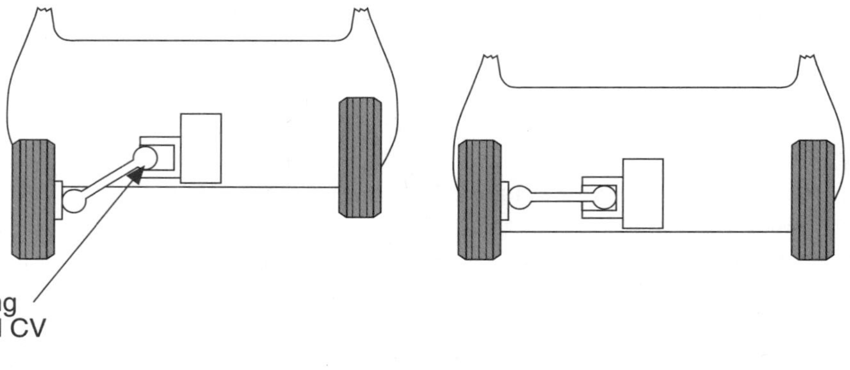

Figure 5-15 Suspension movement and the resulting plunging action of an inboard CV-joint. (Reprinted with the permission of Ford Motor Company)

The outboard joint is a fixed joint. Both joints do not need to plunge if one can do the job. The outboard joint must be able to handle much greater operating angles for steering than would be possible with a plunging joint.

In RWD applications with IRS, one joint on each axle shaft can be fixed and the other plunging, or both can be plunging joints. The operating angles are not as great because the wheels do not have to steer, thus plunging joints can be used at either or both ends of the axle shafts.

CV-Joint Designs

CV-joints are also classified by design. The two basic varieties are the ball type and the tripod type. Both are used as either inboard or outboard joints, and both are available in fixed or plunging designs.

Outboard CV-Joint Designs

Ball-Type Joints

Rzeppa joints are ball-type joints.

The most commonly used type of CV-joint was named after its original designer, A. H. Rzeppa, and is based on a ball-and-socket principle (Figure 5-16). The Rzeppa type of outboard joint has its inner race attached to the axle. The inner race has several precisely machined grooves spaced

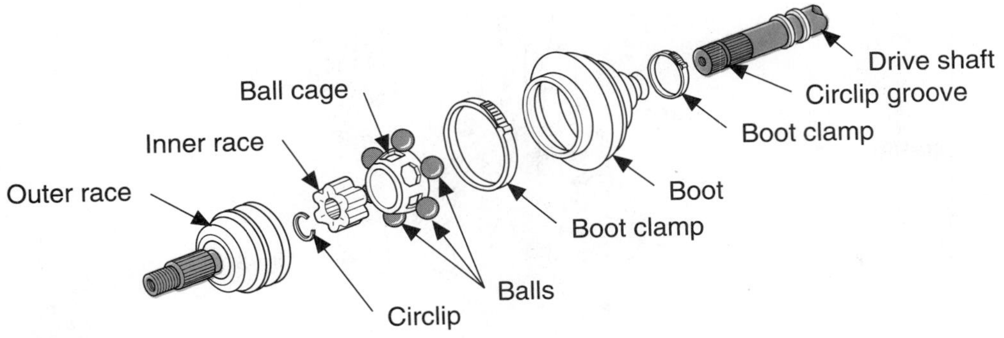

Figure 5-16 Rzeppa CV-joint disassembled.

around its outside diameter. The number of grooves equals the number of ball bearings used by the joint. These joints are designed with a minimum of three to a maximum of six ball bearings. The bearing cage is pressed into the outer housing and serves to keep the joint's ball bearings in place as they ride in the groove of the inner race

When the axle rotates, the inner bearing race and the balls turn with it. The balls, in turn, cause the cage and the outer housing to turn with them (Figure 5-17). The grooves machined in the inner race and outer housing allow the joint to flex. The balls serve both as bearings between the races and the means of transferring torque from one to the other. This type of CV-joint is used on almost every make and model of FWD car, except for most French designs.

If viewed from the side, the balls within the joint always bisect the angle formed by the shafts on either side of the joint regardless of the operating angle (Figure 5-18). This reduces the effective operating angle of the joint by a half and virtually eliminates all vibration problems. The cage helps to maintain this alignment by holding the balls snugly in their windows. If the cage

The ball bearings are often referred to as driving balls or driving elements.

The machined grooves for the ball bearings are often referred to as tracks.

The term *bisect* means to divide by two.

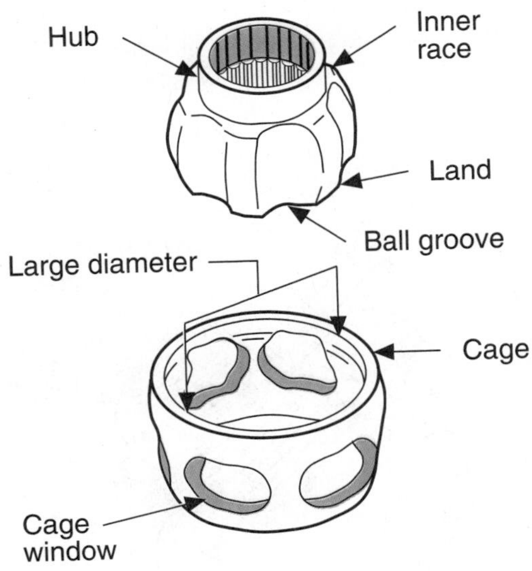

Figure 5-17 The inner race is splined to the axle shaft. The balls, placed between the ball groove and cage window, move the cage with the axle. (Reprinted with the permission of Ford Motor Company)

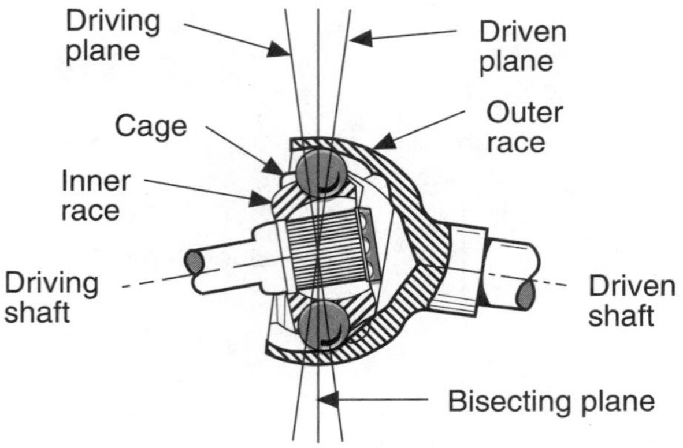

Figure 5-18 In a Rzeppa CV-joint, the balls bisect the angle of the joint.

windows become worn or deformed over time, the resulting play between the ball and window typically results in a clicking noise when turning. It is important to note that the opposing balls in a Rzeppa CV-joint always work together as a pair. Heavy wear in the grooves of one ball almost always results in identical wear in the grooves of the opposing ball.

Another ball-type joint is the disc-style CV-joint, which is used predominantly by Volkswagen as well as many German RWD cars. Its design is very similar to the Rzeppa joint.

Tripod-Type Joints

A tripod is the portion of an inner joint made up of rollers, needle bearings, and three arms.

A trunnion is a pivot or pin on which something can be rotated or tilted.

The outer housing of a tripod joint is called the tulip because of its three-lobed, flowerlike shape.

The fixed tripod CV-joint uses a central hub or tripod that has three trunnions fitted with spherical rollers on needle bearings (Figure 5-19). These spherical rollers or balls ride in the grooves of an outer housing that is attached to the front wheels (Figure 5-20). Because the balls are not held in a set position in the hub, they are free to move back and forth within the hub. This allows for a constant velocity regardless of the movement of the hub as it responds to the steering or suspension system. This type of CV-joint is used on most French cars and has great angular capability. This type of joint is known by three subcategory types that define how they are retained: the Citroen, GKN, and ACI joints (Figure 5-21).

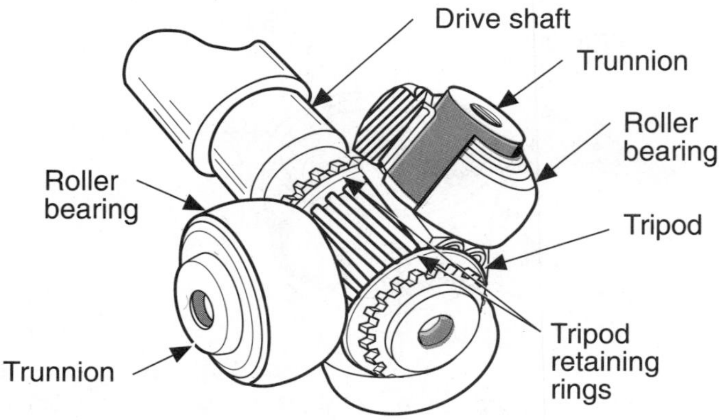

Figure 5-19 A tripod assembly. (Courtesy of Chrysler Corporation)

The hub of a tripod joint is called a spider.

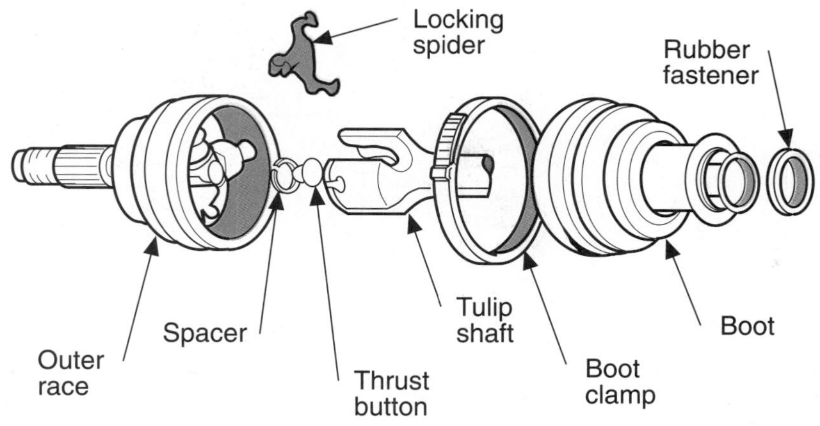

Figure 5-20 Typical tripod CV-joint.

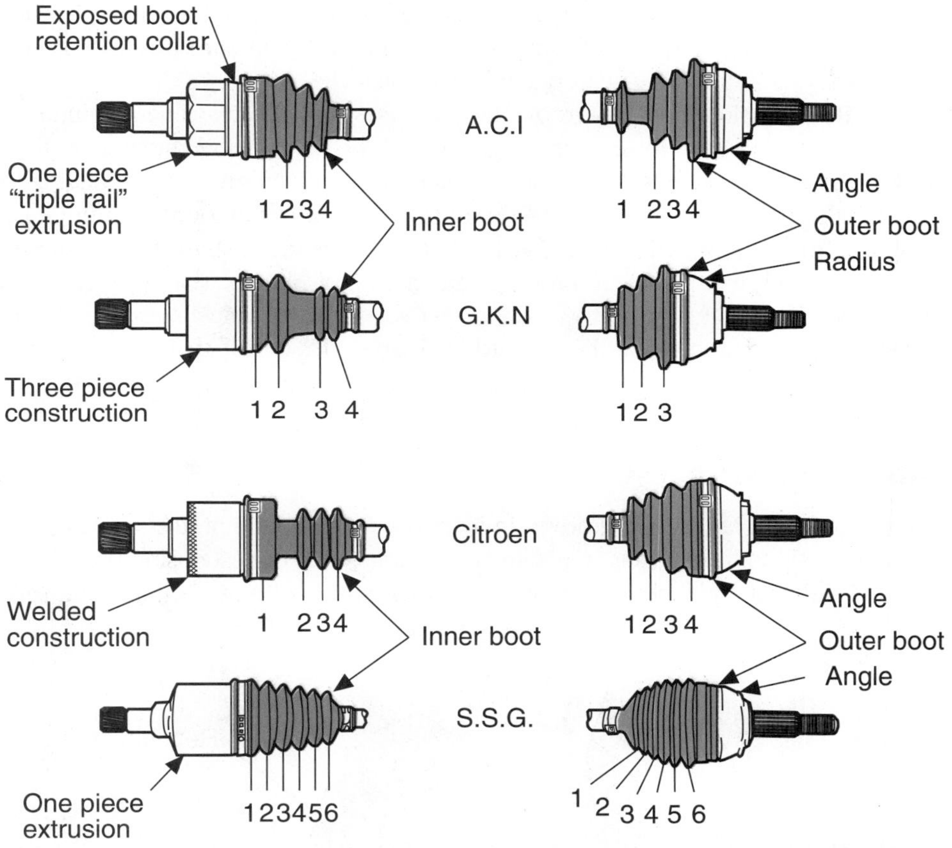

Figure 5-21 The appearance of the joint's boot can identify the design of the joint. (Courtesy of Chrysler Corporation)

Inboard CV-Joint Designs

Ball-Type Joints

A Rzeppa CV-joint can be modified to become a plunging joint, simply by making the grooves in the inner race longer (Figure 5-22). Longer grooves in the inner race of a Rzeppa joint allow the bearing cage to slide in and out. This type of joint is called a double-offset joint and is typically

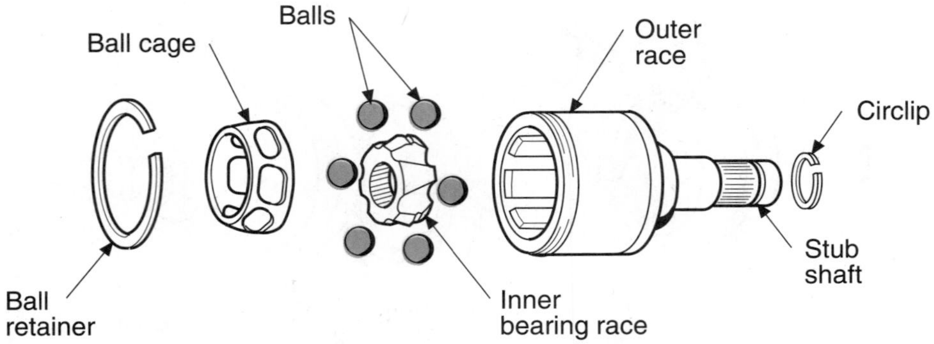

Figure 5-22 A double-offset CV-joint.

The double-offset joint is often listed as a DOJ.

Double-offset joints are commonly found on many Ford, General Motors, Honda, and Subaru cars.

used in applications that require higher operating angles (up to 25 degrees) and greater plunge depth (up to 2.4 inches). This type of joint can be found at the inboard position on some FWD half-shafts as well as on the drive shaft of some 4WD vehicles.

Like the Rzeppa joint, the cross-groove CV-joint uses six balls in a cage and inner and outer races (Figure 5-23). However, instead of the grooves in the races being cut straight, they are cut on an angle. The cross-groove joint has a much flatter design than any other plunging joint. The feature that makes this joint unique is its ability to handle a fair amount of plunge (up to 1.8 inches) in a relatively short distance. The inner and outer races share the plunging motion equally so less overall depth is needed for a given amount of plunge. The cross-groove joint can also handle operating angles of up to 22 degrees. Cross-groove joints are commonly found on German-made cars and are used at the inboard position on FWD half-shafts or at either end of a RWD IRS axle shaft.

Tripod-Type Joints

A plunging tripod joint has longer grooves in its hub than a fixed joint; this allows the spider to move in and out within the housing. On some tripod joints, the outer housing is closed, meaning the roller tracks are totally enclosed within it. On others, the tulip is open and the roller tracks

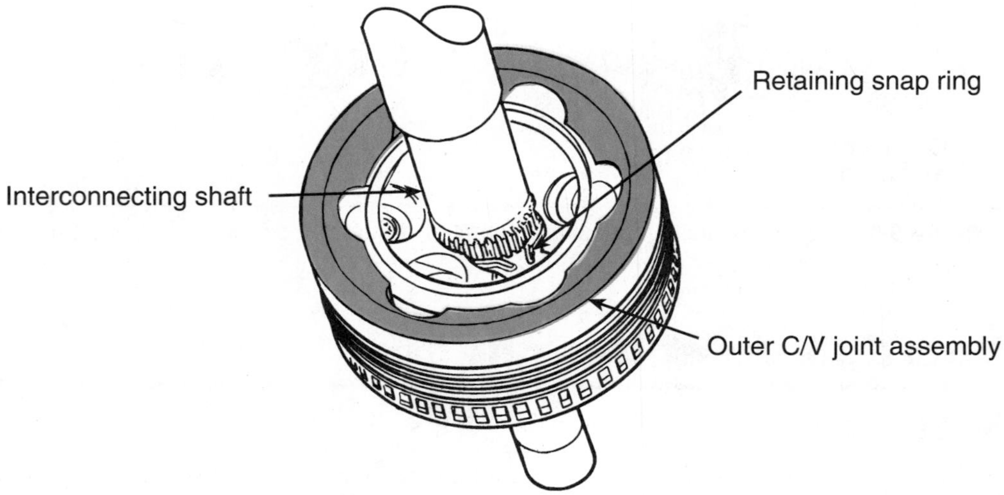

Figure 5-23 Cross-groove CV-joint. (Courtesy of Chrysler Corporation)

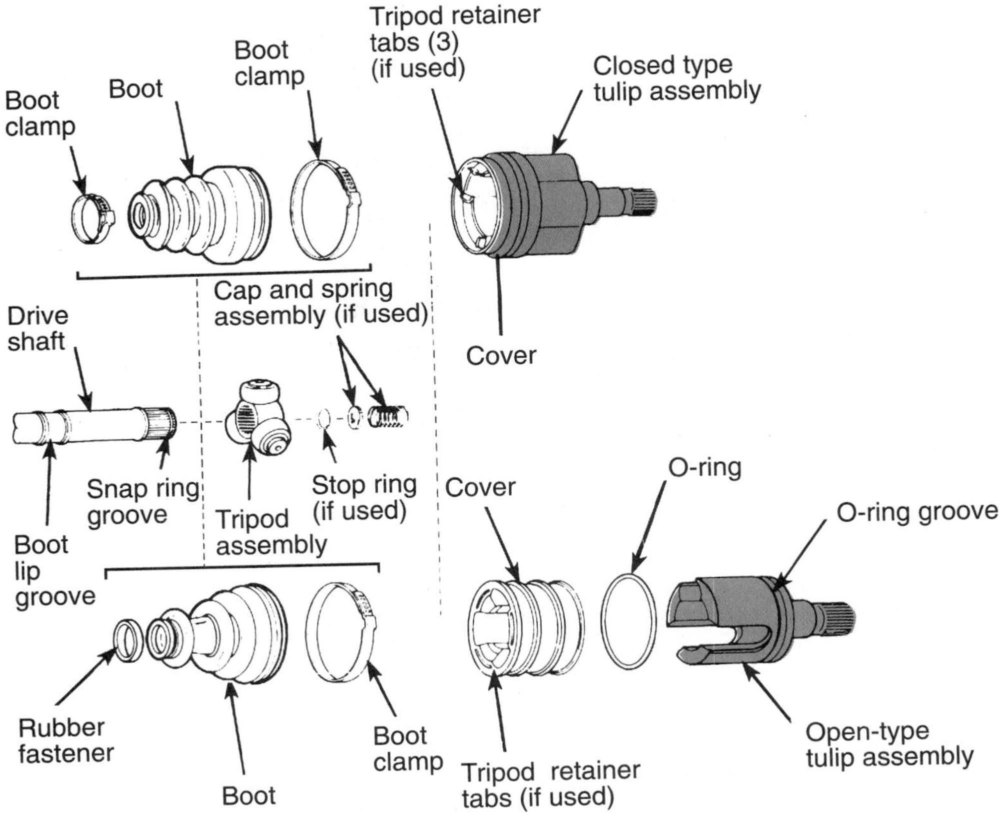

Figure 5-24 Open and closed tripod plunging joints.

are machined out of the housing (Figure 5-24). Plunging tripod-type joints are used on many American and European cars, including some Fords, Chryslers, General Motors, and French cars.

> **WARNING:** Many new vehicles equipped with ABS (antilock brake system) have special toothed rings fitted to the CV-joint housing (Figure 5-25). These rings, called sensor rings, ABS rings, or tone wheels, provide individual wheel-speed information to the ABS computer. Careful inspection and handling procedures are required when CV service is performed to maintain proper ABS operation.

The ABS speed sensors are often called reluctors.

CV-Joint Wear

Regardless of the application, outer joints typically wear at a higher rate than inner joints, because of the increased range of operating angles to which they are subjected. Inner joint angles may change only 10 to 20 degrees as the suspension travels through jounce and rebound. Outer joints can undergo changes of up to 40 degrees in addition to jounce and rebound as the wheels are steered. That combined with more flexing of the outer boots, is why outer joints have a higher failure rate. On an average, nine outer joints are replaced for every inner joint. That does not mean you should overlook the inner joints. They wear too. Every time the suspension travels through jounce and rebound, the inner joints must plunge in and out to accommodate the different arcs between the drive shafts and the suspension. Tripod inner joints tend to develop unique wear patterns on each of the three rollers and their respective tracks in the housing, which can lead to noise and vibration problems.

Jounce is the up and down movement of the car in response to road surfaces.

Rebound is the movement of the suspension system as it attempts to bring the car back to normal heights after jounce.

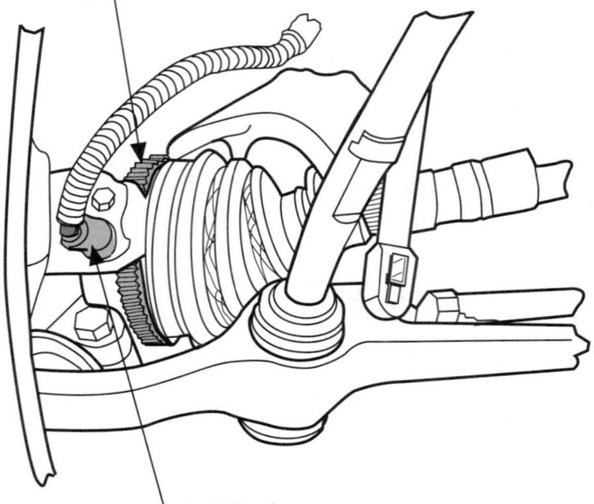

Figure 5-25 ABS speed sensor fitted to outboard joint. (Reprinted with the permission of Ford Motor Company)

FWD Wheel Bearings

The drive axles are supported in the steering knuckle by wheel bearings (Figure 5-26). These bearings allow the axle to rotate evenly and smoothly and keep the axle in the center of the steering knuckle's hub. On some cars the bearing is part of a wheel hub assembly, on others the bearing unit is pressed into the hub, and on a few, the bearings are retained in much the same way as the front wheel bearings on a RWD car.

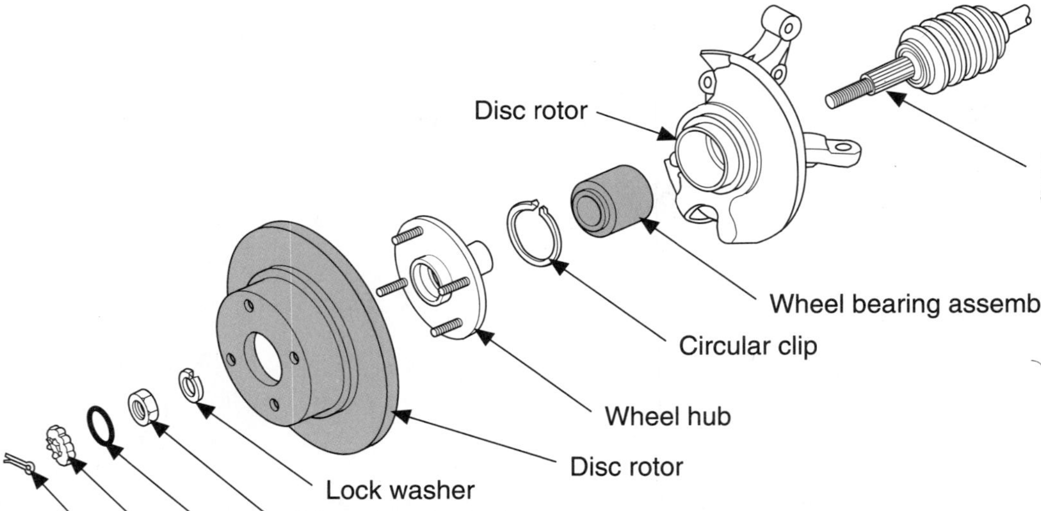

Figure 5-26 A typical FWD wheel bearing and hub assembly. (Courtesy of Nissan Motors)

Maintenance

No periodic service to the drive axles or CV-joints is required. However, if a CV-joint boot is punctured or damaged, it should be replaced immediately; loss of lubricant and entry of dirt and moisture will quickly damage the joint.

Many FWD cars have permanently sealed front-wheel bearings. These do not require periodic adjustment or lubrication. These units are replaced if they are defective. Bearings that require periodic service should be removed, cleaned, inspected, and repacked with fresh grease every 24,000 to 30,000 miles or whenever a brake job is performed or a drive axle removed.

Shop Manual
Chapter 5, page 154

Summary

- ❑ The complete drive axle, including the inner and outer CV-joints is typically called a half-shaft.
- ❑ A FWD car has two short drive shafts, one on each side of the engine, which drive the wheels and adjust for steering and suspension changes.
- ❑ CV-joints are not only used on FWD cars, they are also used on the front axles of many 4WD vehicles. They have also been used on RWD buses and cars that have the engine mounted in the rear.
- ❑ *Torque steer* is a term used to describe a condition in which the car tends to steer or pull in one direction as engine power is applied to the drive wheels.
- ❑ Equal length shafts are used in some vehicles to help reduce torque steer. This is accomplished by making the longer side into two pieces. One piece comes out of the transaxle and is supported by a bearing. The other piece is made to the same length as the shorter side axle.
- ❑ A small damper weight is sometimes attached to one half-shaft to dampen harmonic vibrations in the driveline and to stabilize the shaft as it spins.
- ❑ If the half-shafts are not equal in length, the longer one is usually made thicker than the shorter one, or one axle may be solid and the other tubular. These combinations would allow both axles to twist the same amount while under engine power.
- ❑ Constant velocity joints are used to transfer a uniform torque and a constant speed, while operating through a wide range of angles.
- ❑ Bellows-type neoprene boots are installed over each CV-joint to retain lubricant and to keep out moisture and dirt.
- ❑ Each end of the boot is sealed tightly against the shaft or housing by a retaining clamp or strap.
- ❑ CV-joints come in a variety of types that can be referred to by position (inboard or outboard), by function (fixed or plunging), or by design (ball type or tripod).
- ❑ The CV-joint nearer the transaxle is the inner or inboard joint, and the one nearer the wheel is the outer or outboard joint
- ❑ CV-joints are held onto the axle shafts by three different methods: nonpositive, positive, and single retention.
- ❑ CV-joints are either fixed (meaning they do not plunge in and out to compensate for changes in length) or plunging (one that is capable of in and out movement) joints.
- ❑ The most commonly used type of CV-joint is a Rzeppa, which has its inner race attached to the axle. The inner race has several precisely machined grooves spaced around its outside diameter. The number of grooves equals the number of ball bearings used by the joint. These

Terms to Know

Axle shaft
Bellows
Cross-groove joint
CV-joint
Double-offset joint
Fixed CV-joint
Full-floating axle
Half-shaft
Inboard CV-joint
Intermediate shaft
IRS
Jounce
Outboard CV-joint
Plunging CV-joint
Rebound
Rzeppa CV-joint
Screw boot
Spider
Split boot
Stub axle
Torque steer
Tripod CV-joint
Trunnion
Tulip assembly
Universal joint

joints are designed with a minimum of three to a maximum of six ball bearings. The bearing cage is pressed into the outer housing and serves to keep the joint's ball bearings in place as they ride in the groove of the inner race.

- ❑ The fixed tripod CV-joint uses a central hub or tripod that has three trunnions fitted with spherical rollers on needle bearings. These spherical rollers or balls ride in the grooves of an outer housing that is attached to the front wheels.
- ❑ A Rzeppa CV-joint can be modified to become a plunging joint by making the grooves in the inner race longer. Longer grooves in the inner race of a Rzeppa joint allow the bearing cage to slide in and out. This type is called a double-offset joint.
- ❑ Like the Rzeppa joint, the cross-groove CV-joint uses six balls in a cage and inner and outer races. However, the grooves in the races are cut on an angle rather than straight. The cross-groove joint has a much flatter design than any other plunging joint.
- ❑ A plunging tripod joint has longer grooves in its hub than a fixed joint; this allows the spider to move in and out within the housing.
- ❑ Regardless of the application, outer joints typically wear at a higher rate than inner joints because of the increased range of operating angles to which they are subjected.
- ❑ The drive axles are supported in the steering knuckle by wheel bearings. These bearings allow the axle to rotate evenly and smoothly and keep the axle in the center of the steering knuckle's hub.

Review Questions

Short Answer Essays

1. Define the purpose of a CV-joint.
2. Describe the difference between a fixed and a plunging CV-joint.
3. Explain why CV-joints are preferred over conventional universal joints.
4. Explain the different methods used by automobile manufacturers to offset torque steer.
5. Describe the purpose of the boot on a CV-joint.
6. Describe purposes of the wheel bearing of a FWD car.
7. Describe the major differences between a Rzeppa and a tripod CV-joint.
8. Explain why one type of CV-joint tends to wear much faster than the other.
9. Explain why some half-shafts are fitted with a vibration damper.
10. Explain how a ball-type CV-joint is constructed.

Fill-in-the-Blanks

1. CV-joints come in a variety of types that can be referred to by position
   (__________________ or __________________), by function
   (__________________ or __________________), or by design
   (__________________ or __________________).

2. If the half-shafts are not equal in length, the longer one is usually made ____________ than the other or may be ____________ and the other ____________.

3. The major components of a Rzeppa joint are three to six ____________, an inner ____________, and an outer ____________.

4. The major components of a tripod CV-joint are the ____________, three ____________, and an outer ____________.

5. ____________ CV-joints typically wear at a higher rate than the ____________ ones because of the ____________ range of operating ____________.

6. The type of joint that allows for changes in axle length is the ____________ joint.

7. Half-shafts are also called ____________ ____________ and ____________ ____________.

8. CV-joints are used to transfer ____________ ____________ and a ____________ ____________.

9. CV-joints are held onto the axle shafts by three different methods: ____________, ____________, and ____________.

10. The use of an intermediate shaft allows for half-shafts of ____________ ____________.

ASE Style Review Questions

1. While discussing CV-joints:
Technician A says they are called constant velocity joints because their rotational speed does not change with their operational angle.
Technician B says all FWD and some 4WD vehicles use CV-joints on their front drive axles.
Who is correct?
A. A only
B. B only
C. Both A and B
D. Neither A nor B

2. While discussing the types of CV-joints used on FWD cars:
Technician A says most use a fixed inboard joint.
Technician B says all use one plunging joint on each axle.
Who is correct?
A. A only
B. B only
C. Both A and B
D. Neither A nor B

3. While trying to decide on which is the most commonly used type of CV-joint:
Technician A says the ball type is the most common.
Technician B says the Rzeppa type is the most common.
Who is correct?
A. A only
B. B only
C. Both A and B
D. Neither A nor B

4. While discussing the differences between RWD and FWD front wheel bearings:
Technician A says most FWD cars use one tapered roller bearing whereas RWD cars use two.
Technician B says most FWD front bearings are pressed onto the spindle end of the drive axle.
Who is correct?
A. A only
B. B only
C. Both A and B
D. Neither A nor B

5. While discussing Rzeppa CV-joints:
Technician A says the inner race of the joint is connected to the drive axle.
Technician B says the ball bearings rotate on a trunnion within the outer housing.
Who is correct?
A. A only
B. B only
C. Both A and B
D. Neither A nor B

6. While discussing the types of CV-joints used on FWD cars:
Technician A says plunging joints are most often used as outboard joints.
Technician B says plunging joints connect the axle with the stub axle.
Who is correct?
A. A only
B. B only
C. Both A and B
D. Neither A nor B

7. While explaining torque steer to a customer:
Technician A says the longer axle shaft twists more than the shorter axle shaft.
Technician B says the vibration damper on the half-shaft does not help correct torque steer problems.
Who is correct?
A. A only
B. B only
C. Both A and B
D. Neither A nor B

8. While discussing the purposes of the protective boot on a CV-joint:
Technician A says the boot prevents contamination of the joint's lubricant.
Technician B says the boot prevents the joint's lubricant from flying off when the shaft is rotating.
Who is correct?
A. A only
B. B only
C. Both A and B
D. Neither A nor B

9. *Technician A* says a double-offset joint is similar to a Rzeppa joint.
Technician B says a cross-groove joint is similar to a Rzeppa joint.
Who is correct?
A. A only
B. B only
C. Both A and B
D. Neither A nor B

10. While discussing the differences between the various designs of CV-joints:
Technician A says most joints use the tripod design and all are basically interchangeable.
Technician B says the tripod-type joint uses needle bearings, not ball bearings like the ball type.
Who is correct?
A. A only
B. B only
C. Both A and B
D. Neither A nor B

Drive Shafts and Universal Joints

CHAPTER 6

Upon completion and review of this chapter, you should be able to:

- ❑ Describe the purpose and construction of common RWD drive shaft designs.
- ❑ Describe the purpose and construction of common universal joint designs.
- ❑ Explain the importance of drive shaft balance.
- ❑ Explain the natural speed variations inherent to a drive shaft.
- ❑ Describe the effects of canceling universal joint angles.

On front engined rear-wheel-drive cars, the rotary motion of the transmission's output shaft is carried through the drive shaft to the differential, which causes the rear drive wheels to turn (Figure 6-1).

Although most late model cars are front wheel drive, which are not equipped with a drive shaft and universal joints, the best selling vehicles in America are pick-up trucks, which do have a drive shaft and universal joints. These, plus the many older cars on the road, give many opportunities to technicians trained in the diagnosis and repair of drive shafts and universal joints.

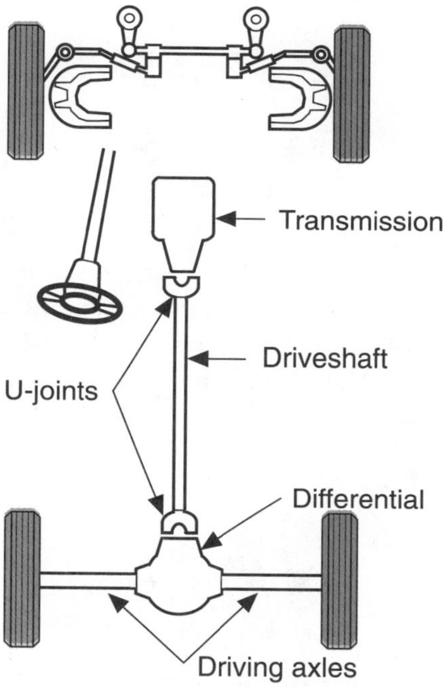

Figure 6-1 RWD drive train.

The drive shaft is normally made from seamless steel tubing with universal joint yokes welded to both ends of the shaft (Figure 6-2). Some drive lines have two drive shafts and four universal joints and use a center support bearing that serves as the connecting link between the two halves (Figure 6-3).

The drive shaft is often called the propeller shaft.

Four-wheel-drive vehicles use two drive shafts, one to drive the front wheels and the other to drive the rear wheels (Figure 6-4). FWD cars, 4WD vehicles equipped with an independent front suspension, and RWD vehicles equipped with independent rear suspension use an additional pair of short drive shafts. These shafts are actually the car's drive axles, which transmit the torque from the differential to each drive wheel.

The short drive shafts or axles used on FWD, 4WD, and RWD with IRS are called half-shafts.

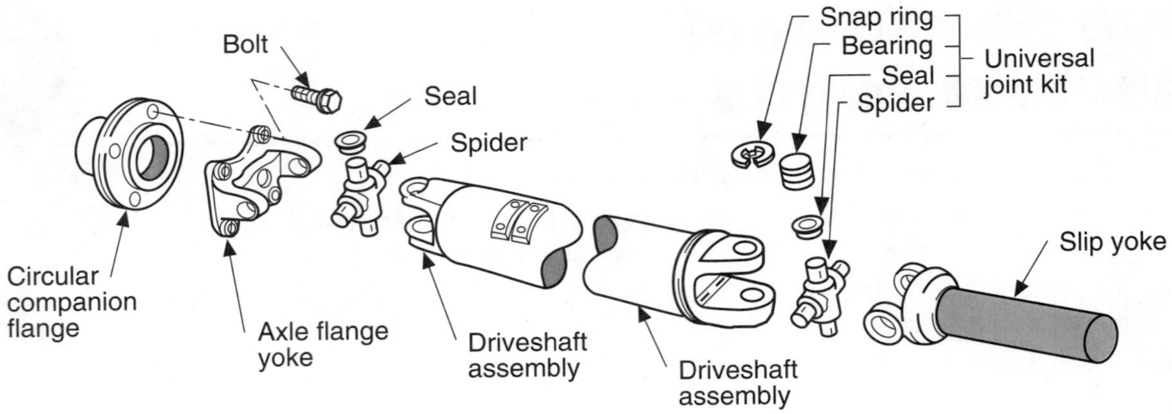

Figure 6-2 Typical drive shaft assembly. (Reprinted with the permission of Ford Motor Company)

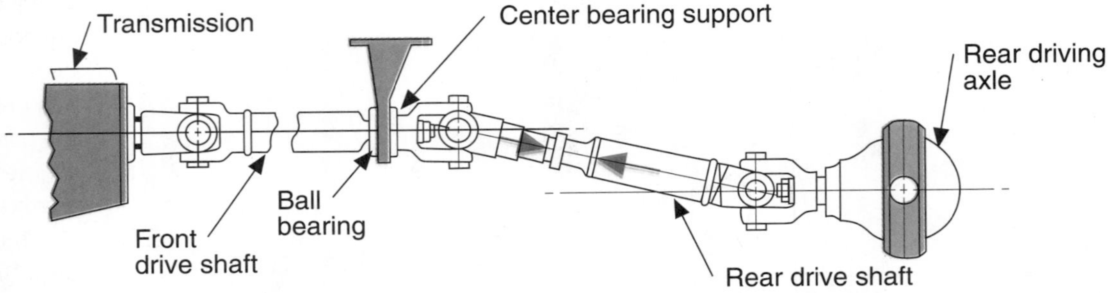

Figure 6-3 A two-piece drive shaft is often used when there is a great distance between the transmission and the rear axle. (Courtesy of Chrysler Corporation)

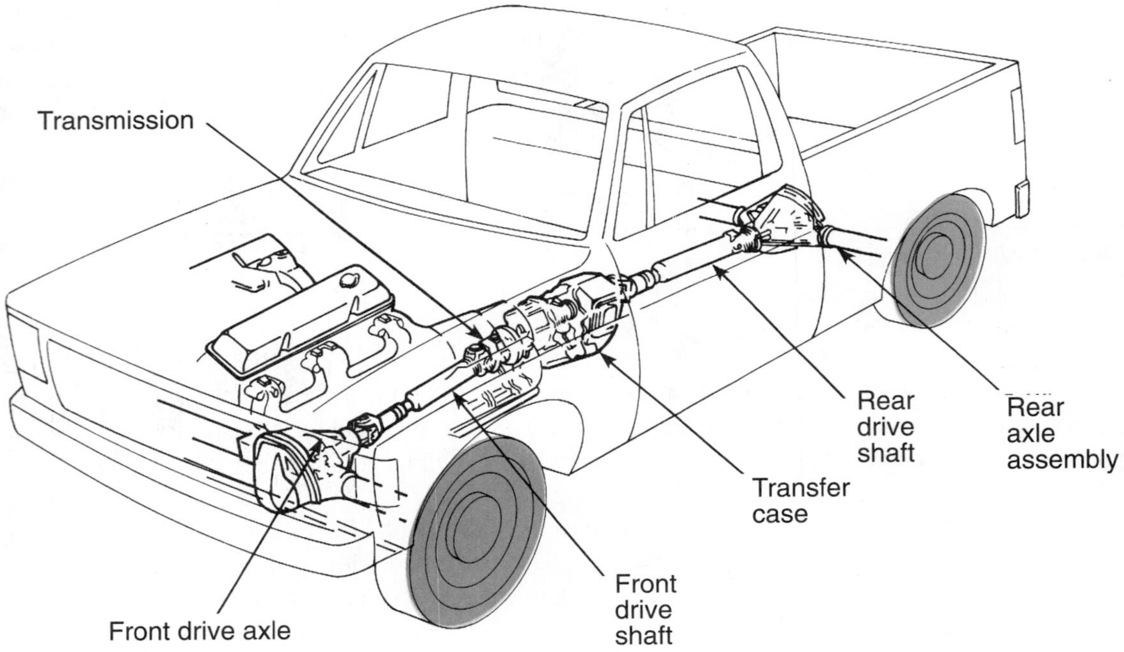

Figure 6-4 4WD truck with two drive shafts.

A BIT OF HISTORY

The Autocar, in 1901, was the first car in the United States to use a drive shaft.

Drive Shaft Construction

Two facts must be considered when designing a drive shaft: the engine and transmission are more or less rigidly attached to the car frame and the rear axle housing, with the wheels and differential, is attached to the frame by springs. As the rear wheels encounter irregularities in the road, the springs compress or expand. This changes the angle of the drive line between the transmission and the rear axle housing. It also changes the distance between the transmission and the differential.

In order for the drive shaft to respond to these constant changes, drive shafts are equipped with one or more universal joints, which permit variations in the angle of the shaft, and a slip joint, which permits the effective length of the drive line to change.

A drive shaft is actually an extension of the transmission's output shaft, as its sole purpose is to transfer torque from the transmission to the drive axle (Figure 6-5). It is usually made from seamless steel tubing with a yoke welded or pressed onto each end. The yokes provide a means of connecting two or more shafts together.

The drive shaft, like any other rigid tube, has natural vibration frequency. This means that if one end of the tube was held tightly, the tube would vibrate at its own frequency when it has been deflected and released or when it rotates. It reaches this natural frequency at its critical

Shop Manual
Chapter 6, page 167

In normal automotive terms, suspension spring compression is called jounce.

The expansion of a suspension spring is suspension rebound.

Critical speed is the rotational speed at which an object begins to vibrate as it turns. This is mostly caused by centrifugal forces.

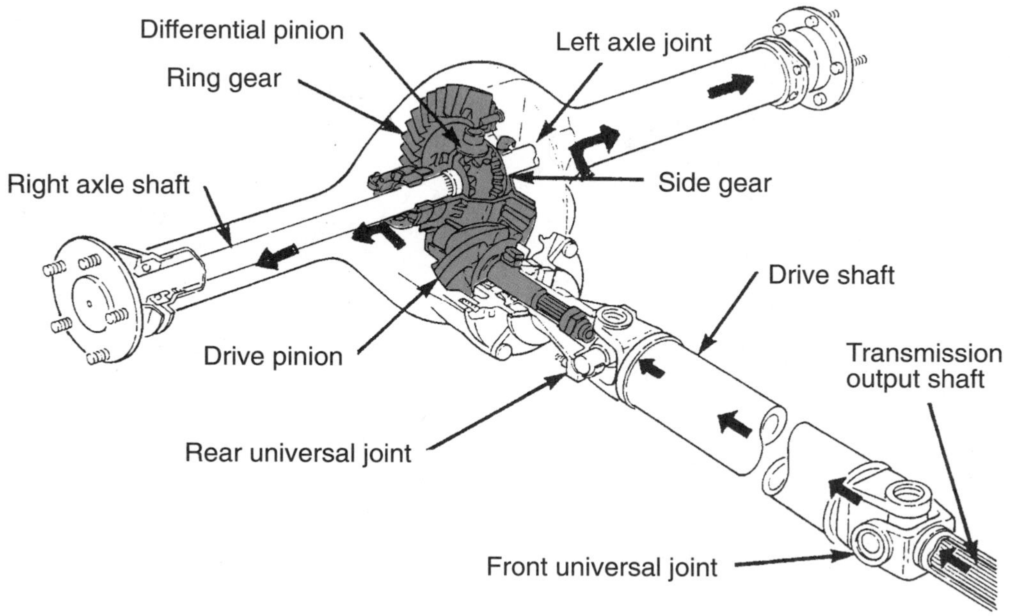

Figure 6-5 Power flows from the transmission to the rear axle. (Reprinted with the permission of Ford Motor Company)

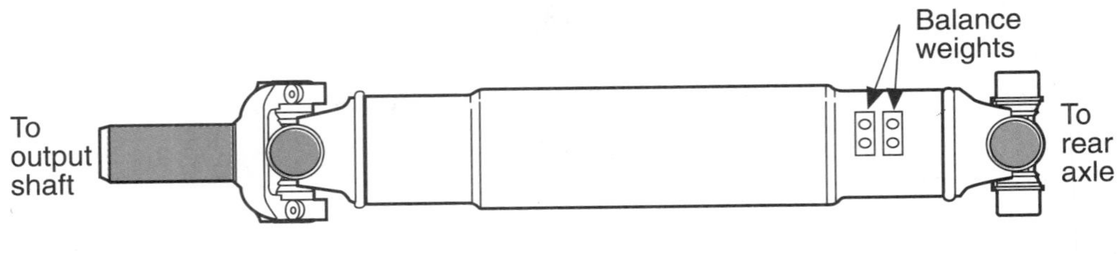

Figure 6-6 Location of the balance weights on a drive shaft. (Courtesy of Oldsmobile Div.—GMC)

A drive shaft rotates at three to four times the speed of the car's tires.

The operating angle of a drive shaft is the amount the drive shaft deviates from the horizontal plane.

speed. The critical speed of a drive shaft depends on the diameter of the tube and the length of the drive shaft. Drive shaft diameters are as large as possible and shafts as short as possible to keep the critical speed frequency above the normal driving range.

Because the drive shaft rotates at high speeds and at varying angles, it must be balanced to reduce vibration. As the drive shaft's length and the operating angle and speed increase, the likelihood of the shaft becoming out of balance also increases. To offset this tendency, several methods are used to balance drive shafts. One of the most common techniques employed by manufacturers is to balance the drive shaft by welding balance weights to the outside diameter of the drive shaft (Figure 6-6).

To reduce the effects of vibrations and the resulting noises, manufacturers have used various methods to construct a drive shaft (Figure 6-7). An example of this is a drive shaft with cardboard liners inserted into the tube, which serve to decrease the shaft's vibration by increasing its strength.

Drive shaft performance has also been improved by placing biscuits between the drive shaft and the cardboard liner. These biscuits are simply rubber inserts that reduce noise transfer within the drive shaft.

Another drive shaft design is the tube-in-tube, in which the input driving yoke has an input shaft that fits inside the hollow drive shaft. Rubber inserts are bonded to the outside diameter of the input shaft and to the inside diameter of the drive shaft. This design reduces the noise

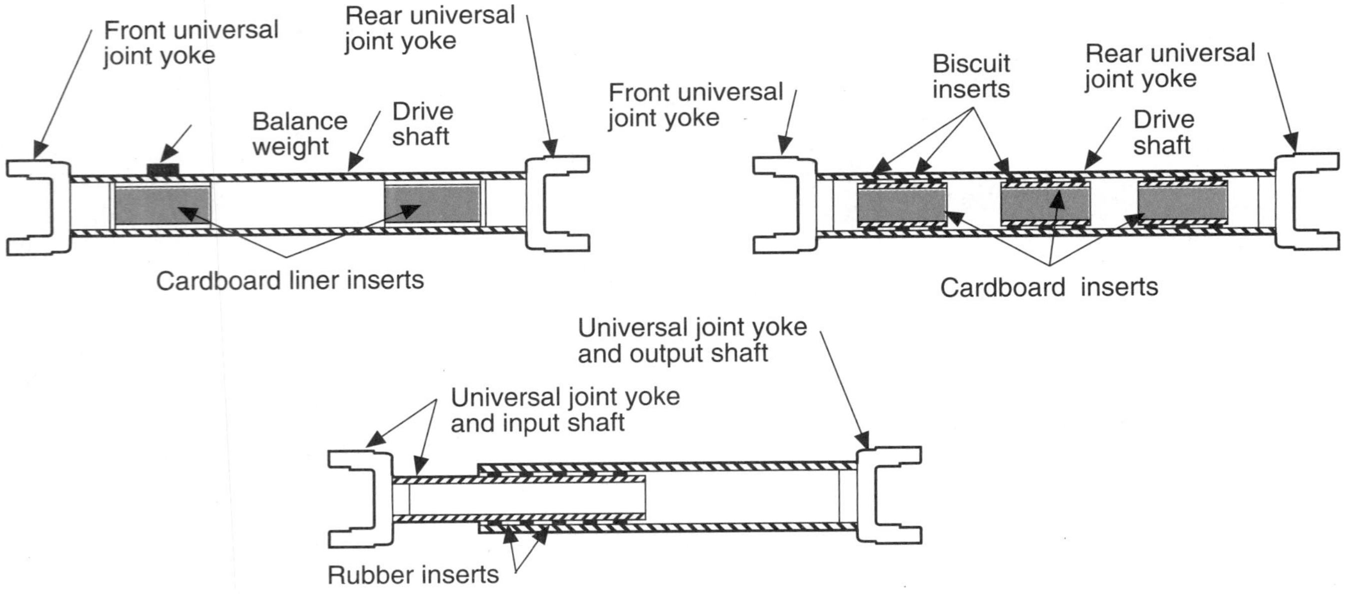

Figure 6-7 Various designs used in drive shaft construction. (Reprinted with the permission of Ford Motor Company)

associated with drive shafts when they are stressed with directional rotation changes. This design also greatly reduces the vibration.

Recently, drive shafts have begun to be made with fiber composites. These composites give the shaft linear stiffness, while the positioning of the fibers provide for torsional strength. The advantages of a fiber composite drive shaft are weight reduction, torsional strength, fatigue resistance, easier and better balancing, and reduced interference from shock loading and torsional problems.

Fiber composites used in the construction of drive shafts are a resin reinforced with fiberglass, graphite, carbon, and/or aluminum.

Types of Drive Shafts

Three types of drive shafts have been used in automobiles. The first type, and the most commonly used, is the Hotchkiss design. This type of drive shaft is readily recognized by its external shaft and U-joints (Figure 6-8). These shafts are either one- or two-piece assemblies consisting of a shaft with U-joints attached to each end. A Hotchkiss drive system can be used with either leaf or coil springs. When it is used with coil springs, additional braces, called control arms, must be used to control the movement of the rear drive axle (Figure 6-9).

Linear stiffness means the shaft will resist deflection regardless of length.

Shop Manual
Chapter 6, page 166

A Hotchkiss drive system actually describes the entire drive shaft and rear axle assembly. This system allows the rear axle to move with the suspension while allowing torque to be transferred from the transmission to the rear axle.

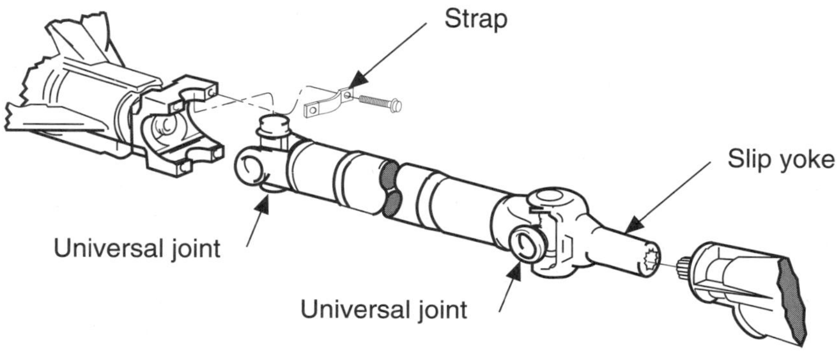

Figure 6-8 Hotchkiss drive. (Courtesy of Oldsmobile Div.—GMC)

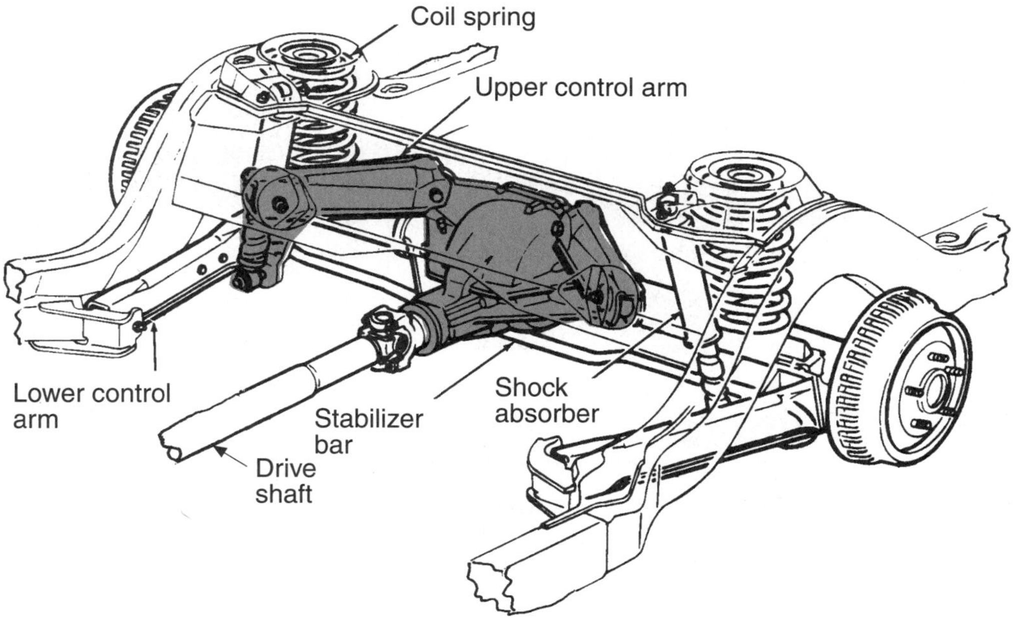

Figure 6-9 Hotchkiss drive with a coil spring suspension.

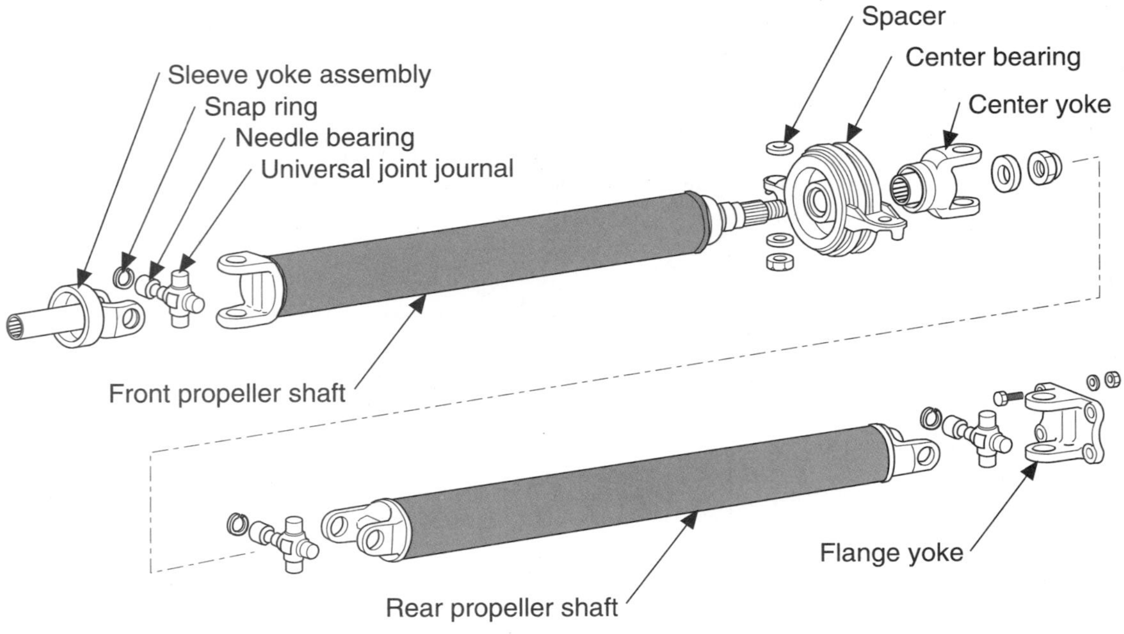

Figure 6-10 A two-piece drive shaft assembly. (Courtesy of Chrysler Corporation)

A two-piece drive shaft is also referred to as a split drive shaft.

A two-piece shaft is used on many long wheelbase vehicles. It uses a third U-joint between the two shaft sections and a center bearing to support the middle of the shaft assembly (Figure 6-10).

The second type of drive shaft is called a torque tube. Vehicles with independent rear suspension and a rear-mounted transaxle—such as the Porsche 924, 928, and 944 models and some Japanese RWD cars—use a torque tube. On these cars, the torque tube is rigidly connected at both ends. The rotating inner drive shaft does not need universal joints because the transaxle location never changes relative to engine location (Figure 6-11). The Chevrolet Chevette used a two-piece drive shaft in which the forward section was a tubular shaft with universal joints at each end and the rear section was a torque tube that was rigidly mounted to the differential.

The third and least commonly used type of drive shaft is the flexible type. This shaft is actually a flexible steel rope, much like an oversized speedometer cable, which does not use universal joints between the engine and the rear-mounted transaxle. The 1961–1963 Pontiac Tempest models used this type of drive shaft.

Universal Joints

Shop Manual
Chapter 6, page172

A drive shaft must smoothly transfer torque while rotating, changing length, and moving up and down. The different designs of drive shafts all attempt to ensure a vibration-free transfer of the engine's power from the transmission to the differential. This goal is complicated by the fact that the engine and transmission are bolted solidly to the frame of the car while the differential is mounted on springs. As the rear wheels go over bumps in the road or changes in the road's surface, the springs compress or expand. This changes the angle of the drive shaft between the transmission

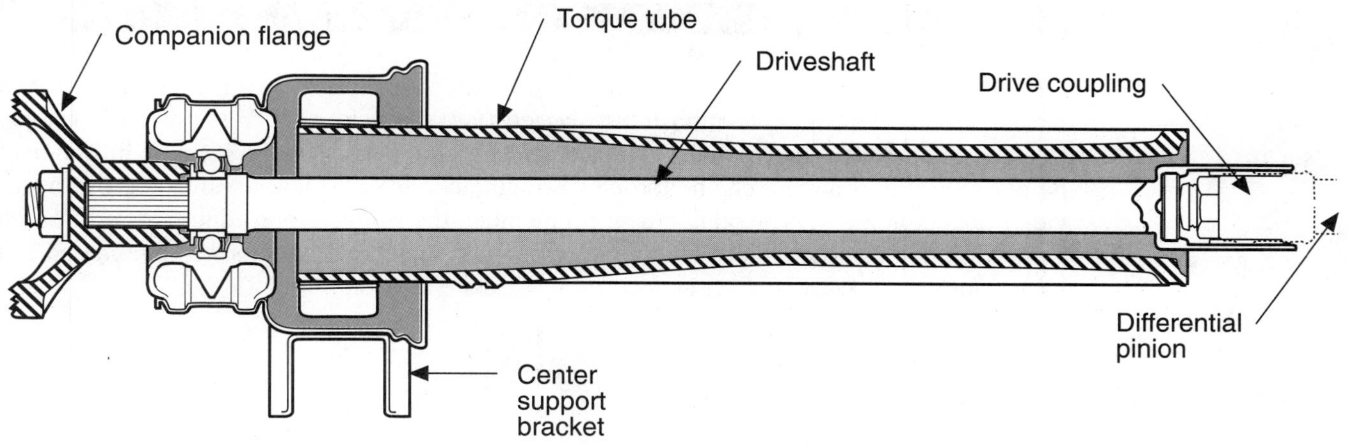

Figure 6-11 A typical torque tube assembly.

and the differential, as well as the distance between the two (Figure 6-12). To allow for these changes, the Hotchkiss-type drive shaft is fitted with one or more U-joints to permit variations in the angle of the drive and a slip joint that permits the effective length of the drive shaft to change.The universal joint is basically a double-hinged joint consisting of two Y-shaped yokes, one on the driving or input shaft and the other on the driven or output shaft, plus a cross-shaped unit called the cross (Figure 6-13). A yoke is used to connect the U-joints together. The four arms of the cross are fitted with bearings in the ends of the two shaft yokes. The input shaft's yoke causes the cross to rotate, and the two other trunnions of the cross cause the output shaft to rotate. When the two shafts are at an angle to each other, the bearings allow the yokes to swing around on their trunnions with each revolution. This action allows two shafts, at a slight angle to each other, to rotate together.

Universal joints allow the drive shaft to transmit power to the rear axle through varying angles that are controlled by the travel of the rear suspension. Because power is transmitted on an angle, U-joints do not rotate at a constant velocity nor are they vibration free.

The joint's cross is often called its spider.

The joint's yoke is a Y-shaped assembly in which two of the joint's arms fit.

The arms of the joint are also called trunnions.

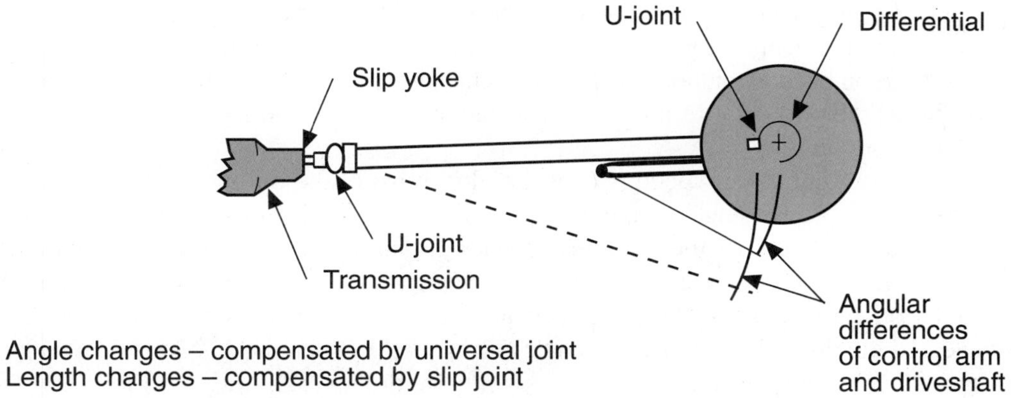

Figure 6-12 Chart showing the changes in the length and angle of a drive shaft. (Reprinted with the permission of Ford Motor Company)

A BIT OF HISTORY

The original universal joint was developed in the sixteenth century by a French mathematician named Cardan. In the seventeenth century, Robert Hooke developed a cross-type universal joint, based on the Cardan. Then in 1902, Clarence Spicer modified Cardan and Hooke's inventions for the purpose of transmitting engine torque to an automobile's rear wheels. By joining two shafts with Y-shaped forks to a pivoting cruciform member, the problem of torque transfer through a connection that also needed to compensate for slight angular variations was eliminated. Both names, Spicer and Hooke, are at times used to describe a Cardan U-joint.

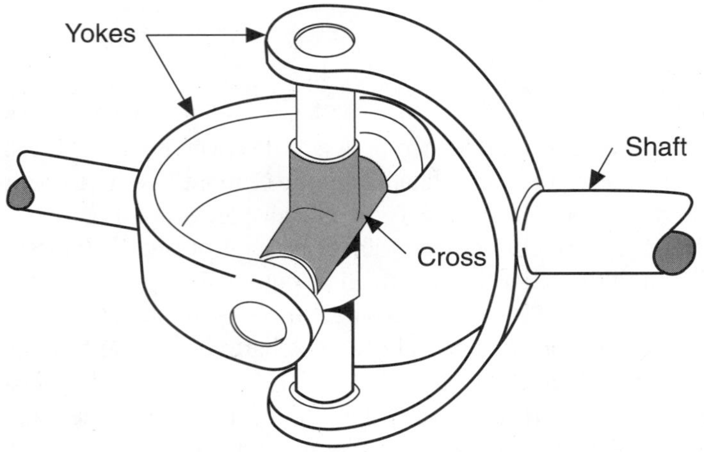

Figure 6-13 A simple universal joint.

Speed Variations

Although simple in appearance, a U-joint is more intricate than it seems. Its natural action is to speed up and slow down twice during each revolution, while operating at an angle. The amount that the speed changes varies according to the steepness of the universal joint's angle.

As a U-joint transmits torque through an angle, its output shaft speed increases and decreases twice on each revolution. These speed changes are not normally apparent, but may be felt as torsional vibration due to improper installation, steep and/or unequal operating angles, and high speed driving.

If a U-joint's input shaft speed is constant, the speed of the output shaft accelerates and decelerates to complete a single revolution at the same time as the input shaft. In other words, the output shaft falls behind, then catches up with the input shaft during this revolution. The greater the angle of the output shaft, the more the velocity will change each shaft revolution.

The installation angle of an object describes how far the object is tilted away from the horizontal plane.

U-joint operating angle is determined by the difference between the transmission installation angle and the drive shaft installation angle (Figure 6-14). When the U-joint is operating at an angle, the driven yoke speeds up and slows down twice during each drive shaft revolution. This acceleration and deceleration of the U-joint is known as speed variation.

These four changes in speed are not normally visible during rotation, but may be understood after examining the action of a U-joint. A universal joint serves as a coupling between two

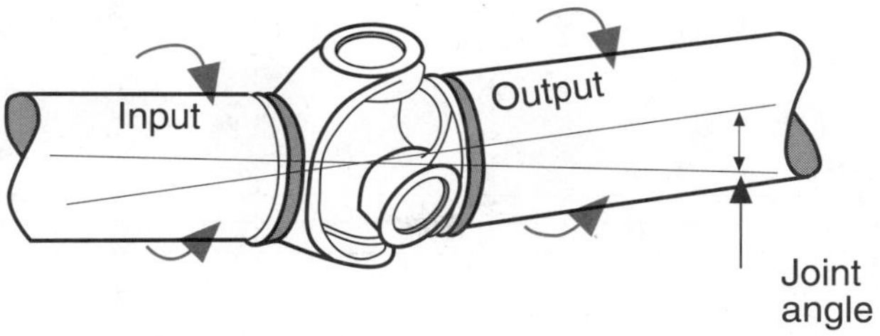

Figure 6-14 Universal joint angles.

shafts that are not in direct alignment. It would be logical to assume that the entire unit simply rotates. This is only true of the joint's input yoke.

The output yoke's rotational path looks like an ellipse because it can be viewed at an angle instead of straight on. This same effect can be obtained by rotating a coin with your fingers. The height of the coin stays the same even though the sides seem to get closer together.

This might seem to be merely a visual effect, however it is more than that. The U-joint rigidly locks the circular action of the input yoke to the elliptical action of the output yoke. The result is similar to what would happen when changing a clock face from a circle to an ellipse.

Like the hands of a clock, the input yoke turns at a constant speed in its true circular path. The output yoke, operating at an angle to the other yoke, completes the path in the same amount of time (Figure 6-15). However, its speed varies, or is not constant, compared to the input.

Speed variation is more easily visualized when looking at the travel of the yokes by 90-degree quadrants (Figure 6-16). The input yoke rotates at a steady or constant speed through a complete 360-degree rotation. The output yoke quadrants alternate between shorter and longer distances of travel than the input yoke quadrants. When one point of the output yoke covers the shorter distance in the same amount of time, it must travel at a slower rate. Conversely, when traveling the longer distance in the same amount of time, it must move faster.

Because the average speed of the output yoke through the four 90-degree quadrants equals the constant speed of the input yoke during the same revolution, it is possible for the two mating yokes to travel at different speeds. The output yoke is falling behind and catching up constantly. The resulting acceleration and deceleration produces a fluctuating torque and torsional vibrations and is characteristic of all Cardan U-joints. The steeper the U-joint angle, the greater the speed fluctuations. Conversely, smaller angles produce less change in speed.

The speed variation of a universal joint is sometimes called speed fluctuation.

An ellipse is merely a compressed form of a circle.

A quadrant is another name for a quarter of something.

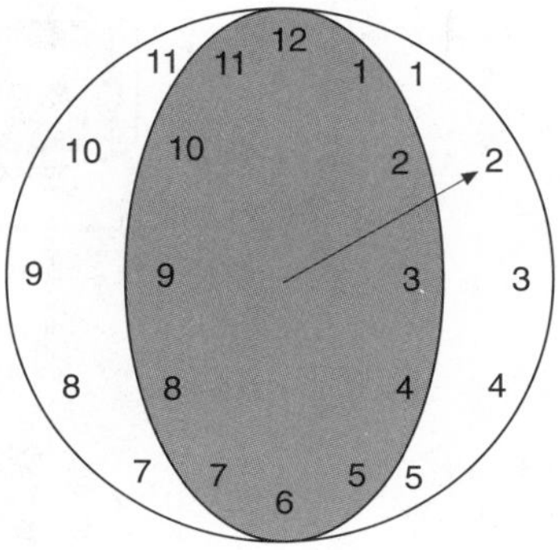

Figure 6-15 The face of a clock can be used to illustrate the elliptical action of the drive shaft's yokes.

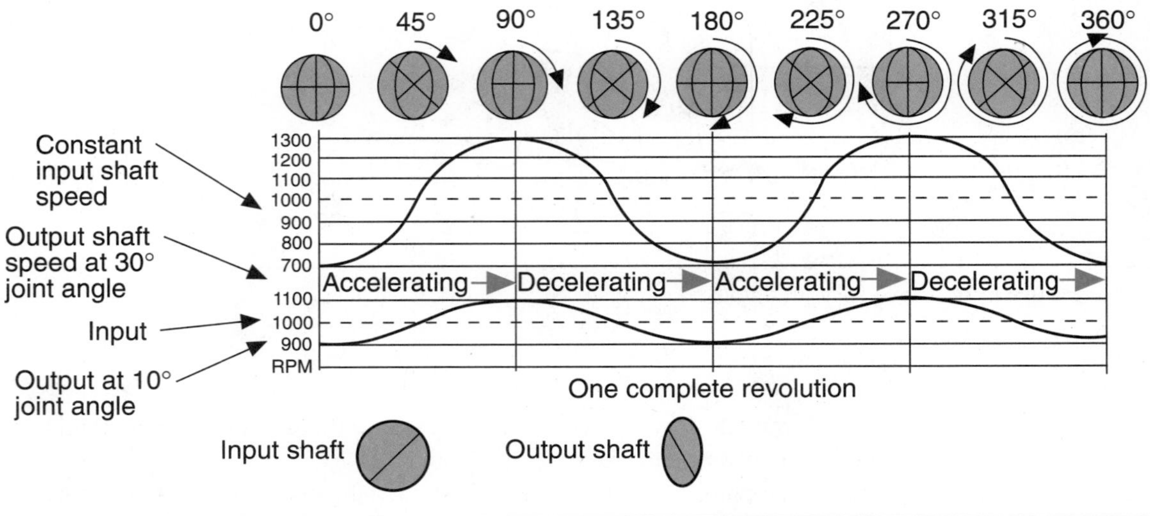

Figure 6-16 A chart showing the speed variations of a drive shaft's yoke.

Phasing of Universal Joints

Shop Manual
Chapter 6, page 169

In-phase describes the condition when two events happen one after the other, regardless of speed.

The torsional vibrations set up by the changes in velocity are transferred down the drive shaft to the next U-joint. At this joint, similar acceleration and deceleration occurs. Because these speed changes take place at equal and reverse angles to the first joint, they cancel out each other whenever both occur at the same angle. To provide for this canceling effect, drive shafts should have at least two U-joints and their operating angles must be slight and equal to each other. Speed fluctuations can be cancelled if the driven yoke has the same point of rotation, or same plane, as the driving yoke. When the yokes are in the same plane, the joints are said to be in-phase (Figure 6-17).

SERVICE TIP: On a two-piece drive shaft, you may encounter problems if you are not careful. The center U-joint must be disassembled to replace the center support

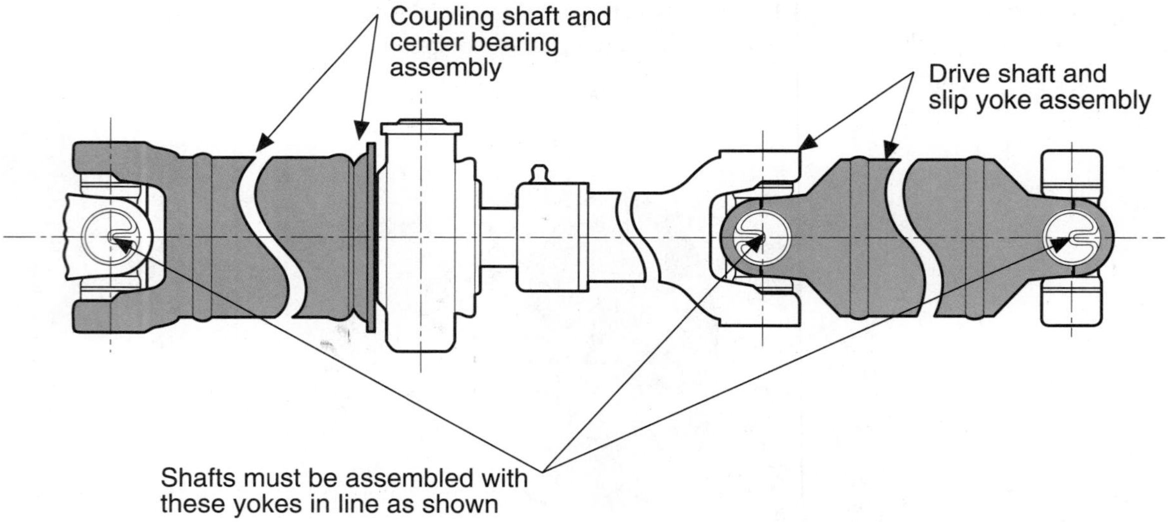

Figure 6-17 The U-joints are in the same plane and are in-phase with each other. (Reprinted with the permission of Ford Motor Company)

bearing. The center driving yoke is splined to the front drive shaft. If the yoke's position on the drive shaft is not indicated in some manner, the yoke could be installed in a position that is out of phase. Manufacturers use different methods of indexing the yoke to the shaft. Some use aligning arrows. Others machine a master spline that is wider than the others. When there are no indexing marks, the technician should always index the yoke to the drive shaft before disassembling the U-joint. This saves time and frustrations during reassembly. Indexing requires only a light hammer and center punch to mark the yoke and drive shaft.

Canceling Angles

Vibrations can be reduced by using canceling angles (Figure 6-18). The operating angle of the front U-joint is offset by the one at the rear of the drive shaft. When the front U-joint accelerates, causing a vibration, the rear U-joint decelerates causing an equal but opposite vibration. These vibrations created by the two joints oppose each other and dampen the vibrations from one to the other. The use of canceling angles provides smooth drive shaft operation.

Shop Manual
Chapter 6, page 166

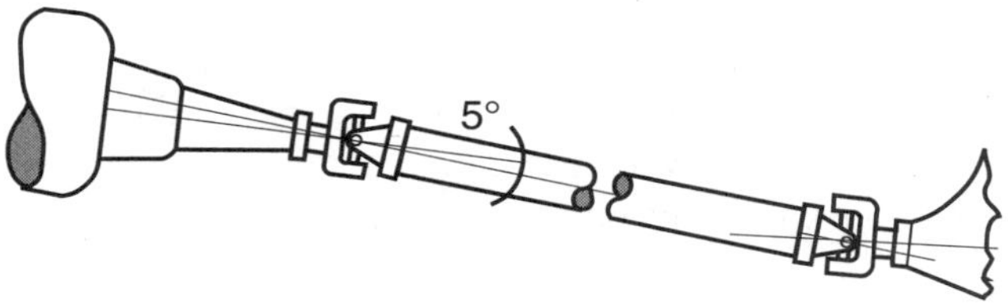

Figure 6-18 Equal U-joint angles reduce the vibrations of the shaft.

The correct operating angle of a U-joint must be maintained in order to prevent drive line vibration and damage. Shimming of leaf springs and the control arms on coil spring suspensions or adjusting the control arm eccentrics allow the operating angle of the drive shaft to be changed. Shimming at the transmission mount can also be done on some vehicles to change universal joint angles.

Shop Manual
Chapter 6, page 183

CAUTION: Extreme care should be taken when working around a rotating drive shaft. Severe injury can result from touching a moving shaft. Never attempt to stop the spinning shaft by hand. It can cause serious physical injury.

Coupled universal joints are commonly called double Cardan constant velocity joints.

Types of Universal Joints

There are three common designs of universal joints: single universal joints retained by either an inside or outside snap ring, coupled universal joints, and universal joints held in the yoke by U-bolts or lock plates.

Shop Manual
Chapter 6, page 172

Single Universal Joints

The single Cardan universal joint's (Figure 6-19) primary purpose is to connect the two yokes that are attached directly to the drive shaft. The joint assembly forms a cross, with four machined trunnions or points equally spaced around the center of the axis. Needle bearings used to reduce friction and provide smoother operation are set into bearing cups. The trunnions of the cross fit into the cup assemblies, which fit snugly into the driving and driven universal joint yokes. U-joint movement takes place between the trunnions, needle bearings, and bearing cups. There should

The Cardan/Spicer universal joint is also known as the cross or four-point joint.

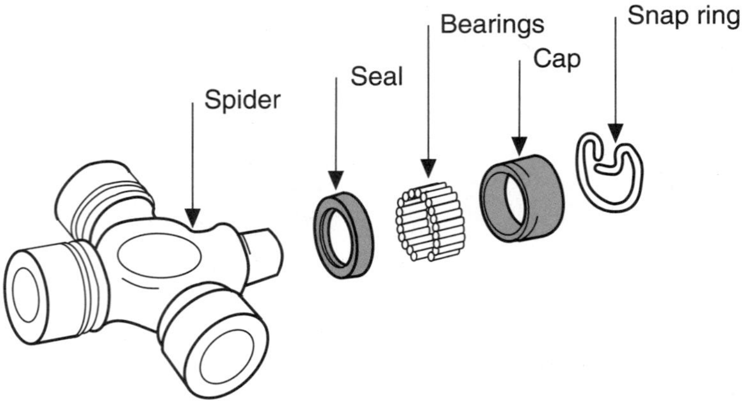

Figure 6-19 A Cardan U-joint. (Courtesy of Oldsmobile Div.—GMC)

be no movement between the bearing cup and its bore in the universal joint yoke. The bearings are usually held in place by snap rings that drop into grooves in the yoke's bearing bores. The bearing caps allow free movement between the trunnion and yoke. The needle bearing caps may also be pressed into the yokes, bolted to the yokes, or are held in place with U-bolts or metal straps.

SERVICE TIP: There are many other methods used to retain the U-joint in its yoke, such as the use of a bearing plate, thrust plate, wing bearing, cap and bolt, U-bolt, and strap (Figure 6-20).

There are other styles of single universal joints. The method used to retain the bearing caps is the major difference between these designs. The Spicer style uses an outside snap ring that fits into a groove machined in the outer end of the yoke (Figure 6-21). The bearing cups for this style are machined to accommodate the snap ring.

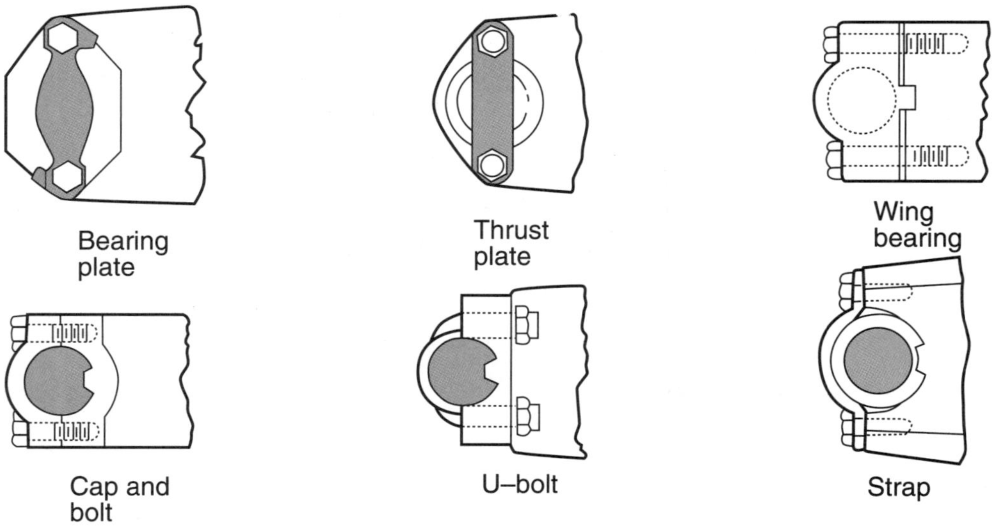

Figure 6-20 Various methods used to retain U-joints in their yokes.

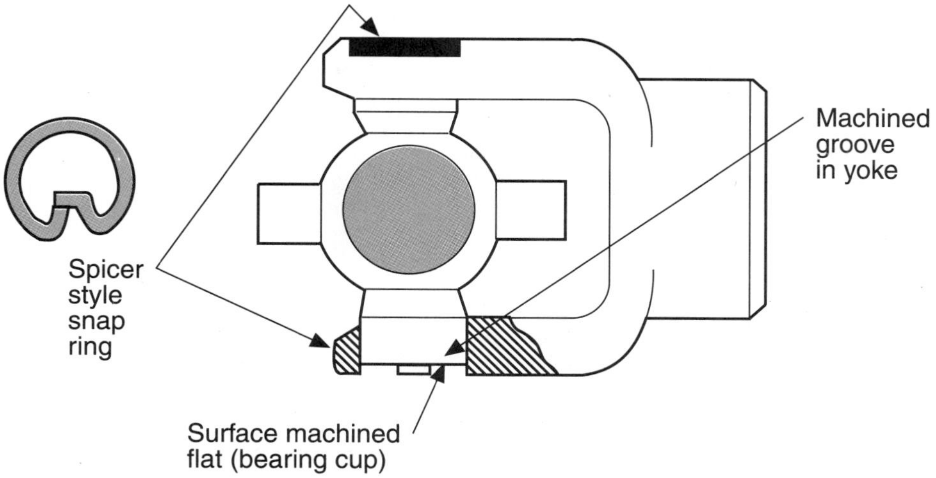

Figure 6-21 Spicer-style U-joint.

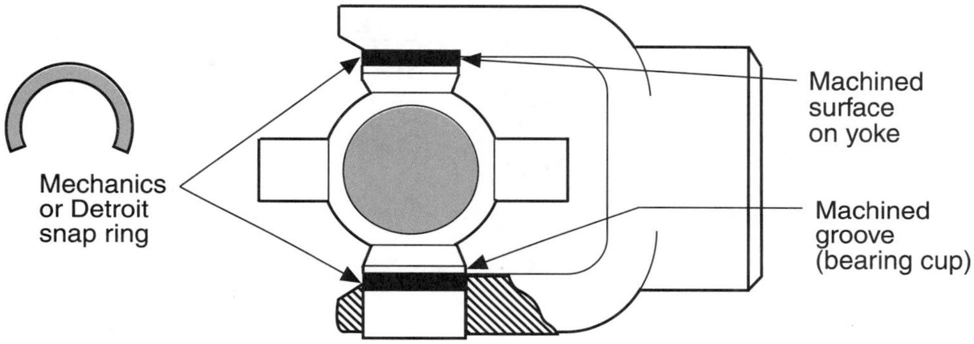

Figure 6-22 Mechanics-style U-joint.

The mechanics-style joint is also called the Detroit/Saginaw style.

The mechanics style uses an inside snap ring or C-clip that fits into a groove machined in the bearing cup on the side closer to the grease seal (Figure 6-22). When installed, the clip rest against the machined portion of the yoke. The snap rings are retained by nylon injected into the retaining ring grooves.

The Cleveland style is an attempt to combine styles of universal joints to obtain more applications from one joint. The bearing cups for this U-joint are machined to accommodate either Spicer- or mechanics-style snap rings. If a replacement U-joint comes with both style clips, use the clips that pertain to your application.

SERVICE TIP: In order to get the correct parts, you should know the type of U-joint and the yoke span originally in the car (Figure 6-23).

Double Cardan Universal Joint

Shop Manual
Chapter 6, page 177

A double Cardan U-joint is used with split drive shafts and consists of two individual Cardan U-joints closely connected by a centering socket yoke and a center yoke that functions like a ball and socket. The ball and socket splits the angle of two drive shafts between two U-joints

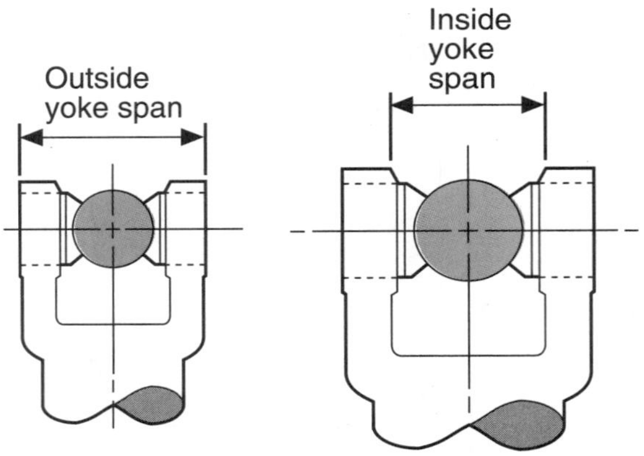

Figure 6-23 You should know both the outside (A) and inside (B) yoke span when ordering a new joint.

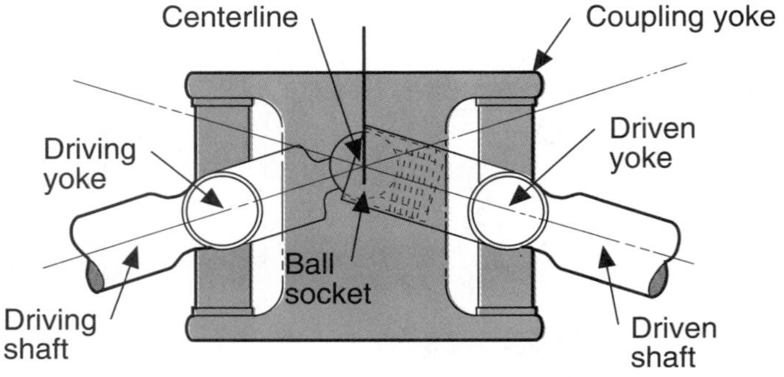

Figure 6-24 The ball and socket of a double Cardan joint splits the angle of the two shafts.

(Figure 6-24). Because of the centering socket yoke, the total operating angle is divided equally between the two joints. Because the two joints operate at the same angle, the normal fluctuations that result from the use of a single U-joint are canceled out. The acceleration or deceleration of one joint is canceled by the equal and opposite action of the other.

Most often installed in front-engined rear-wheel-drive luxury cars, the double Cardan universal joint smoothly transmits torque regardless of the operating angle of the driving and driven members. It is therefore classified as a constant velocity universal joint (Figure 6-25). This joint is used when the U-joint operating angle is too large for a single joint to handle. On some vehicles, the double joint is used at both ends of the drive shaft. On other vehicles, the double joint is used only on the drive end of the drive shaft.

Shop Manual
Chapter 6, page 177

Constant Velocity Joints

CV-joints are primarily used on FWD drive axles; however, some RWD cars with IRS—such as Ford Thunderbird, BMW, and Porsche—use them. The commonly used types of CV-joints are the Rzeppa and the tripod.

Shop Manual
Chapter 6, page 168

Slip Joints

As well as being equipped with universal joints to allow for angle changes, a drive shaft must also be able to change its effective length. As road surfaces change, the drive axle assembly

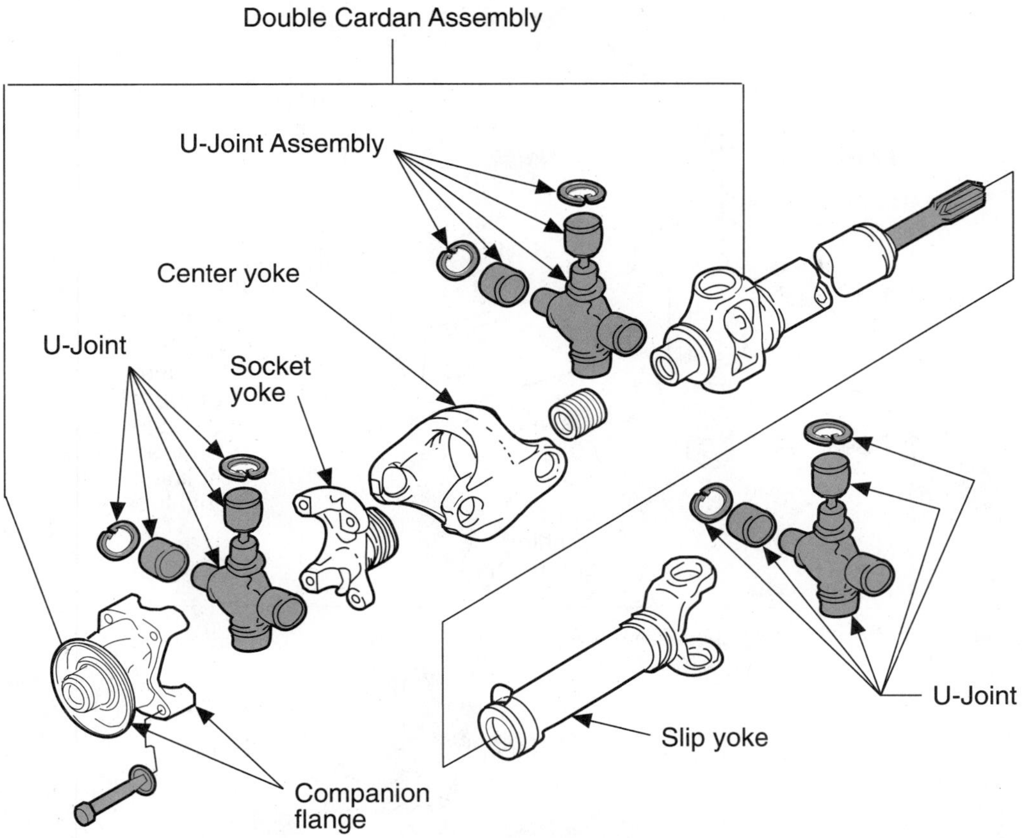

Figure 6-25 A double Cardan joint disassembled. (Reprinted with the permission of Ford Motor Company)

moves up and down with the rear suspension. Because the transmission is mounted to the frame, it does not move with the movement of the suspension and the relative distance between the transmission and rear drive axle changes. Drive shafts use a slip joint at one end of the drive shaft, which allows it to lengthen or shorten. The purpose of the slip joint is similar to the plunging CV-joint used in FWD cars. The slip yoke is either positioned at the center of two-piece designs or at either end of the drive shaft, but is typically fitted to the front U-joint (Figure 6-26).

A slip joint assembly (Figure 6-27) includes the transmission's output shaft, the slip joint itself, a yoke, U-joint, and the drive shaft. The output shaft has external splines that match the internal splines of the slip joint. The meshing of the splines allows the two shafts to rotate together, but permits the ends of the shafts to slide along each other. This sliding motion allows for an effective change in the length of the drive shaft, as the drive axles move toward or away from the car's frame. The yoke of the slip joint is connected to the drive shaft by a U-joint.

Some drive shafts are equipped with plunging-type CV-joints in place of slip and universal joints.

A slip yoke is also called a sliding yoke.

Maintenance

The drive shaft assembly requires very little maintenance. Factory-installed U-joints are normally sealed and require no periodic lubrication. Replacement U-joints, however, are equipped with a grease fitting and should be lubricated on a regular basis. The drive shaft itself requires no maintenance except for an occasional cleaning off of any dirt buildup on the shaft. Dirt and any other buildup on the shaft will affect its balance. While cleaning the shaft, inspect it for damage and missing balance weights.

Shop Manual
Chapter 6, page 167

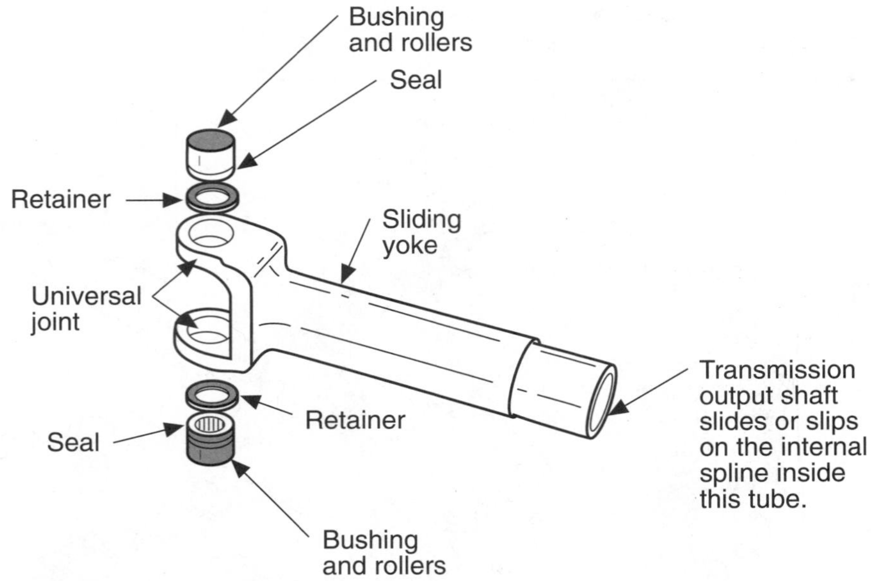

Figure 6-26 Typical slip joint assembly.

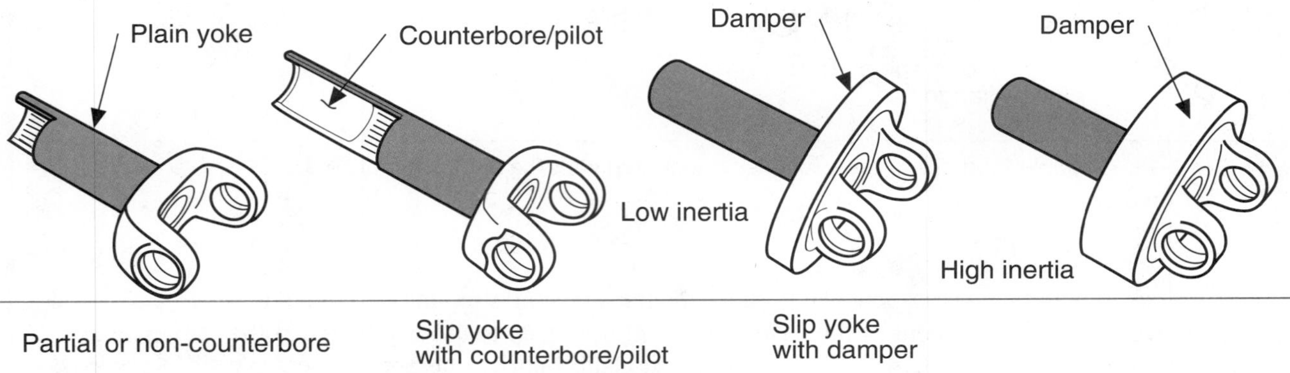

Figure 6-27 Commonly used slip joints (Reprinted with the permission of Ford Motor Company).

Summary

Terms to Know

Balance

Canceling angles

Cardan universal joint

Center hanger bearing

Center Line

Companion flange

Double Cardan universal joint

- ❑ A drive shaft is normally made from seamless steel tubing with universal joint yokes fastened to each end.
- ❑ U-joints are used to allow the angle and the effective length of the drive shaft to change.
- ❑ A drive shaft is actually no more than a flexible extension of the transmission's output shaft.
- ❑ Yokes on the drive shaft provide a means of connecting the shafts together.
- ❑ Balance weights are welded to the outside of a drive shaft to balance it and reduce its natural vibrations.
- ❑ Some drive shafts are internally lined with cardboard or cardboard with rubber inserts to help offset torsional vibration problems.
- ❑ A Hotchkiss drive system has an external drive shaft with at least two universal joints.

- ❑ A torque tube consists of a tubular steel or small diameter solid shaft enclosed in a larger steel tube and is rigidly connected to the transmission and the rear axle housing.
- ❑ The operating angle of a universal joint is determined by the installation angles of the transmission and rear axle housing.
- ❑ U-joints vibrate if the connecting shafts are not on the same plane because one shaft will be accelerating and decelerating at different speeds than the other.
- ❑ Drive shaft vibrations can be reduced by using canceling angles. The operating angle of the front joint is offset by the angle of the rear joint.
- ❑ Speed fluctuations can be canceled if the driven yoke has the same point of rotation, or same plane, as the driving yoke. When the two yokes are in the same plane, the joints are said to be in-phase.
- ❑ There are three common designs of universal joints: single U-joints retained by either an inside or outside snap ring, coupled U-joints, and U-joints held in the yoke by U-bolts or lock plates.
- ❑ A single Cardan joint, the most common type of joint, uses a spider, four machined trunnions, needle bearings, and bearing caps to allow the transmission of power through slight shaft angle changes.
- ❑ A double Cardan joint is called a constant velocity joint because shaft speeds do not fluctuate, regardless of the shaft's angle. These joints are used on two-piece shaft assemblies and are actually two single Cardan joints joined together by a center bearing assembly.
- ❑ The methods used to retain a U-joint in its yoke are the use of a snap ring, C-clip, bearing plate, thrust plate, wing bearing, cap and bolt, U-bolt, and strap.

Terms to Know (continued)

Drive shaft installation angle
Ellipse
Hotchkiss drive
Installation angle
Operating angle
Phasing
Propeller shaft
Sliding yoke
Slip joint
Torque tube
Yoke

Review Questions

Short Answer Essays

1. State the purposes of a universal joint.
2. What methods are used by some auto manufacturers to reduce drive shaft torsional vibration problems?
3. What determines the operating angle of a universal joint?
4. Explain why and when U-joints vibrate.
5. What effect do in-phase universal joints have on the operation of a drive shaft?
6. What is meant by canceling angles?
7. Describe the construction and operation of a single Cardan joint.
8. Describe the construction and operation of a double Cardan joint.
9. Why are some cars equipped with a two-piece drive shaft?
10. What methods are used to retain a U-joint in its yoke?

Fill-in-the-Blanks

1. A drive shaft is normally made from ____________ ____________ tubing with ____________ ____________ ____________ fastened to each end.

2. A drive shaft is actually an extension of the transmission's ____________________

____________________.

3. ____________________ on the drive shaft provide a means of connecting the shafts together.

4. ____________________ ____________________ are welded to the outside of a drive shaft to ____________________ it and reduce its natural ____________________.

5. A slip joint is normally fitted to the ____________________ ____________________ ____________________.

6. A Hotchkiss drive system has an external ____________________ ____________________ with at least ____________________ universal joints.

7. A ____________________ ____________________ is made up of a small diameter solid shaft enclosed in a larger steel tube and is rigidly connected to the transmission and the rear axle housing.

8. There are three common designs of universal joints: ____________________ retained by either an inside or outside snap ring, ____________________ , and ____________________ held in the yoke by U-bolts or lock plates.

9. The three basic styles of Cardan joints are: ____________________ , ____________________ , and ____________________.

10. The type of drive shaft with an input shaft bonded with rubber to the inside of a hollow drive shaft is a ____________________.

ASE Style Review Questions

1. While discussing drive shaft design:
   Technician A says the shorter the shaft is, the less likely it is to become out of balance.
   Technician B says cardboard is often inserted into the shaft to give the shaft more strength.
   Who is correct?
   A. A only
   B. B only
   C. Both A and B
   D. Neither A nor B

2. While discussing the Hotchkiss design of drive shaft:
   Technician A says it consists of a small diameter solid shaft enclosed in a larger tube.
   Technician B says it has a universal joint attached to each end of the shaft.
   Who is correct?
   A. A only
   B. B only
   C. Both A and B
   D. Neither A nor B

3. While discussing the phasing of universal joints:
Technician A says this allows the speed changes at each joint to be canceled by the other joint.
Technician B says this means that both joints are positioned at equal and reverse angles to each other.
Who is correct?
A. A only
B. B only
C. Both A and B
D. Neither A nor B

4. While discussing the different types of universal joints:
Technician A says the most commonly used type is the double cardan.
Technician B says the cardan type is also called the cross and bearing type.
Who is correct?
A. A only
B. B only
C. Both A and B
D. Neither A nor B

5. While discussing slip joints:
Technician A says they allow the drive shaft to maintain its effective length.
Technician B says they normally have external joints on one shaft and internal splines on the other.
Who is correct?
A. A only
B. B only
C. Both A and B
D. Neither A nor B

6. While discussing the different types of drive shafts that have been used to reduce vibrations:
Technician A says the flexible rope design is the most commonly used.
Technician B says the tube-in-tube design is not fitted with universal joints.
Who is correct?
A. A only
B. B only
C. Both A and B
D. Neither A nor B

7. Who is correct?
Technician A says joint speed fluctuations can be canceled if the driven yoke has the same plane of rotation as the driving yoke.
Technician B says joint speed fluctuations can be canceled by putting the joints in-phase.
A. A only
B. B only
C. Both A and B
D. Neither A nor B

8. While discussing slip joints:
Technician A says the yoke normally has external splines.
Technician B says the slip yoke is normally held in place on the transmission's output shaft by a snap ring.
Who is correct?
A. A only
B. B only
C. Both A and B
D. Neither A nor B

9. While discussing double Cardan joints:
Technician A says they are actually two single Cardan joints assembled together at a center bearing support.
Technician B says they are considered constant velocity joints.
Who is correct?
A. A only
B. B only
C. Both A and B
D. Neither A nor B

10. Who is correct?
Technician A says excessive joint operating angles can cause an increase in shaft vibration.
Technician B says a drive shaft can vibrate if the vehicle is driven at high speeds.
A. A only
B. B only
C. Both A and B
D. Neither A nor B

CHAPTER 7

Differentials and Drive Axles

Upon completion and review of this chapter, you should be able to:

- ❑ Describe the purpose of a differential.
- ❑ Identify the major components of a differential and explain their purpose.
- ❑ Describe the various gears in a differential assembly and state their purpose.
- ❑ Describe the various methods used to mount and support the drive pinion shaft and gear.
- ❑ Explain the need for drive pinion bearing preload.
- ❑ Describe the difference between hunting, nonhunting, and partial nonhunting gear sets.
- ❑ Explain the purpose of the major bearings within a differential assembly.
- ❑ Describe the operation of a limited-slip differential.
- ❑ Describe the construction and operation of a rear axle assembly.
- ❑ Identify and explain the operation of the two major designs of rear axle housings.
- ❑ Explain the operation of a FWD differential and its drive axles.
- ❑ Describe the different types of drive axles and the bearings used to support each of them.

Introduction

The drive axle of a RWD vehicle is mounted at the rear of the car. These assemblies use a single housing to mount the differential gears and axles (Figure 7-1). The entire housing is part of the suspension and helps to locate the rear wheels.

Another type of rear drive axle is used with IRS. With IRS the differential is bolted to the chassis and does not move with the suspension. The axles are connected to the differential and drive wheels by U-joints. Because the axles move with the suspension and the differential is bolted to the chassis, a common housing for these parts is impossible.

On most RWD cars, the final drive is located in the rear axle housing. On most FWD cars, the final drive is located within the transaxle. Some FWD cars with longitudinally mounted engines locate the differential and final drive in a separate case that bolts to the transmission (Figure 7-2).

A differential is needed between any two drive wheels, whether in a RWD, FWD, or 4WD vehicle. The two drive wheels must turn at different speeds when the vehicle is in a turn.

RWD final drives normally use a hypoid ring and pinion gearset that turns the powerflow 90 degrees from the drive shaft to the drive axles. A hypoid gearset allows the drive shaft to be positioned low in the vehicle because the final drive pinion gear centerline is below the ring gear centerline (Figure 7-3).

On FWD cars with transversely mounted engines, the powerflow axis is naturally parallel to that of the drive axles. Because of this, a simple set of helical gears in the transaxle serve as the final drive.

The differential is a geared mechanism located between two driving axles. It rotates the driving axles at different speeds when the vehicle is turning, and at the same speed when the vehicle is traveling in a straight line. The drive axle assembly directs drive line torque to its driving wheels, the gear ratio between the pinion and ring gear is used to increase torque and improve driveability, the gears within the differential serve to establish a state of balance between the forces or torques between the drive wheels, and allows the drive wheels to turn at different speeds as the vehicle turns a corner.

Shop Manual
Chapter 7, page 191

Not too long ago, a differential was something that was in the rear axle assembly. Now, with the popularity of FWD vehicles, the differential is part of the transaxle and is most often called the final drive.

Normally, rear axles on RWD vehicles are called live axles because they transmit power.

IRS stands for independent rear suspension.

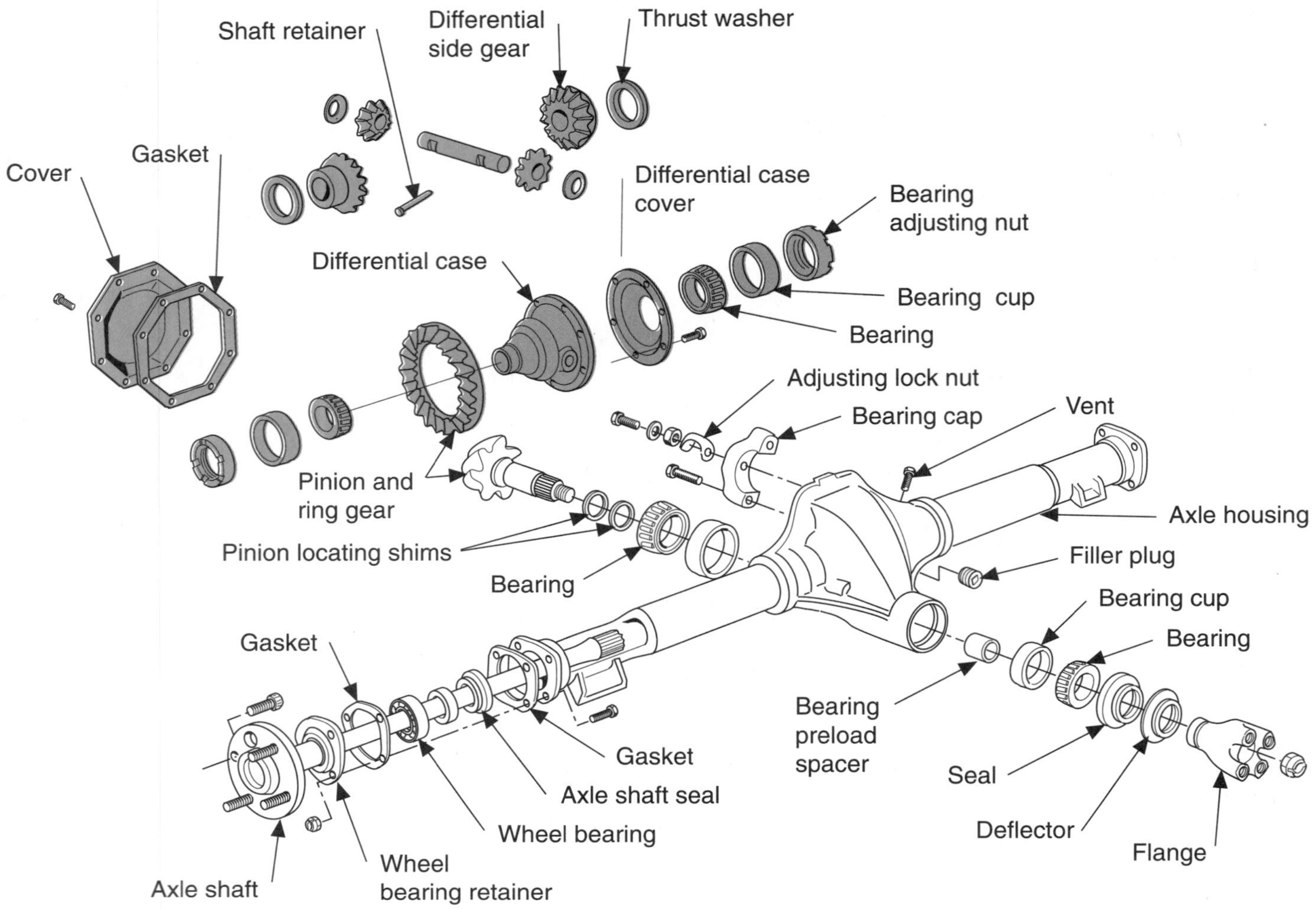

Figure 7-1 Typical rear drive axle assembly. (Reprinted with the permission of Ford Motor Company)

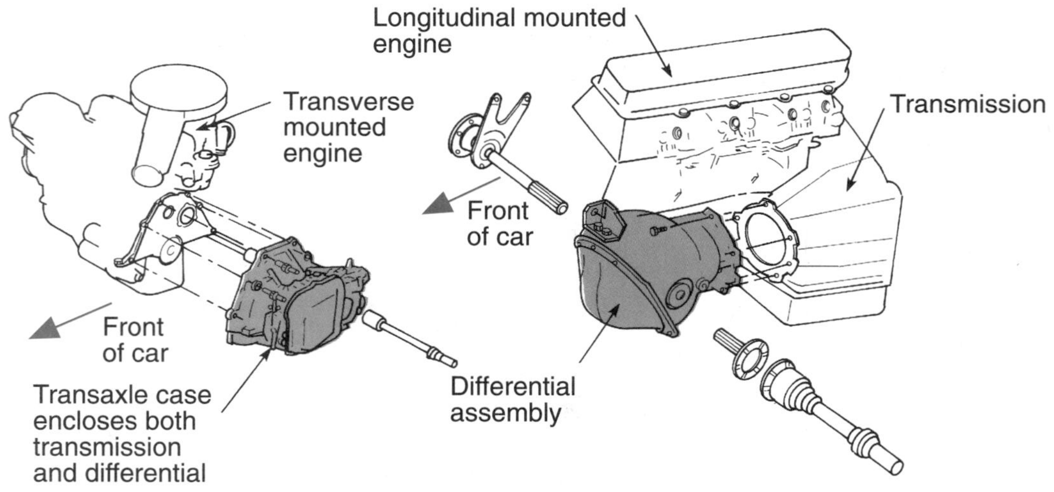

Figure 7-2 Location of differential on FWD vehicles. (Courtesy of Chevrolet)

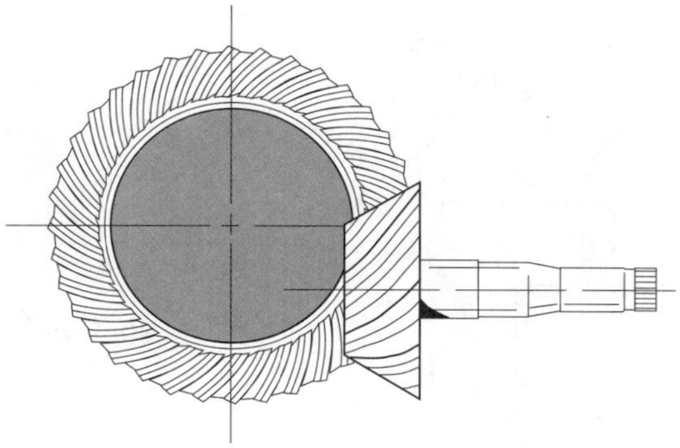

Figure 7-3 A hypoid gearset. (Courtesy of Nissan Motors)

Final drive is the final set of reduction gears the engine's power passes through on its way to the drive wheels.

When engines are placed longitudinally in the car, they are said to have "north/south" placement.

In a ring and pinion gearset, the pinion is the smaller drive gear and the ring gear is the larger driven gear.

Helical gears are gears on which the teeth are at an angle to the gear's axis of rotation.

When engines are mounted transversely in the car, they are said to have "east/west" or sideways placement.

Function and Components

The differential allows for different speeds at the drive wheels when a vehicle goes around a corner or any time there is a change of direction. When a car turns a corner, the outside wheels must travel farther and faster than the inside wheels (Figure 7-4). If compensation is not made for this difference in speed and travel, the wheels would skid and slide, causing poor handling and excessive tire wear. Compensation for the variations in wheel speeds is made by the differential assembly. While allowing for these different speeds, the diffeerential also must continue to transmit torque.

The differential of a RWD vehicle is normally housed with the drive axles in a large casting called the rear axle assembly. Power from the engine enters into the center of the rear axle assembly and is transmitted to the drive axles. The drive axles are supported by bearings and are attached to the wheels of the car. The power entering the rear axle assembly has its direction changed by the differential. This change of direction is accomplished through the hypoid gears used in the differential.

A differential is an assembly of gears located between the axle shafts of the drive wheels. It allows for different speeds between the two drive wheels, multiplies torque, and changes the

A BIT OF HISTORY

Early automobiles were driven by means of belts and ropes around pulleys mounted on the driving wheels and engine shaft or transmission shaft. As there was always some slippage of the belts, one wheel could rotate faster than the other when turning a corner. When belts proved unsatisfactory, automobile builders borrowed an idea from the bicycle and applied sprockets and chains. This was a positive driving arrangement, which made it necessary to provide differential gearing to permit one wheel to turn faster than the other.

The term *differential* means relating to or exhibiting a difference or differences.

Other names commonly used to refer to the differential of RWD vehicles are rear axle, drive axle, third member, center section, or "pumpkin."

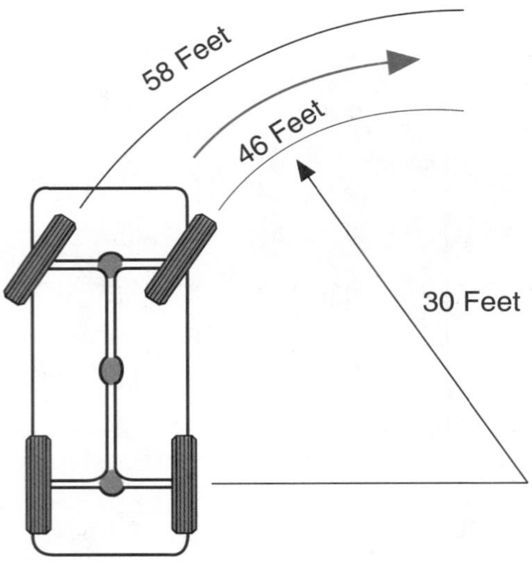

Figure 7-4 Travel of wheels when a vehicle is turning a corner.

direction of the power flow. Power from the drive shaft is transmitted to the axle assembly through the pinion flange. This flange is the connecting yoke to the rear universal joint. Power then enters the differential assembly on the pinion gear (Figure 7-5). The pinion teeth engage the ring gear, which is mounted upright at a 90-degree angle to the pinion. Therefore, as the drive shaft turns, so do the pinion and ring gears.

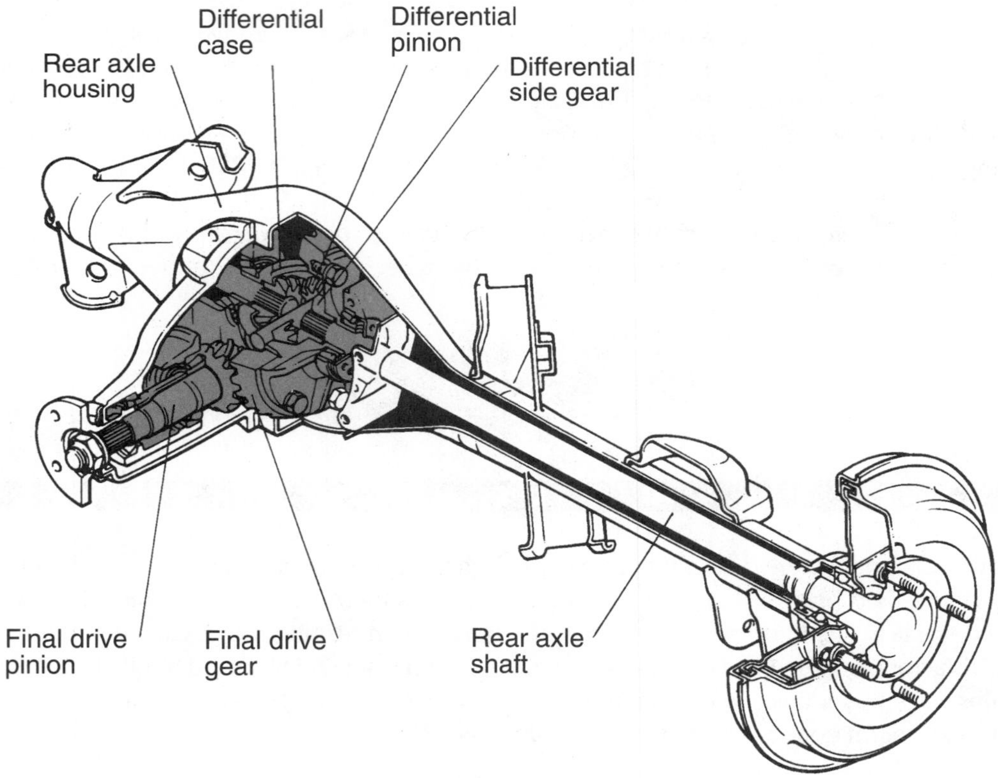

Figure 7-5 Main components of a RWD drive axle. (Courtesy of Chrysler Corporation)

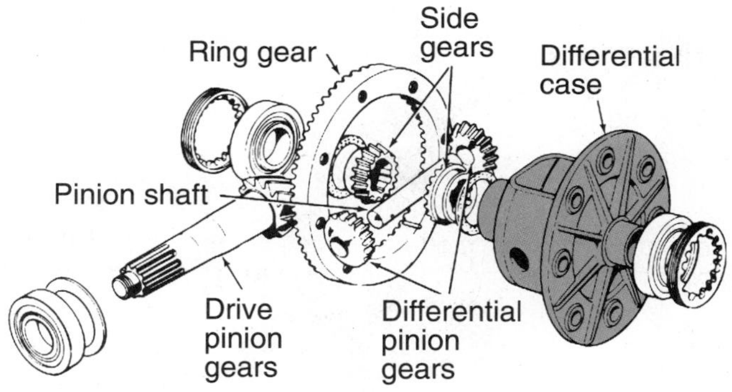

Figure 7-6 Components of a typical differential. (Courtesy of Toyo Kogyo Co., Ltd.)

The ring gear is fastened to the differential case with several hardened bolts or rivets. The differential case is made of cast iron and is supported by two tapered roller bearings and the rear axle housing. Holes machined through the center of the differential housing support the differential pinion shaft. The pinion shaft is retained in the housing case by clips or a specially designed bolt. Two beveled differential pinion gears and thrust washers are mounted on the differential pinion shaft. In mesh with the differential pinion gears are two axle side gears splined internally to mesh with the external splines on the left and right axle shafts (Figure 7-6). Thrust washers are placed between the differential pinions, axle side gears, and differential case to prevent wear on the inner surfaces of the differential case.

Differential Operation

The two drive wheels are mounted on axles that have a differential side gear fitted on their inner ends (Figure 7-7). To turn the power flow 90 degrees, as is required for RWD vehicles, the side gears are bevel gears.

The differential case is located between the side gears and is mounted on bearings so that it is able to rotate independently of the drive axles. Pinion shafts, with small pinion gears, are fitted inside the differential case. The pinion gears mesh with the side gears. The ring gear is bolted

Shop Manual
Chapter 7, page 206

When two beveled gears are meshed, the driving and driven shafts can rotate at a 90-degree angle.

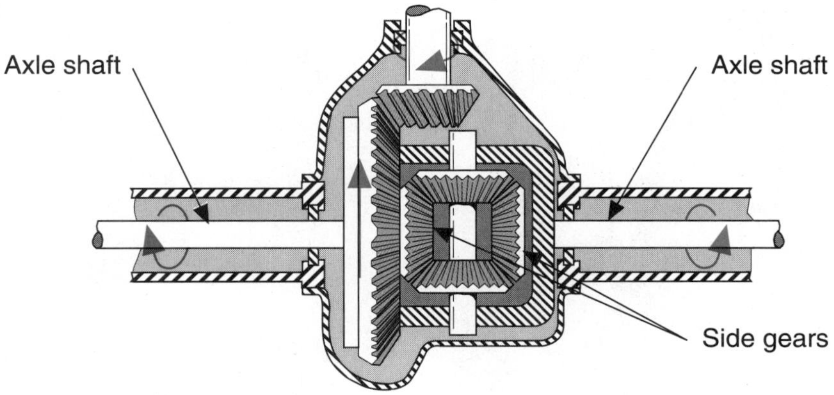

Figure 7-7 Power flow through a RWD differential. (Reprinted with the permission of Ford Motor Company)

The two side gears are placed on the side of the differential case, which is why they are called side gears.

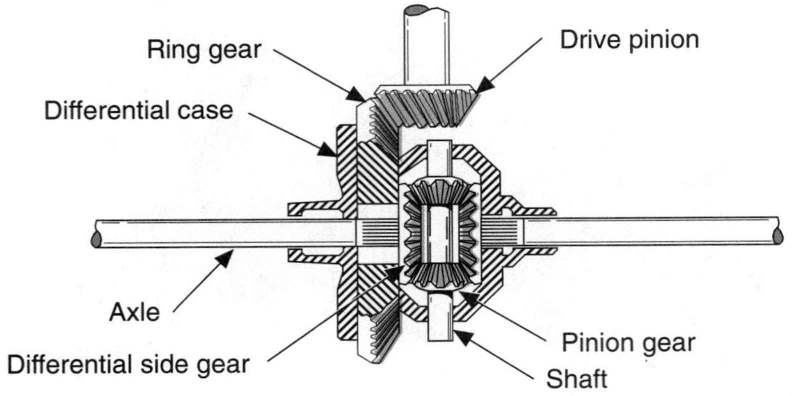

Figure 7-8 Basic differential.

to the flange of the differential case and the two rotate as a single unit. The drive pinion gear meshes with the ring gear and is rotated by the drive shaft (Figure 7-8).

Engine torque is delivered by the drive shaft to the drive pinion gear, which is in mesh with the ring gear and causes it to turn. Power flows from the pinion gear to the ring gear. The ring gear drives the differential case, which is bolted to the ring gear. The differential case extends from the side of the ring gear and normally houses the pinion gears and the side gears. The side gears are mounted so they can slip over splines on the ends of the axle shafts.

The gear ratio in a differential is known as the axle ratio.

The side gears are splined to the axle shafts.

There is a gear reduction between the drive pinion gear and the ring gear, causing the ring gear to turn about one-third to one-fourth the speed of the drive pinion. The pinion gears are located between and meshed with the side gears (Figure 7-9), thereby forming a square of gears inside the differential case. Differentials have two or four pinion gears that are in mesh with the side gears (Figure 7-10). The differential pinion gears are free to rotate on their own centers and can travel in a circle as the differential case and pinion shaft rotate. The side gears are meshed with the pinion gears and are also free to rotate on their own centers. However, because the side gears are mounted at the centerline of the differential case, they do not travel in a circle with the differential case as do the pinion gears.

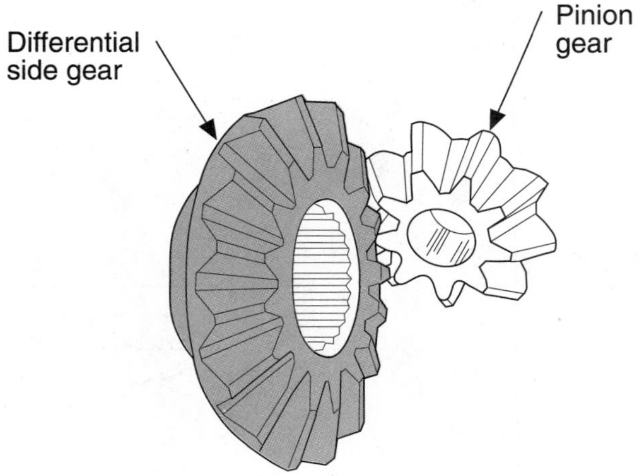

Figure 7-9 Pinion gears in mesh with the side gears.

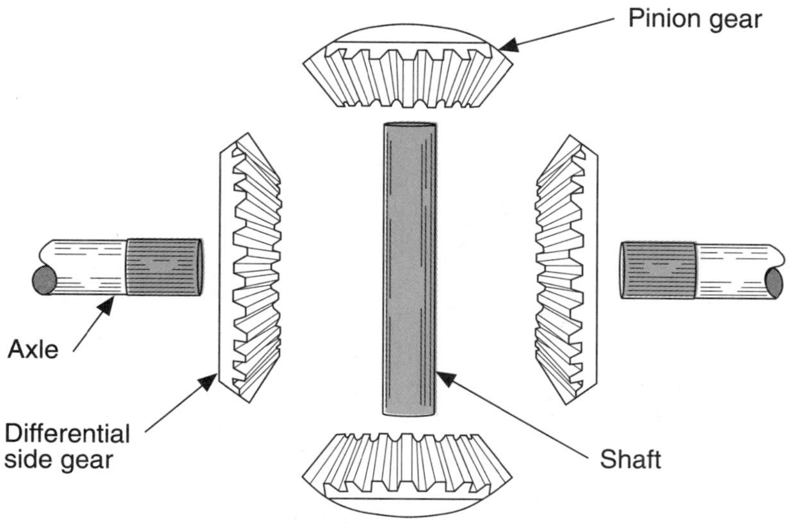

Figure 7-10 Position of side and pinion gears.

The small pinion gears are mounted on a pinion shaft that passes through the gears and the case. The pinion gears are in mesh with the axle side gears, which are splined to the axle shafts. In operation, the rotating differential case causes the pinion shaft and pinion gears to rotate end over end with the case (Figure 7-11). Because the pinion gears are in mesh with the side gears, the side gears and axle shafts are also forced to rotate.

When a car is moving straight ahead, both drive wheels are able to rotate at the same speed. Engine power comes in on the pinion gear and rotates the ring gear. The differential case is rotated with the ring gear. The pinion shaft and pinion gears are carried around by the ring gear and all of the gears rotate as a single unit. Each side gear rotates at the same speed and in the same plane as does the case and they transfer their motion to the axles. The axles

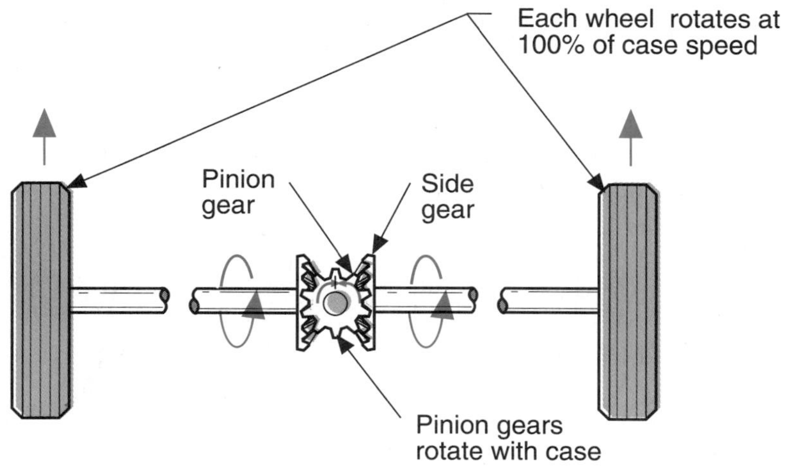

Figure 7-11 Position of pinion gears in the case cause the side gears to rotate. (Reprinted with the permission of Ford Motor Company)

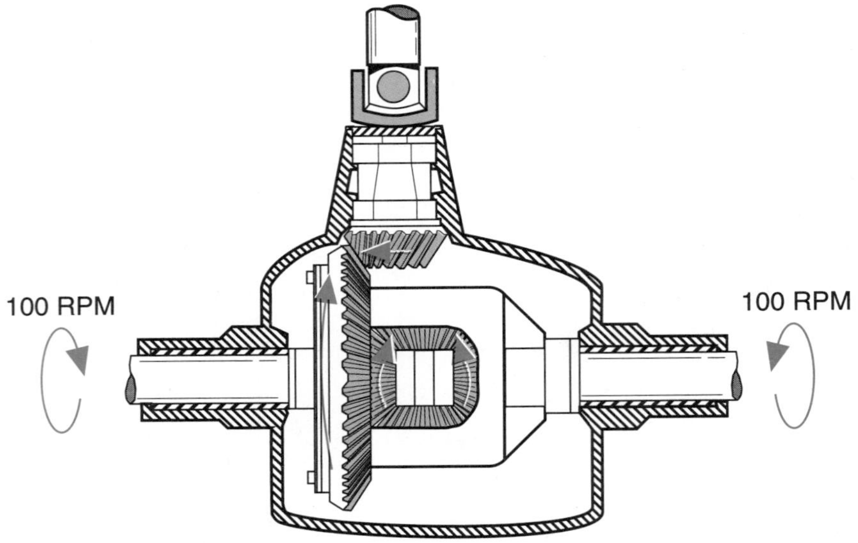

Figure 7-12 Differential action when the vehicle is moving straight.

are thus rotated, and the car moves. Each wheel rotates at the same speed because each axle receives the same rotation (Figure 7-12).

As the vehicle goes around a corner, the inside wheel travels a shorter distance than the outside wheel. The inside wheel must therefore rotate slower than the outside wheel. In this situation, the differential pinion gears will "walk" forward on the slower turning or inside side gear (Figure 7-13). As the pinion gears walk around the slower side gear, they drive the other side gear at a greater speed. An equal percentage of speed is removed from one axle and given to the other (Figure 7-14), however the torque applied to each wheel is equal.

Only the outside wheel is free to rotate when a car is making a very sharp turn; therefore, only one side gear rotates freely. Because one side gear is stationary, the pinion gears now turn on their own centers as they walk around the stationary side gear. As they walk around that side gear, they drive the other side gear at twice their own speed. The moving wheel is now turning at twice the speed of the differential case, but the torque applied to it is only half of the torque applied to the differential case. This increase in wheel speed occurs because of these two actions: the differential pinion gears are rotating end over end with the pinion shaft and the action of the differential pinion gears rotating around the differential pinion shaft.

When one of the driving wheels has little or no traction, the torque required to turn the wheel without traction is very low. The wheel with good traction in effect is holding the axle gear on that side stationary. This causes the pinions to walk around the stationary side gear and drive the other wheel at twice the normal speed but without any vehicle movement. With one wheel stationary, the other wheel turns at twice the speed shown on the speedometer. Excessive spinning of one wheel can cause severe damage to the differential. The small pinion gears can actually become welded to the pinion shaft or differential case.

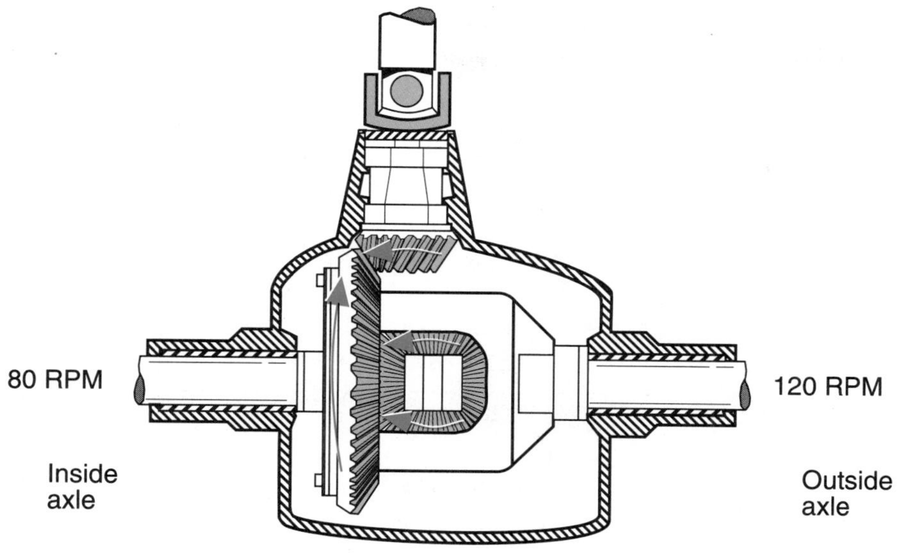

Figure 7-13 Differential action while the vehicle is turning a corner

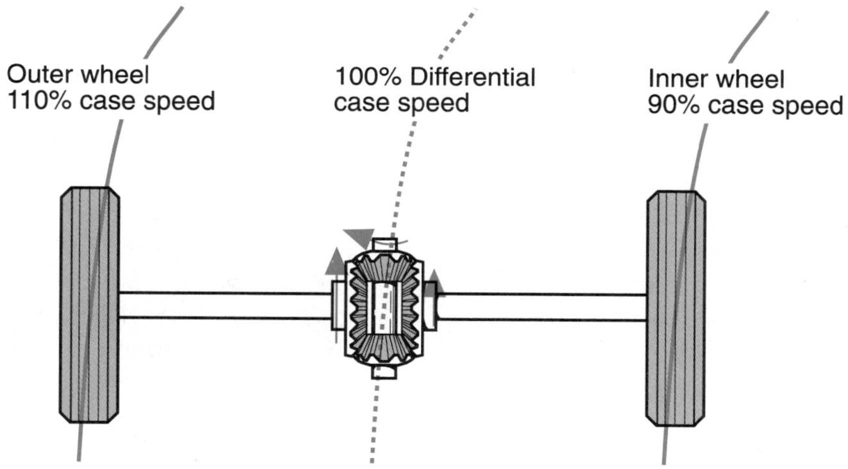

Figure 7-14 Speed differentiation when turning.

Axle Housings

Shop Manual
Chapter 7, page 198

Live rear axles use a one-piece housing with two tubes extending from each side. These tubes enclose the axles and provide attachments for the axle bearings. The housing also shields the parts from dirt and retains the differential lubricant.

In IRS (Figure 7-15) or FWD systems, the housing is in three parts. The center part houses the final drive and differential gears. The outer parts support the axles by providing attachments for the axle bearings. These parts also serve as suspension components and attachment points for

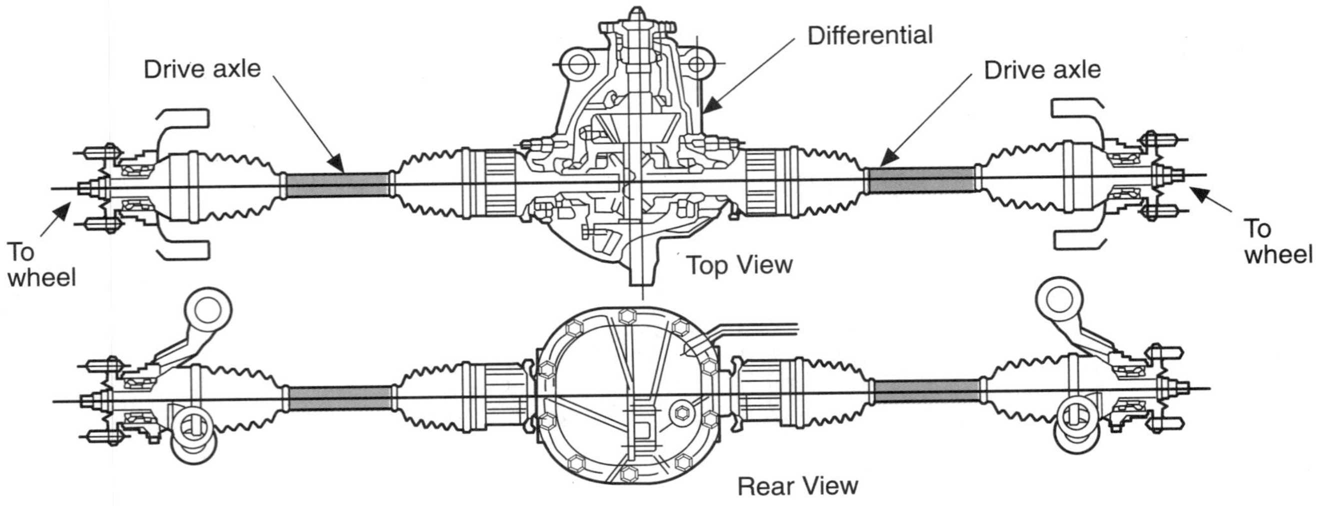

Figure 7-15 Drive axle assembly on a RWD vehicle with IRS. (Reprinted with the permission of Ford Motor Company)

The rear axle housing is sometimes called a banjo because of the bulge in the center of the housing. The bulge contains the final drive gears and differential gears.

In the rear, the outer sections of the housing may be called the uprights and in the front they are usually called the steering knuckle.

Integral carriers are commonly referred to as unitized or Salisbury-type differentials.

the steering gear and/or brakes. In FWD applications, the differential and final drive are either enclosed in the same housing as the transmission or in a separate housing bolted directly to the transmission housing.

Based on their construction, rear axle housings can be divided into two groups, integral carrier or removable carrier. An integral carrier housing attaches directly to the rear suspension. A service cover, in the center of the housing, fits over the rear of the differential and rear axle assembly (Figure 7-16). When service is required, the cover must be removed. The components of the differential unit are then removed from the rear of the housing. Assembly and disassembly of the differential is only possible when it is in the axle housing.

In an integral-type axle housing, the differential carrier and the pinion bearing retainer are supported by the axle housing in the same casting. The pinion gear and shaft is supported by two opposing tapered-roller bearings located in the front of the housing. The differential carrier assembly is also supported by two opposing tapered-roller bearings, one at each side (Figure 7-17).

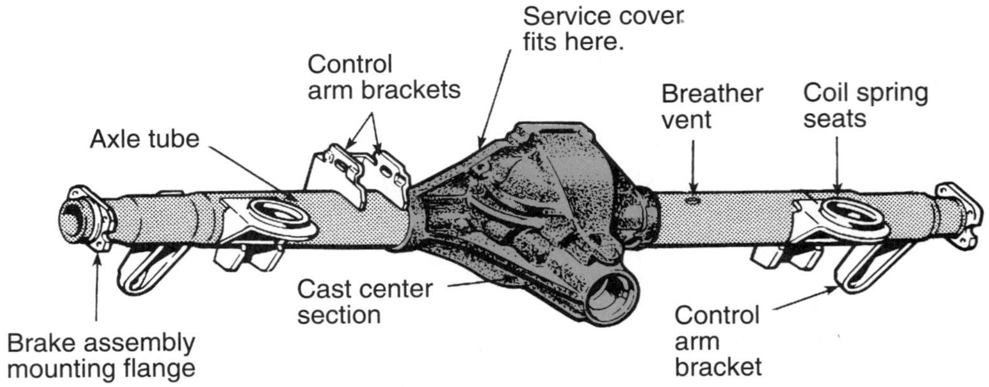

Figure 7-16 Typical integral carrier axle housing. (Reprinted with the permission of Ford Motor Company)

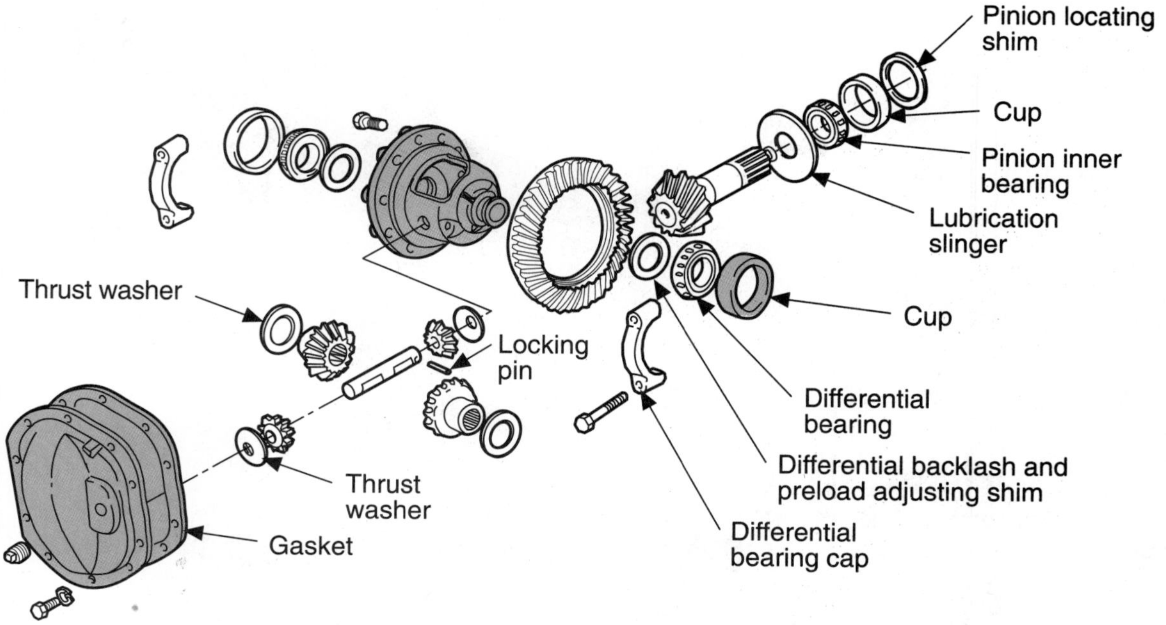

Figure 7-17 Location of bearings in a typical integral housing. (Reprinted with the permission of Ford Motor Company)

The differential assembly of a removable carrier assembly can be removed from the front of the axle housing as a unit. The differential is serviced at a bench and then installed into the axle housing. The differential assembly is mounted on two opposing tapered-roller bearings retained in the housing by removable caps. The pinion gear, pinion shaft, and the pinion bearings are assembled in a pinion retainer, which is bolted to the carrier housing (Figure 7-18).

Removable carriers are often referred to as the third member.

A typical housing has a cast-iron center section with axle shaft tubes pressed and welded into either side. The rear axle housing encloses the complete rear-wheel driving axle assembly. In addition to housing the parts, the axle housing also serves as a place to mount the vehicle's rear suspension and braking system. With IRS, the differential housing is mounted to the vehicle's chassis and does not move with the suspension.

In appearance the two designs of rear axle housing look similar except that the opening for the differential unit on a removable type is at the front and the rear of the housing is solid.

Differential Gears

Two types of gears are currently being used as differential gears: spiral bevel and hypoid (Figure 7-19). Spiral bevel gears are commonly used in heavy duty applications. In a spiral bevel gearset, the centerline of the drive pinion gear intersects the centerline of the ring gear. These designs are noisier than hypoid gears.

Hypoid gearsets are commonly used in RWD passenger car and light truck applications. The pinion gear in a hypoid gearset is mounted well below the centerline of the ring gear. Hypoid gears are quiet running and allow several teeth to absorb the driving force.

This design allows for lower vehicle height and more passenger room inside the vehicle. By lowering the drive pinion gear on the ring gear, the entire drive shaft can be lowered. Lowering the drive shaft allows for a lower drive shaft tunnel, which in turn allows for increased passenger room and a lower ride height

Shop Manual
Chapter 7, page 215

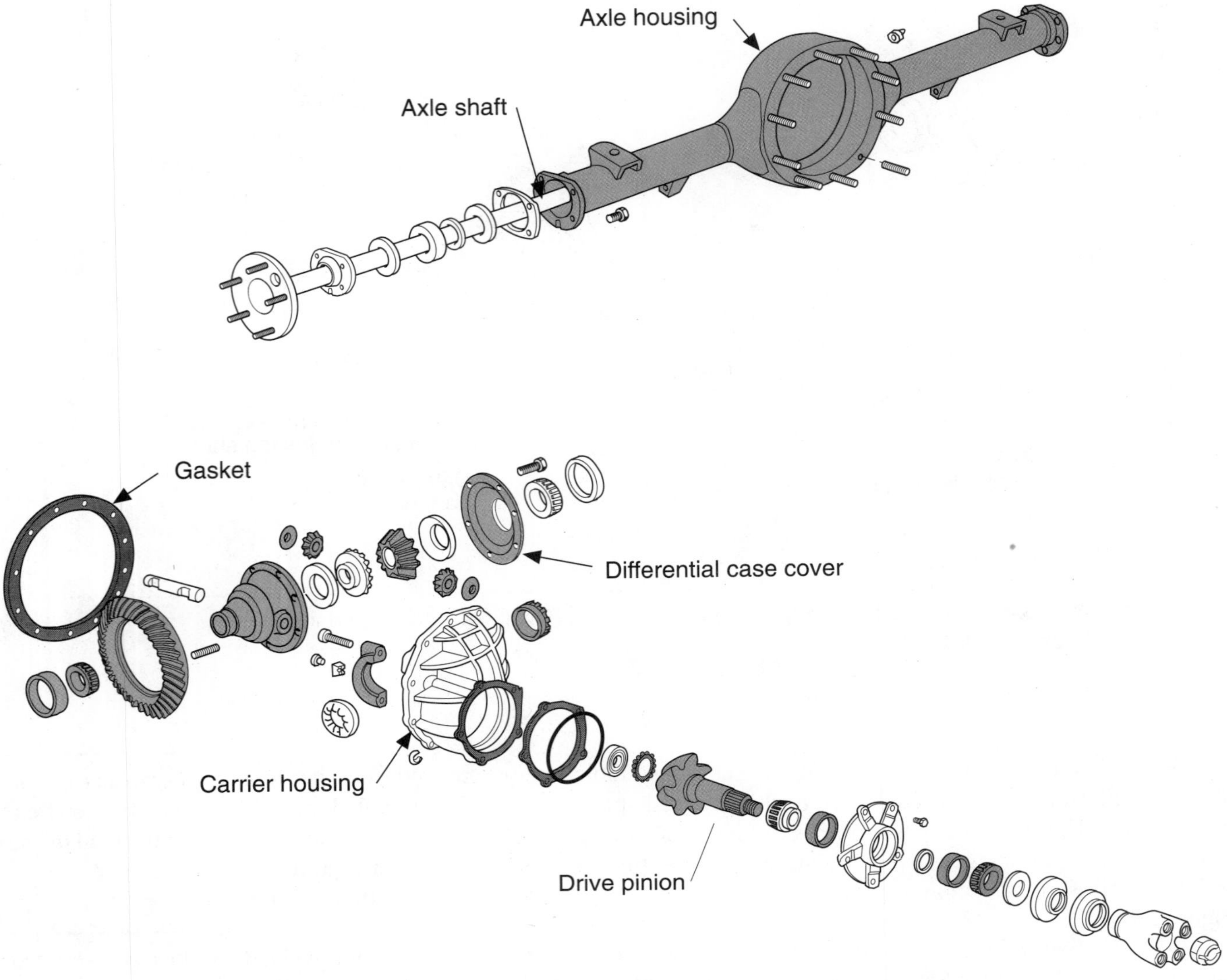

Figure 7-18 Typical removable carrier axle housing. (Reprinted with the permission of Ford Motor Company)

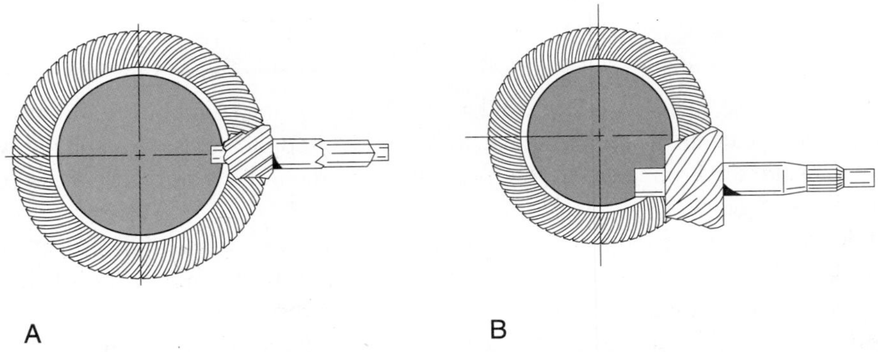

Figure 7-19 Comparison of a spiral bevel and hypoid gearset. (Reprinted with the permission of Ford Motor Company)

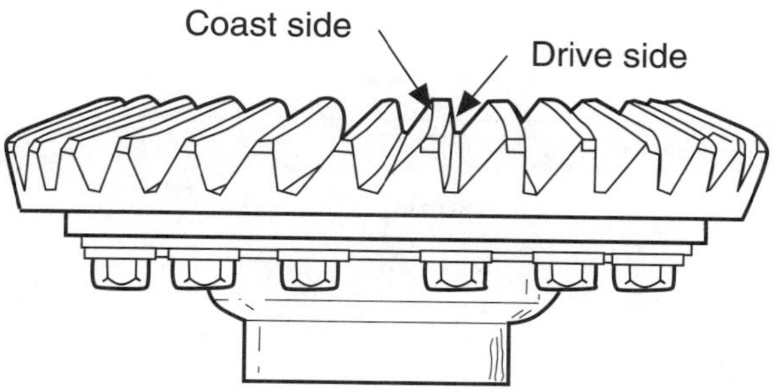

Figure 7-20 The drive and coast side of a ring gear.

The teeth of a hypoid gear are curved to follow the form of a spiral, causing a wiping action while meshing. As the gears rotate, the teeth slide against each other. Because of this sliding action, the ring and pinion gears can be machined to allow for near perfect mating, which results in smoother action and a quiet-running gearset. Because this sliding action produces extremely high pressures between the gear teeth, only a hypoid-type lubricant should be used with hypoid gearsets.

The spiral-shaped teeth result in different tooth contacts as the pinion and ring gear rotate. The drive side of the teeth is curved in a convex shape, and the coast side of the teeth is concave (Figure 7-20). The inner end of the teeth on the ring gear is known as the toe and the outer end of the teeth is the heel (Figure 7-21).

While engine torque is being applied to the drive pinion gear, the pinion teeth exert pressure on the drive side of the ring gear teeth. During coast or engine braking, the concave side of the ring gear teeth exerts pressure on the drive pinion gear.

Upon heavy acceleration, the drive pinion attempts to climb up the ring gear and raises the front of the differential. A rubber bumper between the car's body and differential prevents the front of the differential from contacting the chassis or underbody of the vehicle during heavy acceleration or when the vehicle is heavily loaded and accelerated (Figure 7-22). During other operating conditions, the suspension's leaf springs or the torque arms on coil spring suspensions absorb much of the torque to limit the movement of the axle housing.

The drive side of the teeth is the side that has engine power working on it, whereas the coast side is the side of the teeth that has the forces of the drive wheel on it.

When there is no torque applied either in drive or in coast, the condition is known as float.

The climbing action of the drive pinion is sometimes called wind-up.

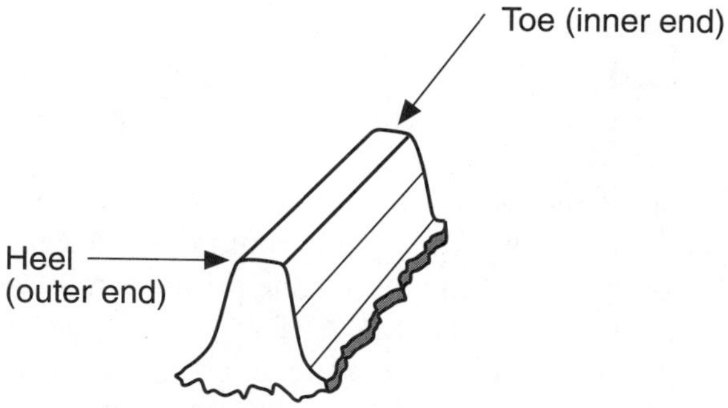

Figure 7-21 The toe and heel of a ring gear's tooth.

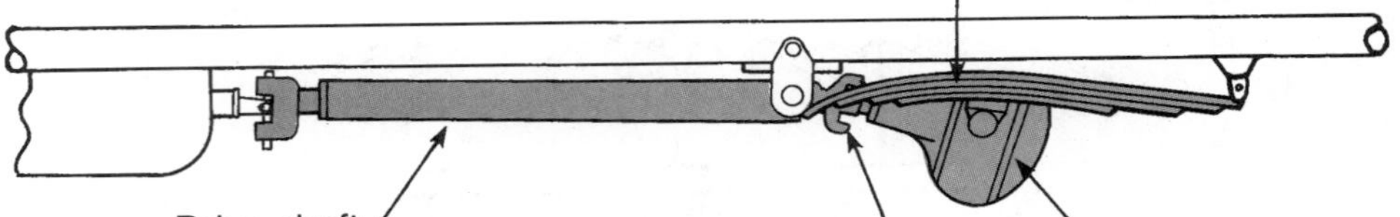

Figure 7-22 Location of rubber bumper to limit axle housing movement.

Gear Ratios

The gear ratio of the pinion and ring gear is often referred to as the final drive ratio.

Gear ratios express the number of turns the drive gear makes compared to one turn of the driven gear it mates with. The ring gear is driven by the pinion gear, therefore to cause torque multiplication, the ring gear is always larger than the pinion. This combination causes the ring gear to turn slower but with greater torque.

Many different final drive ratios are used. A final drive ratio of 2.8:1 is commonly used, especially on cars equipped with an automatic transmission. A 2.8:1 final drive ratio means the drive pinion must turn 2.8 times to rotate the ring gear one time. On cars equipped with a manual transmission, more torque multiplication is often needed, therefore a 3.5:1 final drive ratio is often used. To allow a car to accelerate quicker or to move heavy loads, a final drive ratio of 4:1 can be used. Also, small engine cars with overdrive fourth and fifth gears often use a 4:1 final drive ratio, which allows them to accelerate reasonably well in spite of the engine's low power output. The overdrive in fourth and fifth gear effectively reduces the final drive ratio when the car is moving in those gears. Trucks also use a final drive ratio of 4:1 or 5:1 to provide more torque to enable them to pull or move heavy loads.

It is important to remember that the actual final drive or overall gear ratio is equal to the ratio of the ring and pinion gear multiplied by the ratio of the speed gear the car is operating in. For example, if a car has a final drive ratio of 3:1, the total final drive ratio for each transmission speed is as follows:

The frontal area of a vehicle has much to do with the air resistance on the vehicle's body while it is moving at speed.

	Transmission Ratio	×	**Final Drive Ratio**	=	**Total Final Drive Ratio**
First gear	3:1		3:1		9:1
Second gear	2.5:1		3:1		7.5:1
Third gear	1.5:1		3:1		4.5:1
Fourth gear	1:1		3:1		3:1
Fifth gear	0.75:1		3:1		2.25:1

Vehicles equipped with numerically low gear ratios are said to have high gears, whereas vehicles with numerically high gear ratios are said to have low gears.

Notice that, in this example, the only time the total final drive ratio is the same as the ratio of the ring and pinion gear is when the transmission is in fourth gear, which has a speed ratio of 1:1.

Many factors are considered when a manufacturer selects a final drive ratio for a vehicle. Some of these factors are vehicle weight, engine rpm range, designed vehicle speed, frontal area of the body, fuel economy requirements, engine power output, and transmission type and gear ratios. Cars with final drive ratios around 2.5:1 will take longer to accelerate but will give a higher top speed. At the other end of the scale, a 4.11:1 ratio will give faster acceleration with a lower top speed. Since the 1970s there has been an emphasis on fuel economy, and most cars have been equipped with high gears to allow for lower engine speeds at normal driving speeds.

Determining Final Drive Ratio

To replace a ring and pinion gearset with one of the correct ratio, the ratio of the original set must be known. There are several ways to determine the final drive ratio of a ring and pinion gearset. If a shop manual is available, you can decipher the code found on the assembly or on a tag attached to it (Figure 7-23). Normally a table is given that lists the various codes and the ratios each represents.

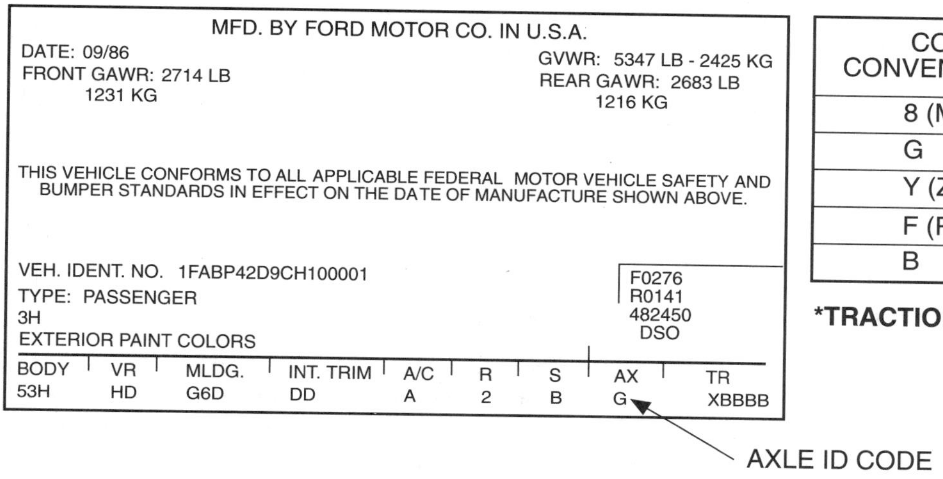

CODE CONVENTIONAL	RATIO
8 (M*)	2:73:1
G	2:26:1
Y (Z*)	3:08:1
F (R*)	3:45:1
B	2:47:1

*TRACTION LOK

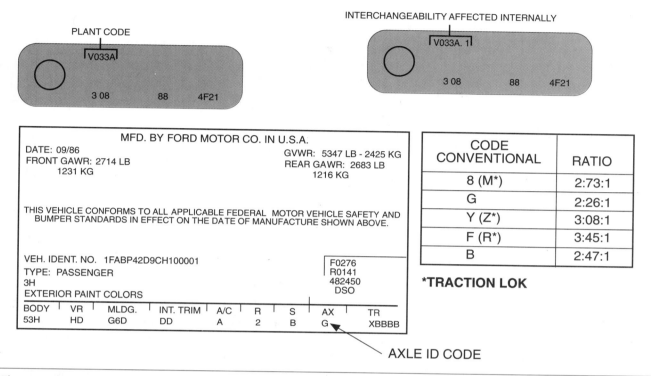

Figure 7-23 Deciphering differential codes from information given on the differential tag or from the VIN. (Reprinted with the permission of Ford Motor Company)

Most axles are shipped with an identification tag bolted to them. These tags contain all of the information needed to identify the axle for diagnosis and service. The tags are located under the housing-to-carrier stud nut or are attached by a cover-to-carrier bolt. Manufacturers also often stamp identification numbers into the axle housing. These codes are normally located on the front side of an axle tube, (Figure 7-24). Always refer to your shop manual to locate and decipher the codes.

SERVICE TIP: If the stamped numbers cannot be found or if the axle tag is not there, refer to the axle code letter or number on the vehicle identification number (VIN) plate (Figure 7-23). This will identify the ratio and type of axle with which the car was originally equipped.

Another way to determine the final drive ratio is to compare the number of revolutions of the drive wheels with those of the drive shaft. While turning both wheels simultaneously, note how many times the drive shaft turns to complete one revolution of the drive wheels. This count represents the ratio of the gears.

The gear ratio can also be determined when the differential is disassembled. Count the number of teeth on both the drive pinion and the ring gear. Divide the ring gear teeth number by the pinion drive number to calculate the final drive ratio.

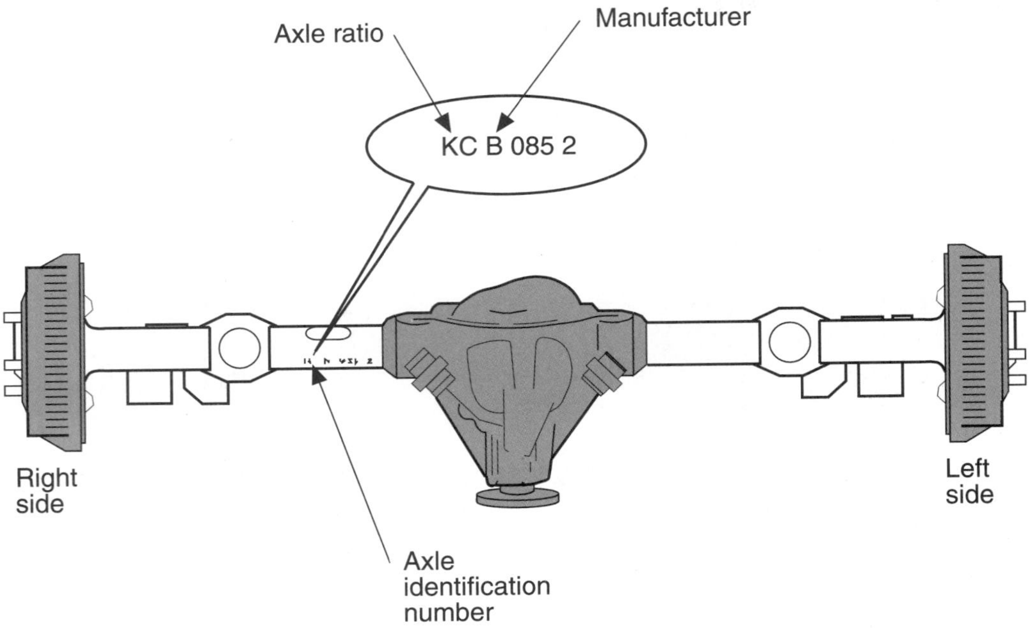

Figure 7-24 Location of differential code on an axle assembly.

Shop Manual
Chapter 7, page 234

The alignment of nonhunting and partial nonhunting gears is often referred to as timing the gears.

Lapping is the process of using a grinding paste to produce a fine finish on the teeth of the two gears that will be in constant contact with each other.

Hunting and Nonhunting Gears

Ring and pinion gearsets are usually classified as hunting, nonhunting, or partial nonhunting gears. Each type of gearset has its own requirements for a satisfactory gear tooth contact pattern. These classifications are based on the number of teeth on the pinion and ring gears.

A nonhunting gearset is one in which any one pinion tooth comes into contact with only some of the ring gear teeth. One revolution of the ring gear is required to achieve all possible gear tooth contact combinations. As an example, if the ratio of the ring gear teeth to the pinion gear teeth is 39 to 13 (or 3.00:1), the pinion gear turns three times before the ring gear completes one turn. One full rotation of the pinion gear will cause its 13 teeth to mesh with one third of the ring gear's teeth. On the next revolution of the pinion gear, its teeth will mesh with the second third of the ring gear's teeth and the third revolution will mesh with the last third of the ring gear. Each tooth of the pinion gear will return to the same tooth space on the ring gear each time the pinion rotates.

A partial nonhunting gearset is one in which any one pinion tooth comes into contact with only some of the ring gear teeth, but more than one revolution of the ring gear is required to achieve all possible gear tooth contact combinations. If the ratio of the ring gear teeth to the pinion gear teeth is 35 to 10 (or 3.5:1), any given tooth of the pinion will meet seven different teeth (two complete revolutions of the pinion gear) of the ring gear before it returns to the space where it started.

When hunting gearsets are rotating, any pinion gear tooth is likely to contact all the ring gear teeth. If the ring gear has 37 teeth and the pinion gear has 9, the gearset has a ratio of 37 to 9 (or 3.89:1). Any given tooth in the pinion gear meets all of the teeth in the ring gear before it meets the first tooth again.

During assembly the nonhunting and partial nonhunting gears must be assembled with the index marks properly aligned (Figure 7-25). When these gearsets were manufactured, they were lapped to ensure the proper mesh and because specified teeth on the pinion will always meet specific teeth on the ring gear, a noisy gear will result if they are not properly aligned. Hunting gears do not need to be aligned because any tooth on the pinion may mesh with any tooth on the ring gear.

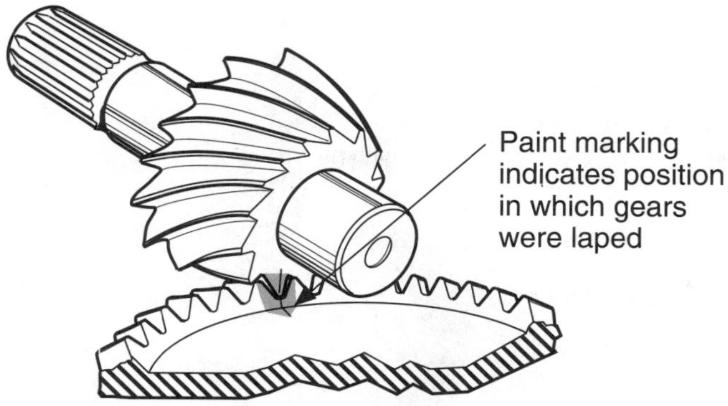

Figure 7-25 Index marks of a ring and pinion gearset. (Reprinted with the permission of Ford Motor Company)

Differential Bearings

At least four bearings are found in all differentials. Two fit over the drive pinion shaft to support it and the other two support the differential case and are usually mounted just outboard of the side gears (Figure 7-26). The drive pinion bearings are typically tapered-roller bearings and the case bearings are normally ball bearings.

Different forces are generated in the differential due to the action of the pinion gear. As the pinion gear turns, it tries to climb up the ring gear and pull the ring gear down. Also, as the pinion gear rotates, it tends to move away from the ring gear and pushes the ring gear equally as

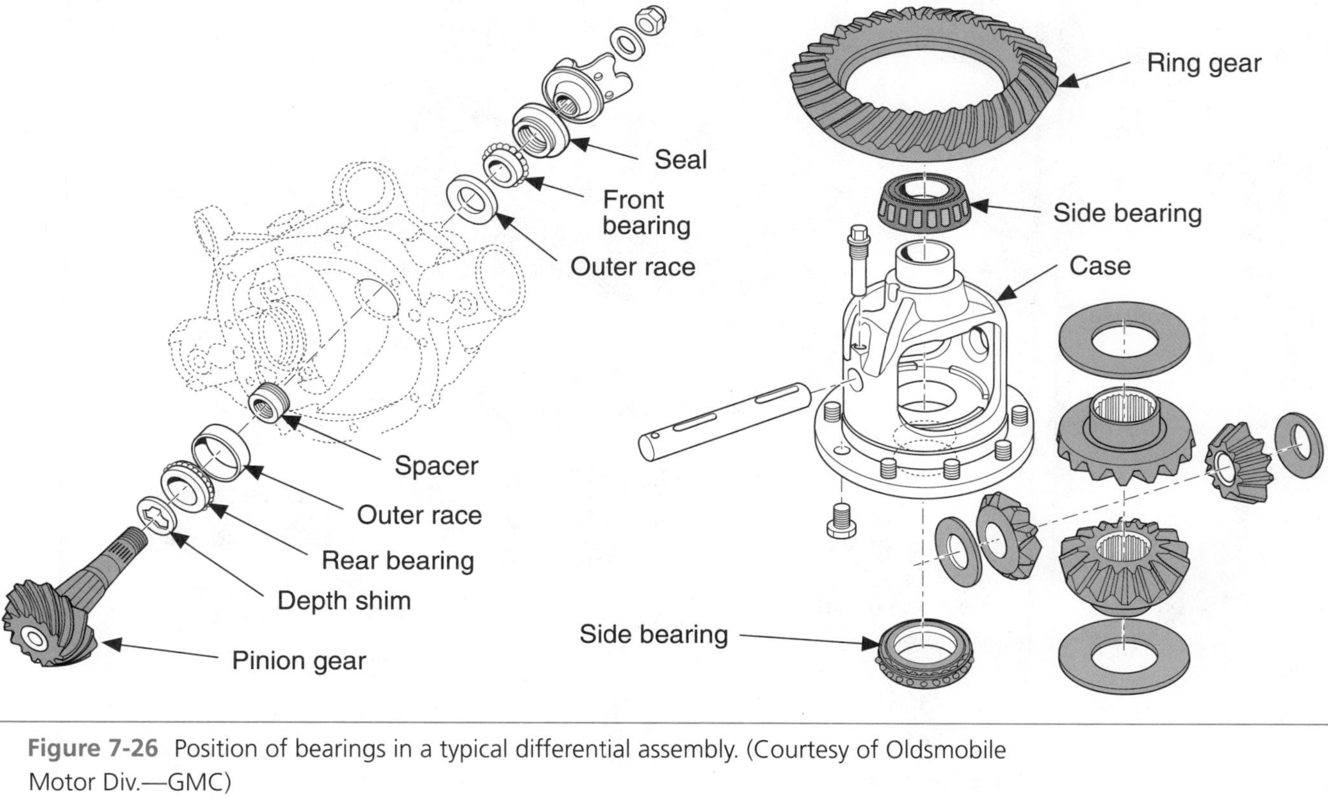

Figure 7-26 Position of bearings in a typical differential assembly. (Courtesy of Oldsmobile Motor Div.—GMC)

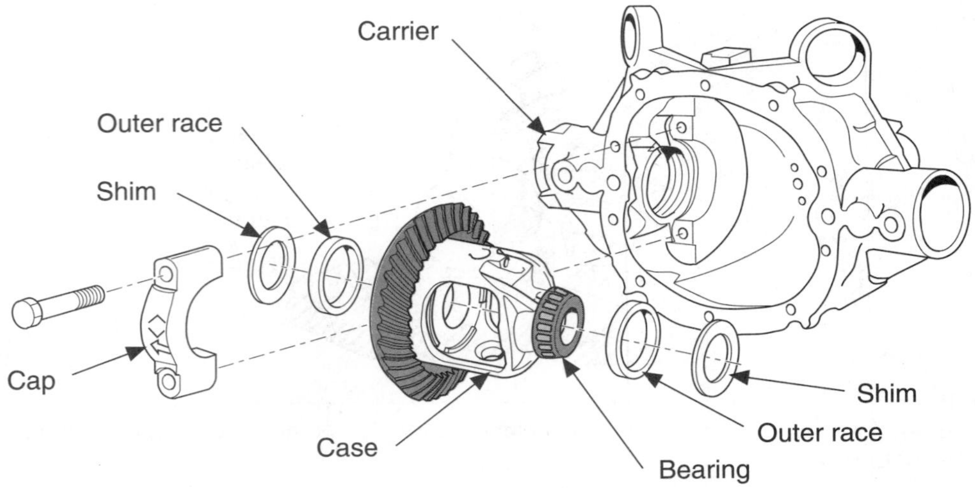

Figure 7-27 Position of differential case side bearings. (Courtesy of Oldsmobile Motor Div.—GMC)

hard in the opposite direction. Because of these forces, the differential must be securely mounted in the carrier housing. The bearings on each end of the differential case support the case and absorb the thrust of the forces (Figure 7-27). The pinion gear and shaft are mounted on bearings to allow the shaft to rotate freely without allowing it to move in response to the torque applied to it. All of these bearings are installed with a preload to prevent the pinion gear and ring gear from moving out of position.

Pinion Mountings

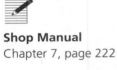

Shop Manual
Chapter 7, page 222

As torque is applied to a pinion gear, the pinion gear rotates and drives the ring gear. As it rotates, three separate forces are produced by its rotation and the torque applied to it. The

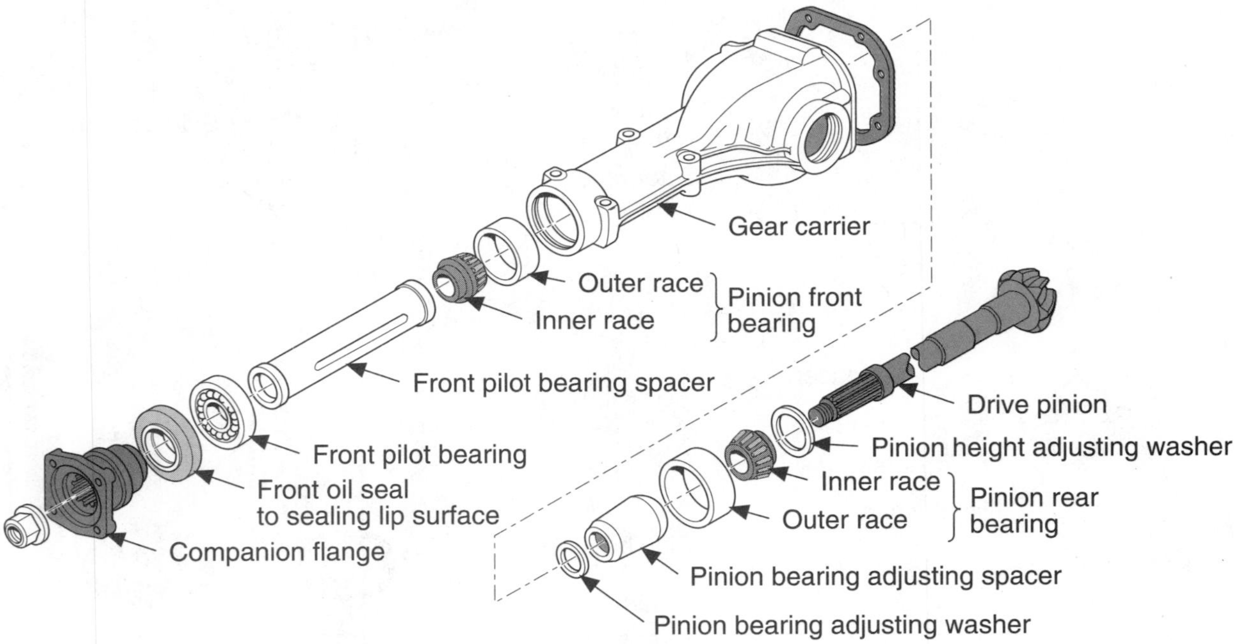

Figure 7-28 Typical straddle-mounted pinion gear.

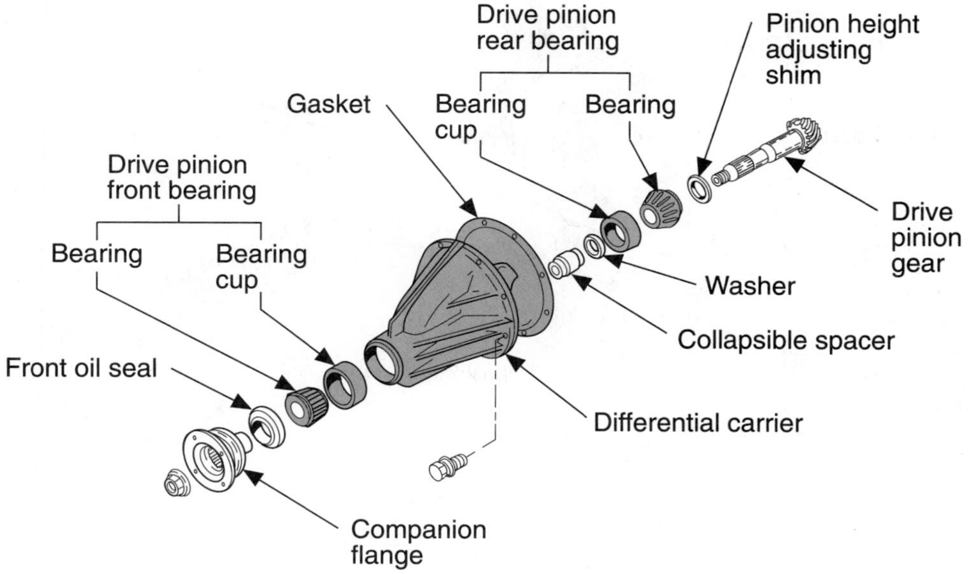

Figure 7-29 Typical overhung-mounted pinion gear. (Reprinted with the permission of Ford Motor Company)

pinion gear tries to screw itself out of the ring gear and move forward. It also tries to climb up on the teeth of the ring gear and it is forced to the side and away from the ring gear. These forces make it necessary to securely mount the pinion gear.

The drive pinion flange is splined to the rear axle's drive pinion gear. The drive pinion gear is placed horizontally in the axle housing and is positioned by one of two types of mounting, straddle or overhung. The straddle-mounted pinion gear is used in most removable carrier-type axle housings. The straddle-mounted pinion has two opposing tapered-roller bearings positioned close together with a short spacer between their inner races and ahead of the pinion gear. A third bearing, usually a straight roller bearing, is used to support the rear of the pinion gear (Figure 7-28).

The rear universal joint of the drive shaft is called the drive pinion flange.

The overhung-mounted pinion also uses two opposing tapered-roller bearings but does not use a third bearing. The two roller bearings must be farther apart than the opposing bearings of a straddle-mounted pinion because a third bearing is not used to support the pinion gear (Figure 7-29). This type of pinion gear mounting can be found on either the removable carrier or integral-type driving axle.

Another name for the collapsible spacer is crush sleeve.

Some pinion shafts are mounted in a bearing retainer that is removable from the carrier housing. This type of pinion assembly utilizes a pilot bearing to support the rear end of the pinion and is equipped with two opposing tapered-roller bearings.

Drive Pinion Bearing Preload

A spacer is placed between the opposing tapered bearings to control the distance between them (Figure 7-30). This spacer also controls the amount of preload or loading pressure applied to the bearings. Preload prevents the pinion gear from moving back and forth in the bearing retainer.

Often a collapsible spacer is used between the two large tapered bearings to provide for proper pinion bearing preload. Some differentials use a solid noncollapsible spacer with selective thickness shims to adjust pinion bearing preload.

Preload is a fixed amount of pressure constantly applied to a component. Preload on bearings eliminates looseness. Preload on limited-slip differential clutches also provides torque transfer to the driven wheel with the least traction.

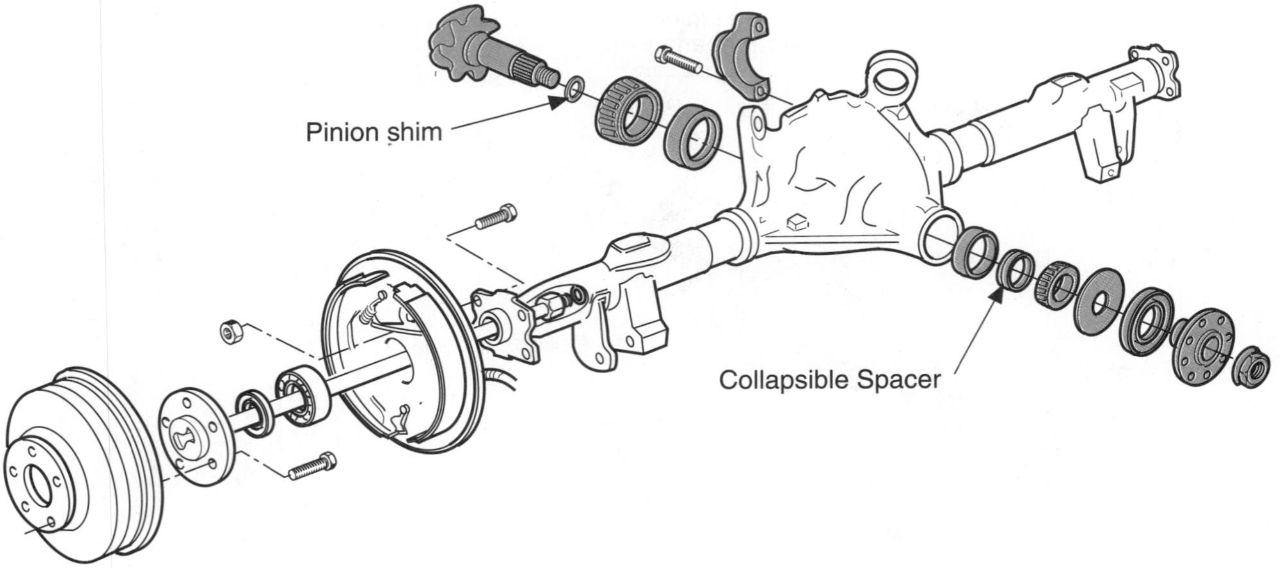

Figure 7-30 Location of pinion bearing spacer. (Reprinted with the permission of Ford Motor Company)

SERVICE TIP: Collapsible spacers should never be reused. After they have been compressed once, they are not capable of maintaining preload when they are compressed again. The spacers should always be replaced when servicing the differential.

Endplay is the amount of axial or end-to-end movement in a shaft due to clearance in the bearings.

When the pinion shaft nut is tightened to specifications, pressure is exerted by the pinion drive flange against the inner race of the front pinion bearing. This applies pressure against the spacer and the rear bearing, which cannot move because it is located against the drive pinion gear. This load on the two pinion bearings assures that there will be no pinion shaft endplay. Any pinion shaft endplay will result in rapid gear and bearing wear.

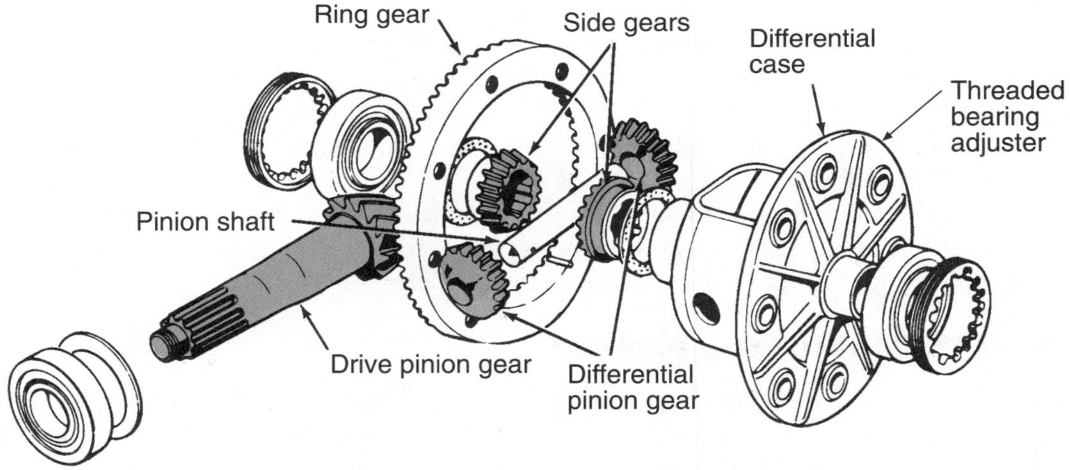

Figure 7-31 Location of bearing adjusting nuts. (Courtesy of Mazda Motors)

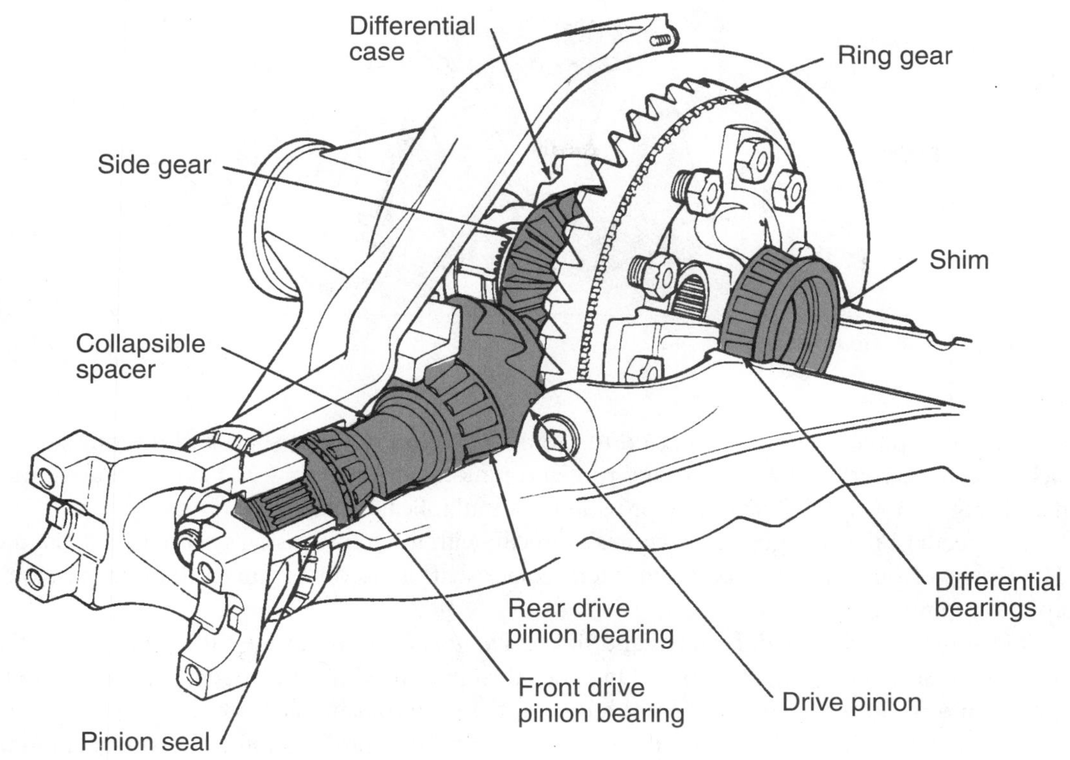

Figure 7-32 Location of bearing selective shims. (Reprinted with the permission of Ford Motor Company)

Differential Case

The differential case is supported in the carrier by two tapered-roller side bearings. This assembly can be adjusted from side to side to provide the proper backlash between the ring gear and-pinion and the required side bearing preload. This adjustment is achieved by threaded bearing adjusters (Figure 7-31) on some units and the placement of selective shims and spacers (Figure 7-32) on others.

Shop Manual
Chapter 7, page 229

Transaxle Final Drive Gears and Differential

Shop Manual
Chapter 7, page 237

Transaxle final drive gears provide the means for transmitting transmission output torque to the differential section of the transaxle.

The differential section of the transaxle has the same components as the differential gears in a RWD axle and basically operate in the same way. The power flow in transversely mounted power trains is in line with the wheels and therefore the differential unit does not need to turn the power 90 degrees.

The drive pinion and ring gears and the differential assembly are normally located within the transaxle housing of FWD vehicles. There are four common configurations used as the final drives on FWD vehicles: helical, planetary, hypoid, and chain drive. The helical, planetary, and chain final drive arrangements are usually found in transversely mounted power trains. Hypoid final drive gear assemblies are normally used with longitudinal power train arrangements.

The ring gear in a transaxle is sometimes referred to as the differential drive gear.

Backlash is the clearance or play between two gears in mesh. It is the amount one of the gears can be moved without moving the other (Figure 7-33).

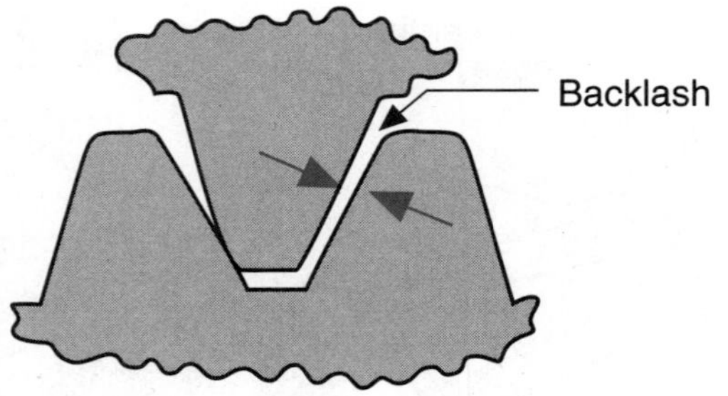

Figure 7-33 Gear backlash.

The drive pinion gear is connected to the transmission's output shaft and the ring gear is attached to the differential case. Like the ring and pinion gearsets in a RWD axle, the drive pinion and ring gear of a FWD assembly provide for a multiplication of torque.

The teeth of the ring gear usually mesh directly with the transmission's output shaft (Figure 7-34). However on some transaxles, an intermediate shaft is used to connect the transmission's output to the ring gear.

On some models, the differential and final drive gears operate in the same lubricant as the transmission section of the transaxle. On other designs, the differential section is separately enclosed and is lubricated by a different lubricant than the transmission section. These designs require positive sealing between the differential unit and the transmission to keep the different lubricants from mixing. All transaxles use seals between the differential and the drive axles to prevent dirt from entering the transaxle and to prevent lubricant from leaking past the attachment point of the drive axles.

Helical Final Drive Assembly

Helical gears are gears with teeth that are cut at an angle or are spiral to the gear's axis of rotation.

Helical (Figure 7-35) final drive assemblies use helical gearsets that require the centerline of the pinion gear to be at the centerline of the ring gear. The pinion gear is cast as part of the main shaft and is supported by tapered-roller bearings. The pinion gear is meshed with the ring gear to provide the required torque multiplication. Because the ring is mounted on the differential case, the case rotates in response to the pinion gear.

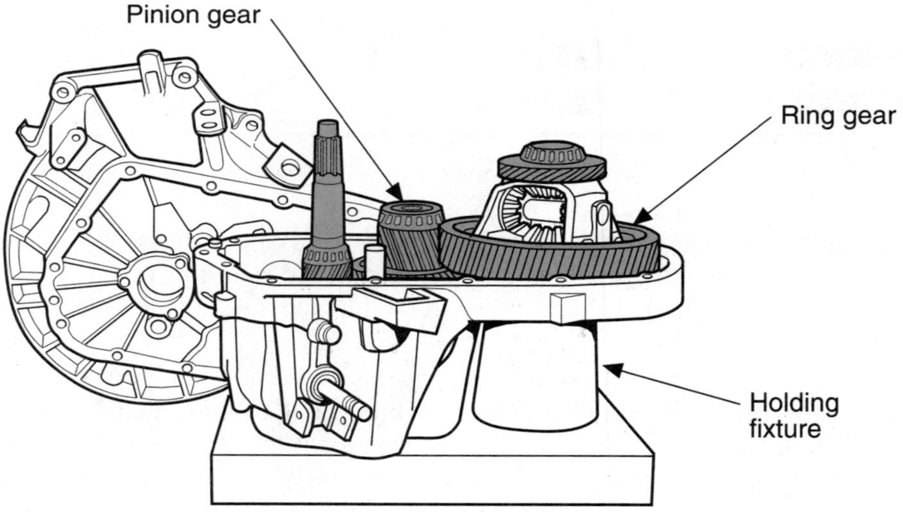

Figure 7-34 Typical ring and pinion gearset in a transaxle.

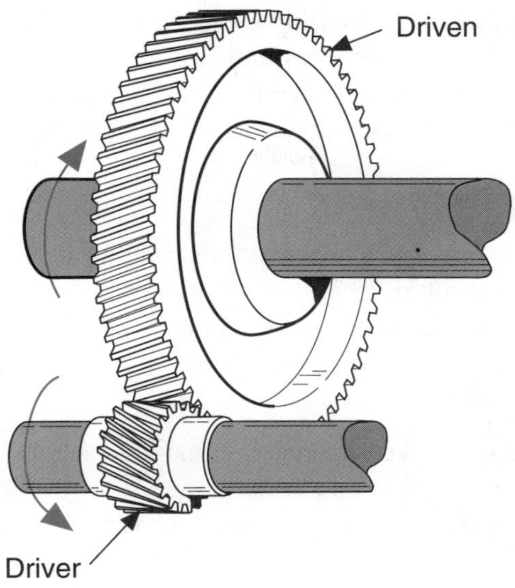

Figure 7-35 Helical gearset.

Planetary Final Drive Assembly

The ring gear of a planetary final drive assembly has lugs around its outside diameter. These lugs fit into grooves machined inside the transaxle housing. These lugs and grooves hold the ring gear stationary. The transmission's output shaft is splined to the planetary gearset's sun gear. The planetary pinions are in mesh with both the sun gear and ring gear and form a simple planetary gearset (Figure 7-36). The planetary carrier is constructed so that it also serves as the differential case.

Another name for the ring gear in a planetary gear set is an internal gear.

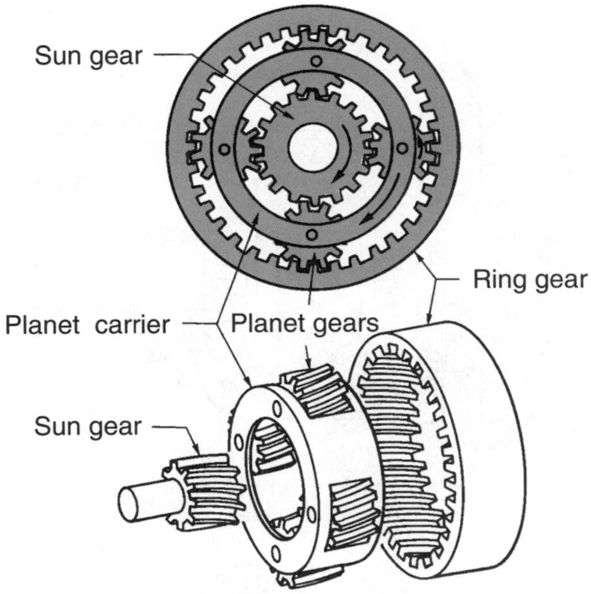

Figure 7-36 Planetary final drive gearset.

In operation, the transmission's output drives the sun gear, which, in turn, drives the planetary pinions. The planetary pinions walk around the inside of the stationary ring gear. The rotating planetary pinions drive the planetary carrier and differential housing. This combination provides maximum torque multiplication from a simple planetary gearset.

Hypoid Final Drive Assembly

Hypoid gears have the advantage of being quiet and strong because of their thick tooth design. And due to their strength, hypoid-type gears can be used with large engines that are longitudinally mounted in vehicles. This type of final drive unit is identical to those used in RWD vehicles.

Chain Drive Assembly

Commonly used in automatic transaxles, chain drive final drive assemblies use a chain to connect a drive sprocket—connected to the transmission's output shaft—to the driven sprocket, which is connected to the differential case. This design allows for remote positioning of the differential within the transaxle housing.

Limited Slip Differentials

An open differential is a standard-type differential.

Many names are used for limited slip differentials, including Posi-Traction, Traction-Lok, and Posi-units.

An open differential is built with a combination of interlocking gears that eliminates tire scrubbing, as the outer tire has further distance to travel during cornering. Although this differential is the easiest on the car's tires and suspension, it has one major disadvantage—a lack of traction. Power, is for the most part, transferred to the wheel with the least resistance or traction.

When a car is stuck in mud or snow, one drive wheel spins while the other remains stationary. In this example, the differential is transferring all of the torque and rotary motion to the drive wheel with the least resistance. Resistance, in this case, means traction. Applying torque to the wheel without traction does little good while trying to move the car.

A limited-slip differential assembly provides more driving force to the wheel with traction when one wheel begins to spin. With the addition of clutches to the differential case (Figure 7-37), differential action can be restricted so that if one drive wheel has no traction, the other

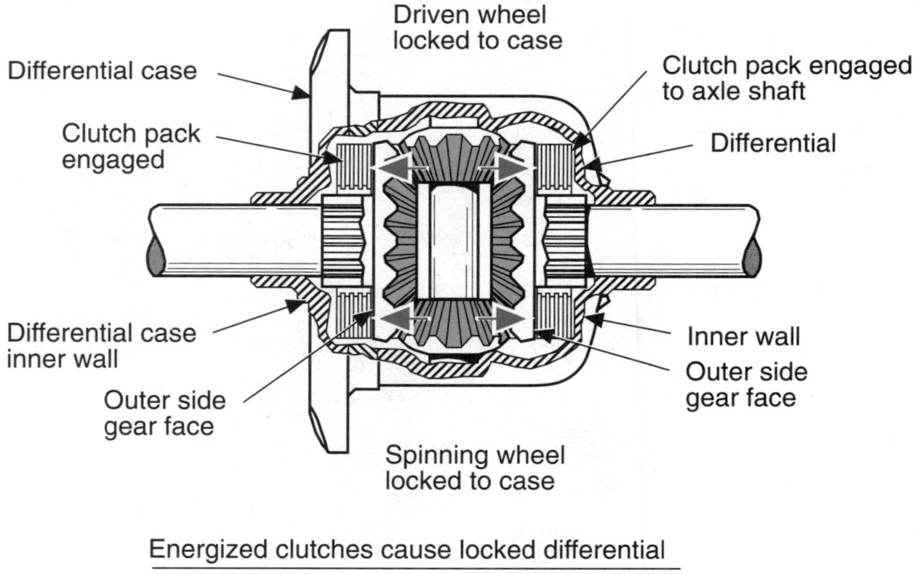

Figure 7-37 Action of clutches in a limited-slip differential.

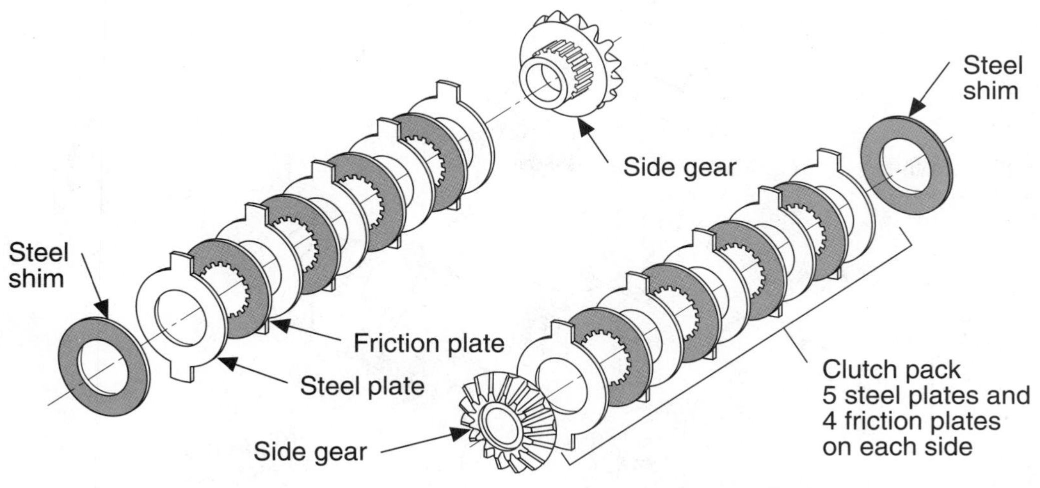

Figure 7-38 The clutch assembly in a typical limited slip differential assembly. (Reprinted with the permission of Ford Motor Company)

wheel that has some traction will at least receive some torque. This is the theory behind the limited slip differential.

Limited-slip differentials are used on sports cars for increased traction while cornering and on off-road vehicles where the drive wheels are constantly losing traction. Power flows in the same way as in an open differential. Most limited-slip differentials transfer at least 20% of the available torque to the wheel with traction. Limited-slip differentials merely limit the amount of differential action between the side gears through the use of these clutches.

Limited-slip differential cases are similar to open differential cases except for a large internal recess around the area of each side gear. This recess accepts either a clutch pack, a cone clutch, or a viscous clutch assembly, depending on design.

Clutch Pack

Shop Manual
Chapter 7, page 243

Often limited-slip differentials use two sets of multiple disc clutches to control differential action. Each clutch pack consists of a combination of steel plates and friction plates. The plates are stacked on the side gear hub and are housed in the differential case. A preload spring applies an initial force to the clutch packs (Figure 7-38).

The friction plates are splined to each side gear's hub. The ears of the steel plates are fitted into the case so that the clutch packs are always engaged. The discs rotate with the side gear and the plates with the differential case.

A clutch pack consists of a complete set of alternating clutch plates and discs.

The clutch assembly consists of a multiple plate clutch, a center block, preload springs, and a preload plate (Figure 7-39). The clutch assembly is always engaged due to pressure constantly being applied to it by the preload springs. Under normal driving conditions, the clutch brake slips as the torque generated by differential action easily overcomes the capacity of the clutch assembly. This allows for normal differential action when the vehicle is turning. During adverse road conditions, where one or both wheels may be on a low friction surface such as snow, ice, or mud, the friction between the clutch plates will transfer a portion of usable torque to the wheel with the most traction.

The clutch packs are mounted behind each of the axle's side gears and springs between the side gears force the gears against the clutches. Although the springs allow enough slippage to permit driving around a curve, during slippery conditions they keep the side gears against the clutches with enough pressure to make those gears spin at the same speed. If one wheel begins to slip, the friction of the clutches ensures that the slipping wheel does not receive all of the engine's torque.

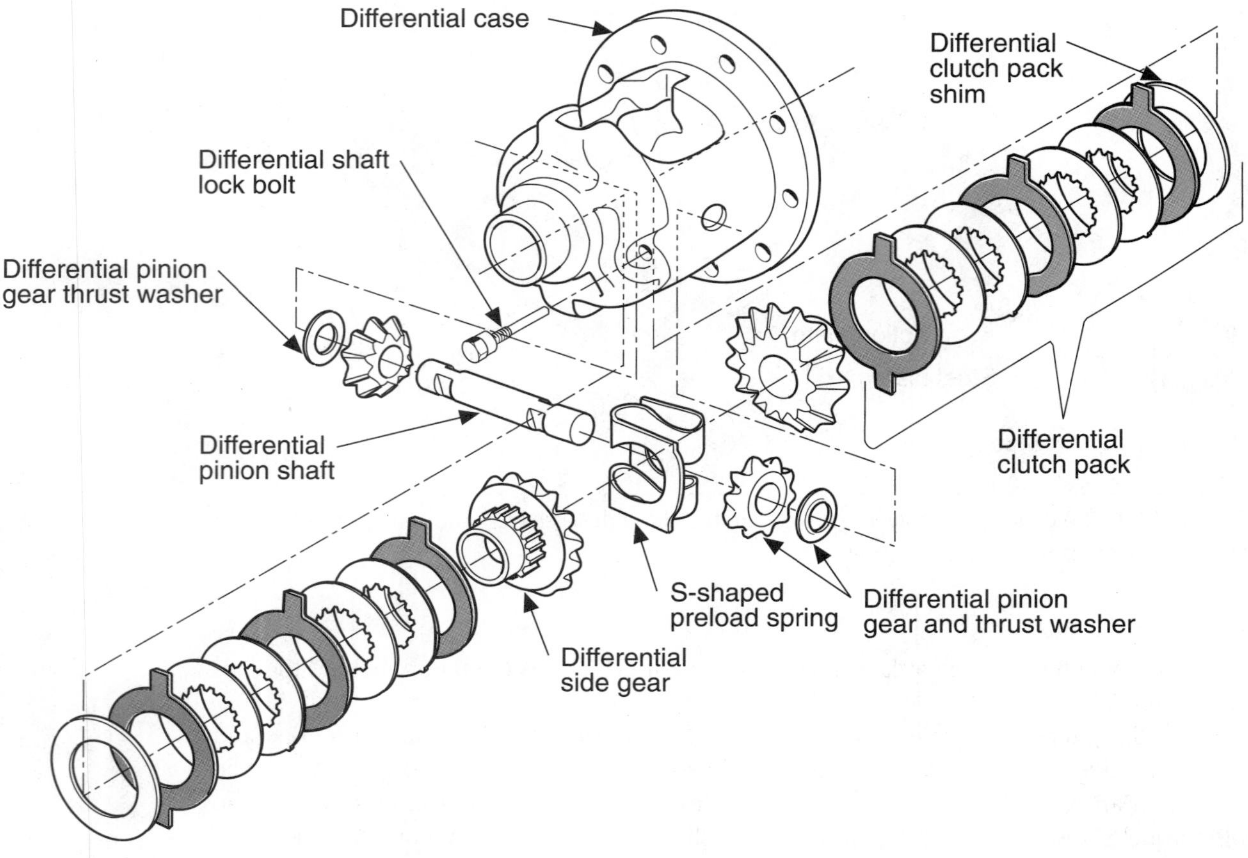

Figure 7-39 Typical limited-slip differential assembly. (Reprinted with the permission of Ford Motor Company)

A preload spring or springs provide the necessary clutch apply pressure to provide drive to both axles and wheels during unequal traction on the drive wheels. The spring tension is low enough to allow clutch slippage on the inner drive axle when turning corners.

Cone Clutches

Shop Manual
Chapter 7, page 245

Cone clutches are also used (Figure 7-40) to limit differential action. A cone clutch is simply a cone covered with frictional material that fits inside an internal cone in the differential case. When the two cones are pressed together, friction allows them to rotate as one unit.

One cone is splined to the axle shaft and the other is attached to the side gear (Figure 7-41). The operation of a cone clutch is the same as that of a unit equipped with a clutch pack.

Viscous Clutch

Shop Manual
Chapter 7, page 193

When speed differential increases, the viscous torque also increases.

Some late model cars use a viscous clutch in their limited-slip differentials (Figure 7-42). These units rely on the resistance generated by a high-viscosity silicon oil to transmit power between the wheels. When there is no rotational difference between the left- and right-side wheels, power is distributed evenly to both wheels in the same way as an open differential. However, when one wheel is on a slippery surface or in hard cornering or when the loading on the outside rear wheel is considerably higher than that on the inside wheel, a rotational speed difference will exist. This speed differential causes the silicon oil to shear, generating viscous torque. This torque effectively reduces the speed differential and reduces the spinning of the lightly loaded wheel, increasing traction and cornering ability.

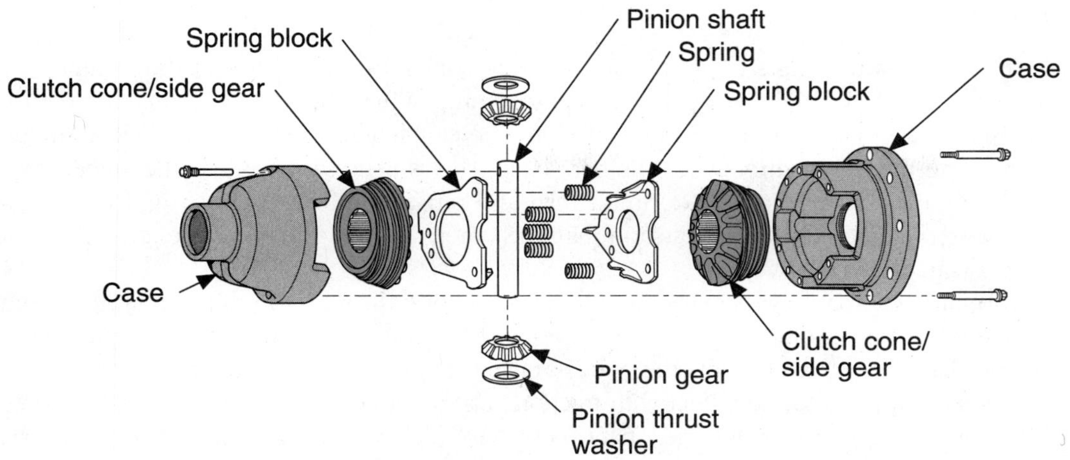

Figure 7-40 Typical cone clutched limited-slip differential. (Courtesy of Oldsmobile Motor Div.—GMC)

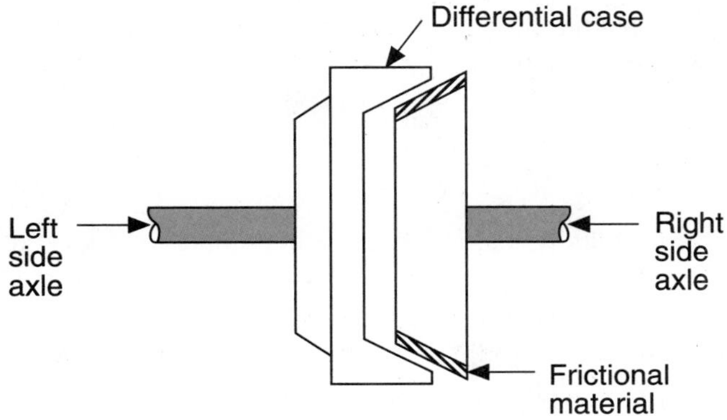

Figure 7-41 Basic construction of a cone-type clutch.

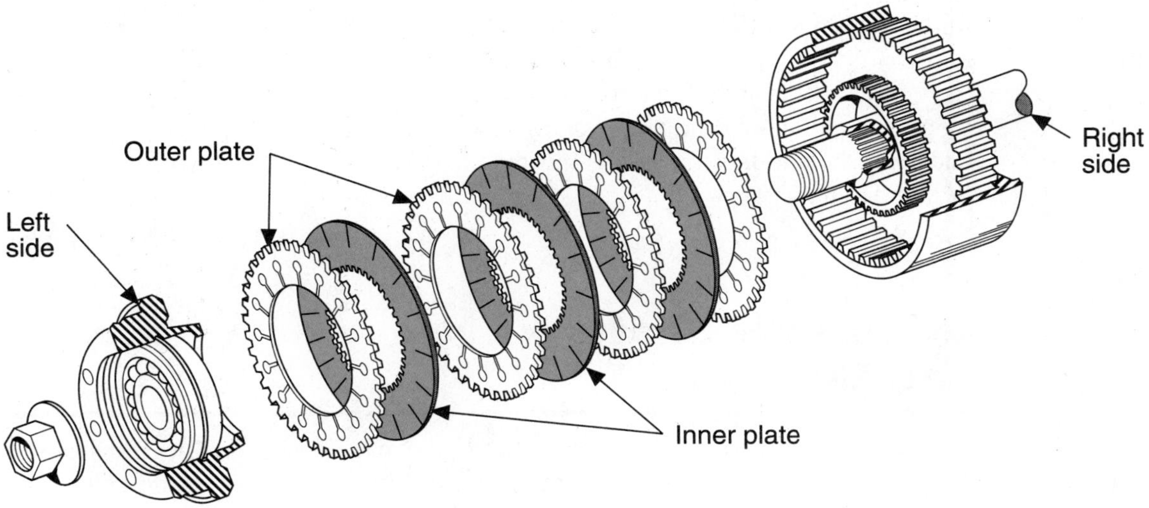

Figure 7-42 Typical viscous clutch assembly.

Operation

When a vehicle is moving straight ahead, the axle shafts are linked to the differential case through the clutch and each wheel gets equal torque. While the vehicle is making a turn, depending on the direction the vehicle is turning, one clutch assembly slips a sufficient amount to allow a speed differential between the two axles. This is necessary because the wheels must move through two different arcs during a turn and must therefore spin at slightly different speeds. When one wheel has less traction than the other, a larger portion of the torque goes to the wheel with the most traction.

Normally, each axle gets an equal amount of torque through the differential. However when one wheel slips, some of that wheel's torque is lost through the pinion gears spinning on the pinion shaft. The clutch on the other wheel remains applied and some of the torque from the slipping side is applied to the wheel with traction. The amount of torque applied to the wheel with traction is determined by the frictional capabilities of its clutch assembly. Power is delivered to that wheel only until the torque overcomes the frictional characteristics of the clutch assembly, at which time it begins to slip. The friction between the clutch plates and discs will transfer a portion of the engine's torque to the wheel with the most traction. This action limits the maximum amount of torque that can be applied to the wheel with traction.

WARNING: A car equipped with a limited-slip differential should never be driven with a minitype spare tire. Two different tire sizes will cause differentiation in the unit and may cause excessive damage to the clutches.

Locked Differential

Locked differentials are often called "lockers."

A common design of a locked differential is a Detroit Locker, which provides maximum traction even with one wheel off the ground.

Another type of special traction differential is a locked differential. In a locked differential, the pinion and side gears are welded together and then welded to the differential case or hydraulic pressure is used to clamp the clutch plates against the side gears, which locks them mechanically. When locked, all differential action is stopped. Therefore, both drive wheels turn at the same speed and with the same torque at all times. Locked differentials are used primarily on race cars and off-road vehicles. They provide superior traction while cornering and traveling off the road, but they absorb horsepower and increase tire wear and fuel consumption. These disadvantages result from the fact that the engine must provide enough power to allow the drive wheels to slide when both wheels have traction and the vehicle is turning.

A spool assembly is basically a ring gear mounted to an empty differential case splined to the axles and offers no differentiating at all. It will scrub the tires, even in the slightest of turns.

Some trucks and off-the-road equipment use differentials that can be locked and unlocked by pressing a button. The button activates an air pump, which applies pressure on the clutches and locks them to the side gears. This type of system gives the advantages of both an open and locked differential unit.

Drive Axle Shafts and Bearings

Shop Manual
Chapter 7, page 245

Located within the hollow horizontal tubes of the axle housing are the axle shafts (Figure 7-43). The purpose of an axle shaft is to transmit the driving force from the differential side gears to the drive wheels. Axle shafts are heavy steel bars splined at the inner end to mesh with the axle side

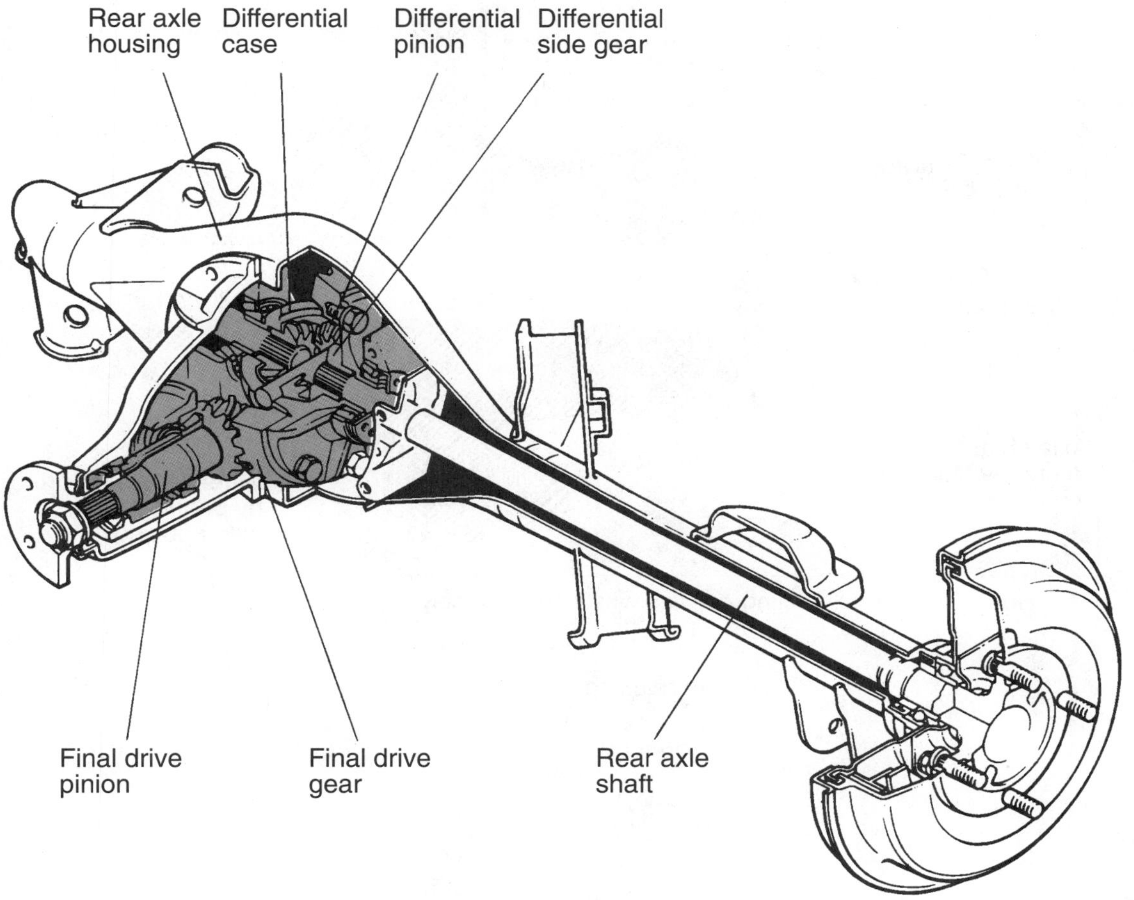

Figure 7-43 Position of drive axles within a RWD axle housing. (Courtesy of Chrysler Corporation)

gear in the differential. The driving wheel is bolted to the wheel flange at the outer end of the axle shaft. The drive wheels rotate to move the vehicle forward or reverse.

The drive axles in a transaxle usually have two CV-joints to allow independent front-wheel movement and steering of the drive wheels. These CV-joints also allow for lengthening and shortening of the drive axles as the wheels move up and down.

The purpose of the axle shaft is to transfer driving torque from the differential assembly to the vehicle's driving wheels. There are two types of axles: the "dead" axle, which supports a load, and the vehicle's wheels and "live" axles, which support and drive the vehicle.

Dead axles are found on trailers and are the type of axle found in the rear of FWD vehicles.

There are basically three designs by which axles are supported in a live axle: full-floating, three-quarter floating, and semifloating. These refer to where the axle bearing is placed in relation to the axle and the housing. The bearing of a full-floating axle is located on the outside of the housing (Figure 7-44). This places all of the vehicle's weight on the axle housing with no weight on the axle.

Three-quarter and semi-floating axles are supported by bearings located in the housing and thereby carry some of the weight of the vehicle (Figure 7-45). Most passenger cars are equipped with three-quarter or semi-floating axles. Full-floating axles are commonly found on heavy-duty trucks.

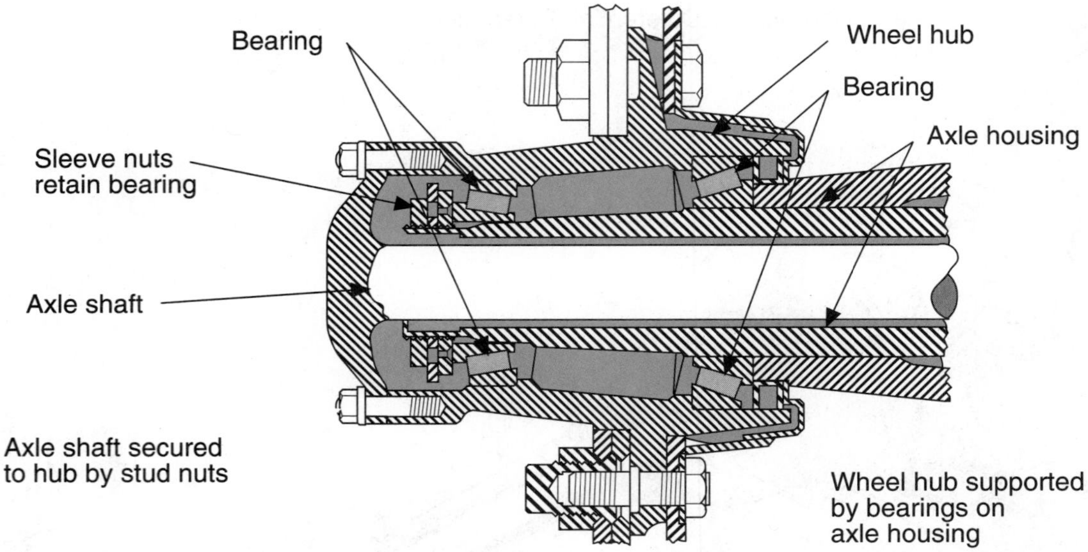

Figure 7-44 A full-floating axle and wheel hub assembly.

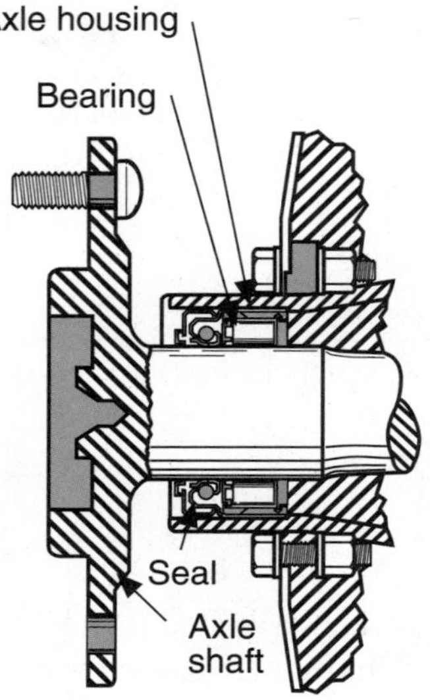

Figure 7-45 Locating the axle bearings in the axle housing places some of the vehicle's weight on the bearing. (Courtesy of Oldsmobile Div.—GMC)

Thrust loads are loads placed on a part that is parallel to the center of the axis of rotation.

Radial loads are loads applied at 90 degrees to an axis of rotation.

The axle shaft bearing supports the vehicle's weight and reduces rotational friction. With semifloating axles, radial and thrust loads are always present on the axle shaft bearing when the vehicle is moving. Radial bearing loads act at 90 degrees to the axle's center of axis. Radial loading is always present whether or not the vehicle is moving. Thrust loading acts on the axle bearing parallel with the center of axis. It is present on the driving wheels, axle shafts, and axle bearings when the vehicle turns corners or curves.

Three designs of axle shaft bearings are used on semifloating axles: ball-type, straight-roller, and tapered-roller bearings. The load on a bearing that is of primary concern is the axle's end thrust. When a vehicle moves around a corner, centrifugal force acts on the vehicle's body, causing it to lean to the outside of the curve. The vehicle's chassis does not lean because of the tire's contact with the road's surface. As the body leans outward, a thrust load is placed on the axle shaft and axle bearing. Each type of axle shaft handles axle shaft end thrus differently.

The end-to-end movement of the axle is controlled by a C-type retainer on the inner end of the axle shaft or by a bearing retainer and retainer plate at the outer end of the axle shaft.

When the centrifugal force created by turning is great enough to cause the chassis to lean, the wheels on the outside of the vehicle will leave the ground.

Ball-type Axle Bearings

An axle with ball-type axle bearings has the axle shaft and bearing held in place inside the axle housing by a stamped metal bearing retainer plate (Figure 7-46). The plate is bolted to the axle housing. The bearing is held in place on the axle shaft by a retaining ring, which is pressed onto the axle shaft.

Shop Manual
Chapter 7, page 247

Body lean is also called body roll.

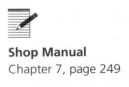

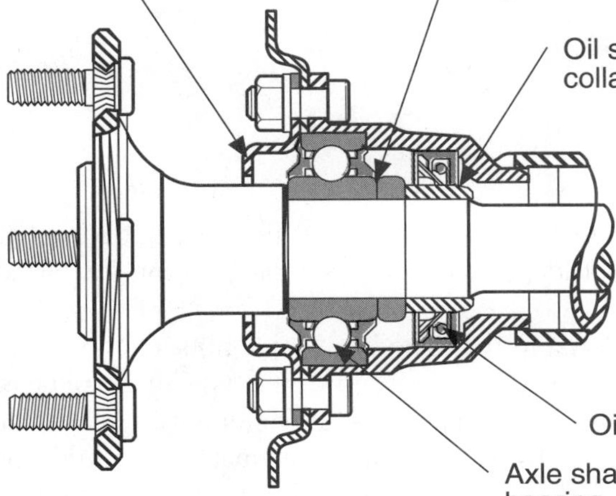

Figure 7-46 An axle shaft with a ball-type bearing.

The operation of the ball-type bearing is designed to absorb radial load as well as the axle shaft end thrust. Because both bearing loads are taken at the bearing, there is no axle shaft end thrust absorption or adjustment designed into the rear axle housing.

To seal in the lubricant, an oil seal collar and oil seal is used. The oil seal collar is a machined sleeve or finished portion of the axle on which the lips of the seal ride. The oil seal retains the gear lubricant inside the axle housing. The axle seal prevents the lubricant from leaking into the brakes.

Straight-Roller Axle Bearings

The straight-roller bearing uses the axle shafts as its inner race (Figure 7-47). The outer bearing race and straight rollers are pressed into the axle tubes of the rear axle housing. The inner end of the axle shaft at the differential has a groove machined around its outside diameter where the C-type retainer fits.

Shop Manual
Chapter 7, page 249

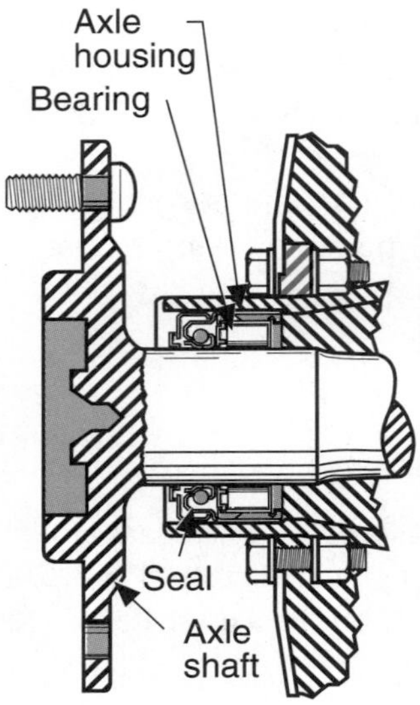

Figure 7-47 An axle shaft with a straight roller-type bearing. (Courtesy of Oldsmobile Div.—GMC)

The bearing is lubricated by hypoid lubricant from the differential area of the axle housing. The grease seal, located outside the axle shaft bearing, prevents the lubricant from leaking out of the housing.

When the vehicle takes a turn, the body and axle housing move outward and the axle shaft moves inward on the bearings. The inner end of the axle shaft contacts the differential pinion shaft. The axle shaft end thrust exerted against the differential pinion shaft moves the differential housing and differential side bearing assembly against the integral housing axle tube, which absorbs the axle shaft end thrust. There is no end thrust adjustment designed into the rear axle housing.

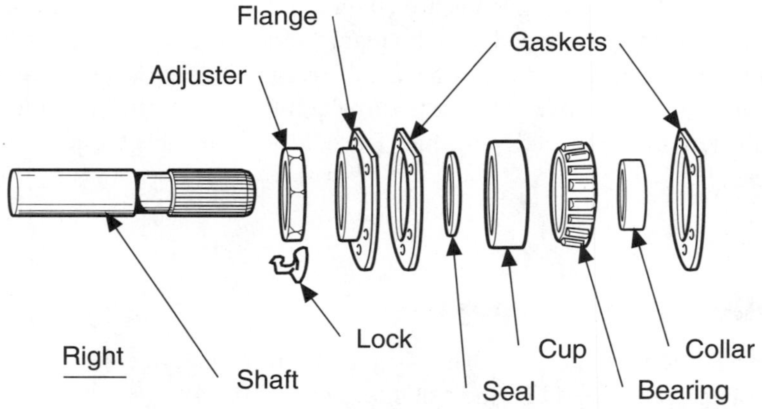

Figure 7-48 Tapered-roller bearing assembly. (Courtesy of Chrysler Corporation)

Tapered-Roller Axle Bearings

Shop Manual
Chapter 7, page 249

The tapered-roller bearing and axle shaft assembly are held inside the axle housing by a flange, which is bolted to the axle housing (Figure 7-48). The inside of the flange may be threaded to receive the adjuster or it is machined to accept adjustment shims.

The axle shaft is designed to float slightly based on in and out movements of the axles. As the axle shaft moves inward, it contacts a thrust block that separates both axle shafts at the center of the differential. The inward moving axle shaft contacts the thrust block, which passes the thrust force onto the opposite axle shaft. There the axle shaft end thrust becomes an outward moving force, which causes the opposite tapered-roller axle bearing to seat in its bearing cup. Axle end thrust adjustments can be made by a threaded adjuster or thin metal shims placed between the brake assembly plate and the axle housing.

The tapered-roller bearing is lubricated prior to installation in the axle housing. A seal and two gaskets keep the hypoid lubricant and foreign matter out of the bearing operating area. A collar holds the rear axle bearing in place on the axle shaft.

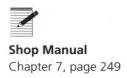

A BIT OF HISTORY

The use of axle half-shafts and a transverse tube to position the rear wheels of a vehicle was patented in 1894 by De Dion-Bourton. This design of axle is known as the De Dion axle (Figure 7-49).

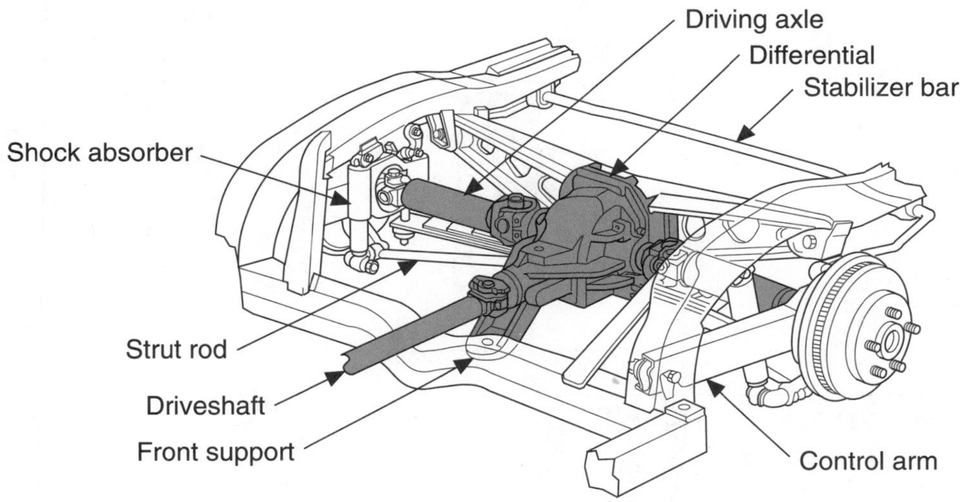

Figure 7-49 A De Dion or IRS axle assembly.

IRS Axle Shafts

The drive axles on most newer IRS systems use two U- or CV-joints per axle to connect the axle to the differential and the wheels (Figure 7-50). They are also equipped with linkages and control arms to limit camber changes. The axles of an IRS system are much like those of a FWD system. The outer portion of the axle is supported by an upright or locating member that is also part of the suspension.

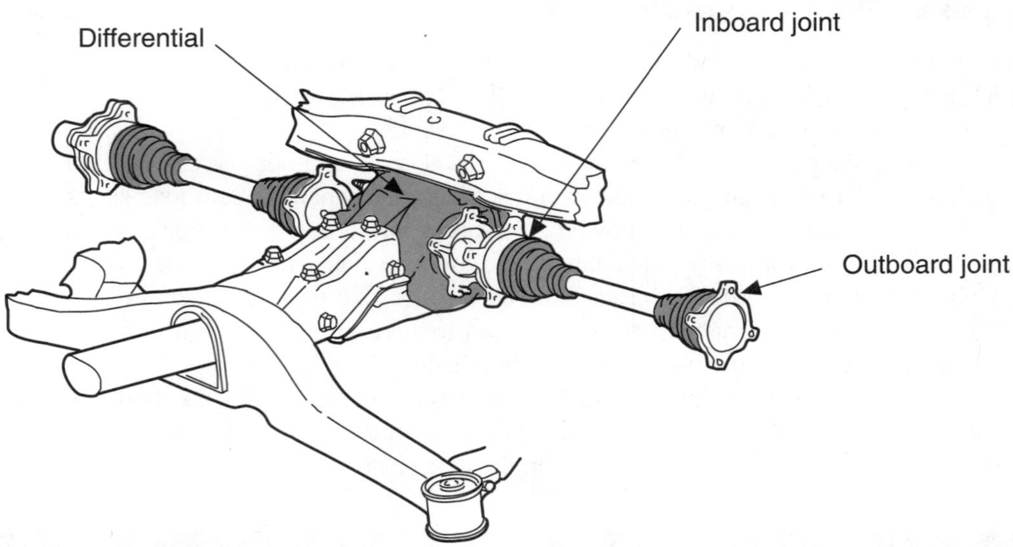

Figure 7-50 A typical IRS drive axle assembly. (Courtesy of Toyota Motors)

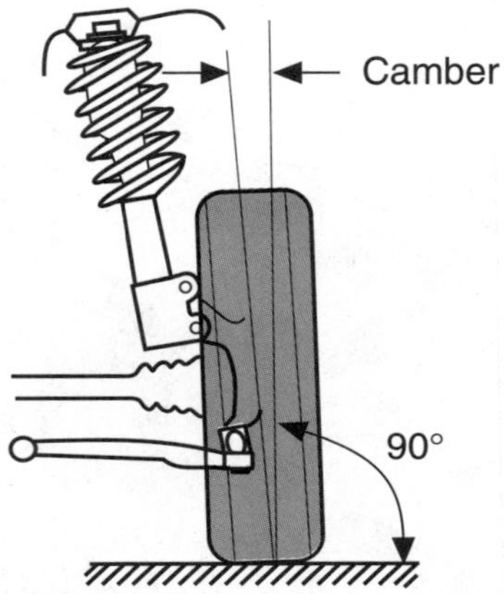

Figure 7-51 Wheel camber. (Courtesy of Nissan Motors)

Camber is a suspension alignment term used to define the amount that the centerline of a wheel is tilted inward or outward from the true vertical plane of the wheel (Figure 7-51). If the top of the wheel is tilted inward, the camber is negative. If the top of the wheel is tilted outward, the camber is positive.

Swing axles are a form of IRS. In a swing axle car, the differential is also bolted to the chassis but the axles are U-jointed only to where they meet the differential and not at the wheel end. This makes the wheels move in an arc as they move up and down with the suspension. The swinging of the wheels in an arc causes large camber changes and raises and lowers the rear of the car. These motions make the car difficult to drive in some situations, which is why swing axles are not found in many new cars. Swing axles were popular from the late 1940s through the 1960s because they were inexpensive to make and gave a softer ride than a solid live axle.

Maintenance

Hypoid gears require hypoid gear lubricant of the extreme pressure type. Gear lube viscosity is generally SAE 75 to 90. Limited-slip differentials require special limited-slip lubricant, which provides the required coefficient of friction for the clutch discs or cones, as well as proper lubrication.

Transaxles and some RWD differentials may require a lower viscosity oil, such as ATF. Also some transaxles may require separate lubricants for the transmission and differential. Always refer to manufacturers' recommendations for lubricants and frequency of change for every unit.

Summary

- ❑ The drive axle of a RWD vehicle is mounted at the rear of the car. It is a single housing for the differential gears and axles. It also is part of the suspension and helps to locate the rear wheels.
- ❑ The final drive is the final set of reduction gears the engine's power passes through on its way to the drive wheels.
- ❑ A differential is needed between any two drive wheels, whether in a RWD, FWD, or 4WD vehicle because the two drive wheels must turn at different speeds when the vehicle is in a turn.
- ❑ RWD final drives use a hypoid ring and pinion gearset, which turns the power flow 90 degrees from the drive shaft to the drive axles. A hypoid gearset also allows the drive shaft to be positioned low in the vehicle.
- ❑ The differential rotates the driving axles at different speeds when the vehicle is turning and at the same speed when the vehicle is traveling in a straight line.
- ❑ The differential is normally housed with the drive axles in the rear axle assembly. Power from the engine enters into the rear axle and is transmitted to the drive axles, which are attached to the wheels of the car.
- ❑ The differential allows for different speeds between the two drive wheels, multiplies torque, and changes the direction of the power flow through a series of gears.
- ❑ The pinion gear meshes with the ring gear, which is fastened to the differential case. The pinion shafts and gears are retained in the differential case and mesh with side gears splined to the drive axles.
- ❑ When both driving wheels are rotating at the same speed, the differential pinions do not rotate on the differential pinion shaft and the differential assembly rotates as one and the driving wheels, axles, and axle side gears rotate at the same speed.
- ❑ When the vehicle turns, the drive wheels rotate at different speeds because the differential case forces the pinion gears to walk around the slow turning axle side gear. This action causes the outside axle side gear to reach a higher speed than the inside wheel. The amount of differential action taking place depends on how sharp the corner or curve is. Differential action provides control on corners and prolongs drive wheel tire life.
- ❑ Live rear axles use a one-piece housing with two tubes extending from each side. These tubes enclose the axles and provide attachments for the axle bearings. The housing shields all parts from dirt and retains the differential lubricant.
- ❑ Rear axle housings can be divided into two groups, integral carrier or removable carrier.
- ❑ An integral carrier housing has a service cover that fits over the rear of the differential and rear axle assembly. The assembly and disassembly of the differential is only possible when it is in the axle housing.
- ❑ The differential assembly of a removable carrier assembly can be removed from the front of the axle housing as a unit and is serviced at a bench and then installed into the axle housing.
- ❑ The types of gears currently used as differential gears are the helical, spiral bevel, and hypoid gears.

Terms To Know

Axle shaft tubes
Axle shaft end thrust
C-clip
Center section
Clutch packs
Cone clutch
Differential case
Differential housing
Differential pinion gears
Differential pinion shaft
Differential ring gear
Differential side gears
Drive pinion
Drive pinion flange
Drive pinion gear
Extreme-pressure lubricant
Forward coast side
Forward drive side
Full-floating rear axle
Heel
Hunting gears
Hypoid gear lubricant
Integral carrier
Lapping
Matched gearset code
Nonhunting gears
Pinion carrier
Pinion gear

Terms to Know (continued)

Removable carrier housing

Ring gear

Semifloating rear axle

Side clearance

Side gears

Swing axles

Three-quarter floating axle

- ❑ Hypoid gearsets are commonly used in RWD passenger car and light trucks because they are quiet running and require a lower hump in the floor of the vehicle's body.
- ❑ With hypoid gears, the drive side of the teeth is curved in a convex shape, whereas the coast side of the teeth is concave. The inner end of the teeth on a hypoid ring gear is known as the toe and the outer end of the teeth as the heel.
- ❑ The gear ratio of the pinion and ring gear is often referred to as the final drive ratio.
- ❑ Gear ratios express the number of turns the drive gear makes compared to one turn of the driven gear it mates with.
- ❑ Ring and pinion gearsets are usually classified as hunting, nonhunting, or partial nonhunting gears.
- ❑ A nonhunting gearset is one in which any one pinion tooth comes into contact with only some of the ring gear teeth and is identified by a .00 gear ratio.
- ❑ A partial nonhunting gearset is one in which any one pinion tooth comes into contact with only some of the ring gear teeth, but more than one revolution of the ring gear is required to achieve all possible gear tooth contact combinations. These gears are identified by a .50 ratio.
- ❑ When hunting gearsets are rotating, any pinion gear tooth is likely to contact all the ring gear teeth.
- ❑ During assembly the nonhunting and partial nonhunting gears must be assembled with the timing marks properly aligned. Hunting gears do not need to be aligned because any tooth on the pinion may mesh with any tooth on the ring gear.
- ❑ Lapping is the process of using a grinding paste to produce a fine finish on the teeth of the two gears that will be in constant contact with each other.
- ❑ At least four bearings are found in all differentials. Two fit over the drive pinion shaft to support it while the other two support the differential case.
- ❑ The drive pinion gear is placed horizontally in the axle housing and is positioned by one of two types of mounting, either straddle or overhung.
- ❑ The straddle-mounted pinion gear is used in most removable carrier-type axle housings and uses two opposing tapered-roller bearings positioned close together with a short spacer between their inner races and a third bearing to support the rear of the pinion gear.
- ❑ The overhung-mounted pinion uses two opposing tapered-roller bearings but not a third bearing.
- ❑ A spacer is placed between the opposing pinion shaft bearings to control the amount of preload applied to the bearings. Preload prevents the pinion gear from moving back and forth in the bearing retainer.
- ❑ Preload is a fixed amount of pressure constantly applied to a component to prevent it from loosening up.
- ❑ The differential section of the transaxle has the same components as the differential gears in a RWD axle and basically operate in the same way, except that the power flow in transversely mounted power trains does not need to turn 90 degrees.
- ❑ The final drive gears and differential assembly are normally located within the transaxle housing of FWD vehicles.
- ❑ There are four common configurations used as the final drives on FWD vehicles: helical, planetary, hypoid, and chain drive. The helical, planetary, and chain final drive arrangements are usually found in transversely mounted power trains. Hypoid final drive gear assemblies are normally used with longitudinal power train arrangements.
- ❑ A limited-slip unit provides more driving force to the wheel with traction when one wheel begins to spin by restricting the differential action.

- ❑ Limited-slip differentials use either a clutch pack, a cone clutch, or a viscous clutch assembly.
- ❑ The purpose of the axle shaft is to transfer driving torque from the differential assembly to the vehicle's driving wheels.
- ❑ There are basically three ways drive axles are supported by bearings in a live axle: full-floating, three-quarter floating, and semifloating.
- ❑ There are three designs of axle shaft bearings used on semifloating axles: ball-type, straight-roller, and tapered-roller bearings.
- ❑ Hypoid gears require hypoid gear lubricant of the extreme pressure type. Gear lube viscosity is generally SAE 75 to 90.
- ❑ Limited-slip differentials require special limited-slip lubricant, which provides the required coefficient of friction for the clutch discs or cones, as well as proper lubrication.
- ❑ Transaxles and some RWD differentials may require a lower viscosity oil, such as ATF. Also some transaxles may require separate lubricants for the transmission and differential.

Review Questions

Short Answer Essays

1. Define final drive.
2. Explain why a differential is needed.
3. List the reasons why hypoid gears are used in RWD final drives.
4. List the three major functions of a differential.
5. Describe the components of a differential and state their location.
6. Explain the major differences between an integral carrier housing and a removable carrier housing.
7. Explain the differences between hunting, nonhunting, and partial nonhunting gears.
8. Describe the difference between a straddle-mounted pinion and an overhung pinion.
9. Explain the purpose and general operation of a limited-slip differential.
10. List the different ways drive axles are supported in an axle housing and explain the major characteristics of each one.

Fill-in-the-Blanks

1. The differential's ____________________ gear meshes with the ____________________ gear, which is fastened to the differential case, which houses the ____________________ shafts and gears, which are in mesh with the ____________________ gears, which are splined to the drive axles.

2. Rear axle housings can be divided into two groups: ____________________ carrier and ____________________ carrier.

3. The types of gears currently used as differential gears are the ____________________ , ____________________ ____________________ , and ____________________ .

4. The drive side of a hypoid's gear teeth is curved in a ________________ shape, and the coast side of the teeth is ________________ . The inner end of the teeth on a hypoid ring gear is known as the ________________ and the outer end of the teeth as the ________________ .

5. Gear ratios express the number of turns the ________________ gear makes compared to one turn of the ________________ gear it mates with.

6. Ring and pinion gearsets are usually classified as ________________ , ________________ , or ________________ gears.

7. A straddle-mounted pinion gear is usually mounted on ________________ bearings, whereas an overhung-mounted pinion is mounted on ________________ .

8. The four common configurations used as the final drives on FWD vehicles are: ________________ , ________________ , ________________ , and ________________ .

9. Limited-slip differentials use either a ________________ , a ________________ , or a ________________ to limit the action of the differential.

10. There are three designs of axle shaft bearings used on semifloating axles: ________________ , ________________ , and ________________ .

ASE Style Review Questions

1. Who is correct?
   Technician A says when a car is making a turn, the outside wheel must turn faster than the inside wheel.
   Technician B says a locked differential causes the car to slide around a turn.
   A. A only
   B. B only
   C. Both A and B
   D. Neither A nor B

2. While discussing the torque multiplication factor of the differential:
   Technician A says all of the gears in a differential affect torque multiplication.
   Technician B says there is a gear reduction as the power flows from the pinion to the ring gear.
   Who is correct?
   A. A only
   B. B only
   C. Both A and B
   D. Neither A nor B

3. Who is correct?
Technician A says that when a car is moving straight ahead, all differential gears rotate as a unit.
Technician B says that when a car is turning a corner, the inside differential side gear rotates slowly on the pinion, causing the outside side gear to rotate faster.
A. A only
B. B only
C. Both A and B
D. Neither A nor B

4. While discussing the mounting of the drive pinion shaft:
Technician A says most pinion gears are mounted on a long bushing.
Technician B says pinion gears are mounted on two tapered-roller bearings.
Who is correct?
A. A only
B. B only
C. Both A and B
D. Neither A nor B

5. While discussing gear ratios:
Technician A says they express the number of turns the driven gear makes compared to one turn of the drive gear.
Technician B says the gear ratio of a differential unit expresses the number of teeth on the ring gear compared to the number of teeth on the pinion gear.
Who is correct?
A. A only
B. B only
C. Both A and B
D. Neither A nor B

6. While discussing final drive gear ratios:
Technician A says lower gear ratios allow for better acceleration.
Technician B says higher gear ratios allow for improved fuel economy but lower top speeds.
Who is correct?
A. A only
B. B only
C. Both A and B
D. Neither A nor B

7. While discussing different types of ring and pinion gearsets:
Technician A says that with a nonhunting gearset, each tooth of the pinion will return to the same tooth space on the ring gear each time the pinion rotates.
Technician B says that when a hunting gearset rotates, any pinion gear tooth is likely to contact each and every tooth on the ring gear.
Who is correct?
A. A only
B. B only
C. Both A and B
D. Neither A nor B

8. While discussing limited-slip differentials:
Technician A says these differentials are mostly used to increase acceleration.
Technician B says these differentials limit the amount of differential action between the side gears.
Who is correct?
A. A only
B. B only
C. Both A and B
D. Neither A nor B

9. While discussing the different designs of rear axles:
Technician A says the bearings for full-floating shafts are located within the axle tubes of the rear axle housing.
Technician B says the names used to classify the different designs actually define the amount of vehicle weight that is supported by the axles.
Who is correct?
A. A only
B. B only
C. Both A and B
D. Neither A nor B

10. Who is correct?
Technician A says limited-slip differentials require a special lubricant.
Technician B says all differential units require a special hypoid compatible lubricant.
A. A only
B. B only
C. Both A and B
D. Neither A nor B

CHAPTER 8

Four-Wheel-Drive Systems

Upon completion and review of this chapter, you should be able to:

- ❑ Describe the different designs of four-wheel-drive systems and their applications.
- ❑ Compare and contrast the components of part- and full-time four-wheel-drive systems.
- ❑ Describe the operation of various transfer case designs and their controls.
- ❑ Identify the differences in operation and construction between manual and automatic locking front wheel hubs.
- ❑ Discuss the purpose of an interaxle differential and the design variations used by the industry.
- ❑ Discuss the purpose, operation, and application of a viscous coupling in four-wheel-drive systems.
- ❑ Identify the suspension requirements of vehicles equipped with four wheel drive.

Many light trucks have been produced through the years with four wheel drive; these vehicles represent many opportunities for automotive technicians versed in the diagnostics and service of these special systems. Recently, powering all four wheels of performance cars has become more commonplace and provides even more opportunities for skilled technicians.

The primary focus of this chapter is on the transfer cases and related systems used for four-wheel-drive on light trucks (Figure 8-1), automobiles, and off-the-road vehicles. Although the principles of operation are the same for all four-wheel-drive units, the components, location, and controls of the various systems differ according to manufacturer and application.

SERVICE TIP: Whenever diagnosing, servicing, or repairing a four-wheel-drive performance car, light truck, or off-the-road vehicle, refer to the appropriate service manual for specific information and service procedures for that vehicle.

With four wheel drive, engine power can flow to all four wheels. This action can greatly increase a vehicle's traction when traveling in adverse conditions and can also improve handling as side forces generated by the turning of a vehicle or by wind gusts will have less of an effect on a vehicle that has power applied to the road on four wheels.

The common acronym for four wheel drive is 4WD.

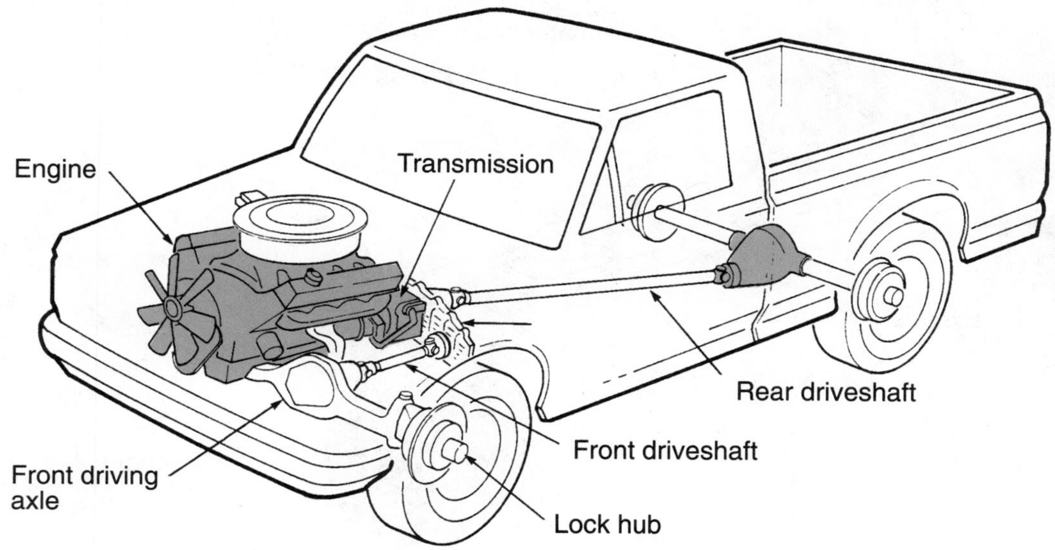

Figure 8-1 Typical arrangement of 4WD components. (Reprinted with the permission of Ford Motor Company)

A BIT OF HISTORY:

The first known gasoline-powered four-wheel-drive automobile was the Spyker, which was built in the Netherlands in 1902.

Four-Wheel-Drive Design Variations

Shop Manual
Chapter 8, page 271

Four wheel drive is most useful when a vehicle is traveling off the road or in deep mud or snow. However, some high performance cars are equipped with four wheel drive to improve the handling characteristics of the car. Nearly all of these cars are front-wheel-drive models converted to four wheel drive. Normally, FWD cars are modified by adding a transfer case, a rear drive shaft, and a rear axle with a differential (Figure 8-2). Although this is the typical modification, some cars are equipped with a center differential (Figure 8-3) in place of the transfer case. This differential unit allows the rear and front wheels to turn at different speeds.

The center differential is commonly referred to as the interaxle differential.

Due to the many names manufacturers give their 4WD systems, it is often difficult to clearly define the difference between 4WD and AWD (all wheel drive). In this text, all references to 4WD systems will refer to those systems that use a separate transfer case and allow the driver to select or feature automatic controls that select the transfer of the engine's power to either two or four wheels. If the transfer case is manually controlled, this is accomplished by a shifter, electrical switches, or through locking wheel hubs. These locking hubs can be either manual or automatic.

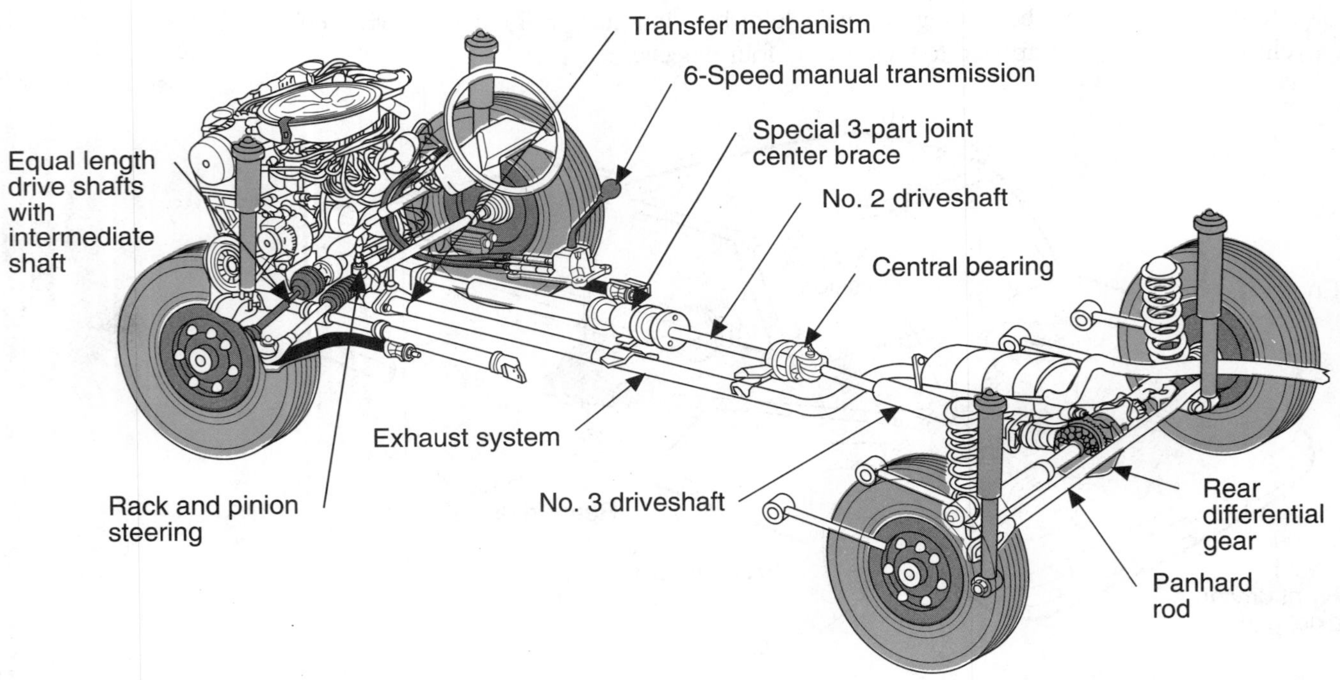

Figure 8-2 An AWD system based on a FWD platform. (Courtesy Of American Honda Motor Co., Inc.)

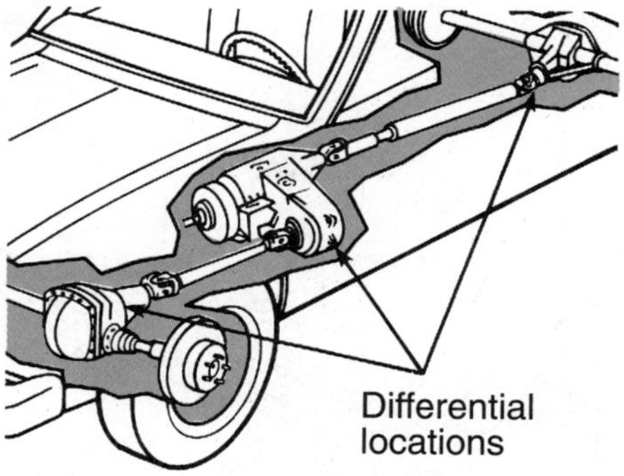

Figure 8-3 Location of front, rear, and center differentials.

Normally AWD systems do not use a transfer case; rather, they use a center differential, viscous coupling, or transfer clutch assembly to transmit engine power to the front and rear axles (Figure 8-4). AWD systems do not allow the driver to select between two or four wheel drive; they always operate in 4WD.

Four-wheel-drive vehicles designed for off-the-road use are normally RWD vehicles equipped with a transfer case, a front drive shaft, and a front differential and drive axles. Many

The designation of 4 x 4 is used with vehicles to indicate the number of wheels on the ground and the number of wheels that can be driven. For example, the typical car has four wheels, two of which can deliver power, therefore they could be designated as 4 x 2 vehicles. A 4WD vehicle is designated as a 4 x 4 because it has four wheels and power can be applied to all four wheels.

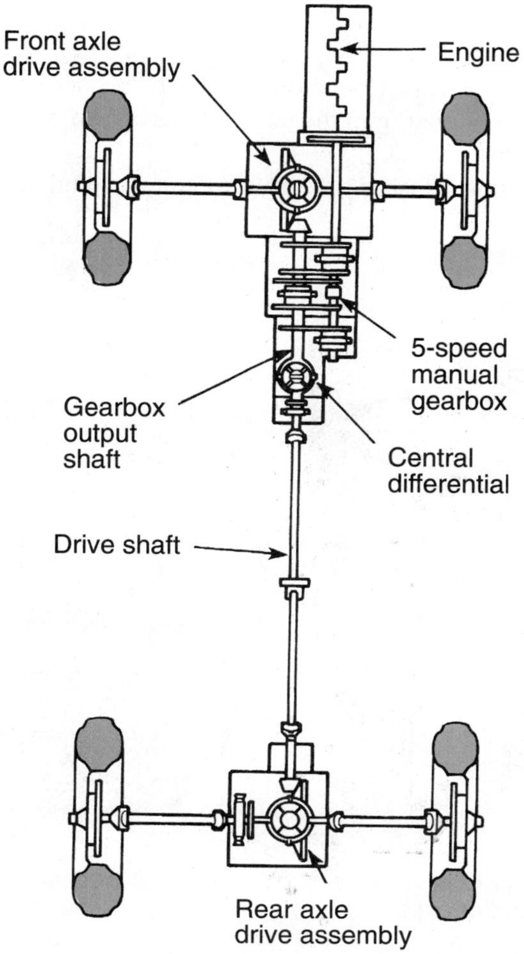

Figure 8-4 A typical AWD powertrain. (Courtesy of Porsche/Audi)

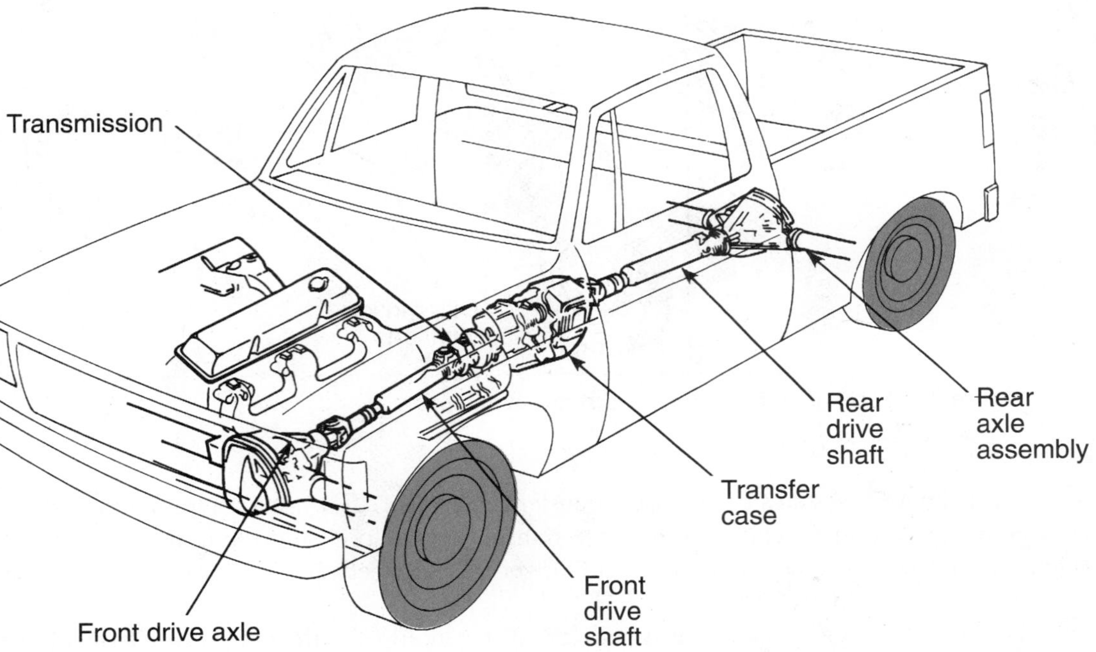

Figure 8-5 Location of major components of a typical 4WD vehicle.

4WD vehicles use three drive shafts. One short drive shaft connects the output of the transmission to the transfer case. The output from the transfer case is then sent to the front and rear axles through separate drive shafts.

The typical 4WD system consists of a front-mounted, longitudinally positioned engine;

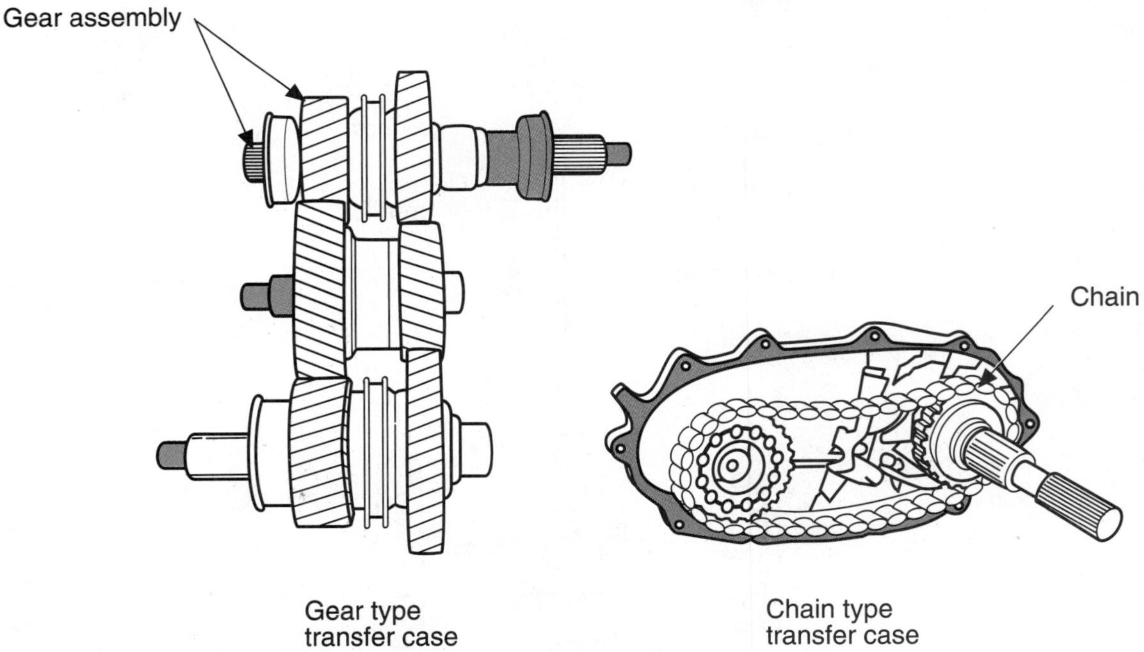

Figure 8-6 Typical gear and chain drive assemblies. (Reprinted with the permission of Ford Motor Company)

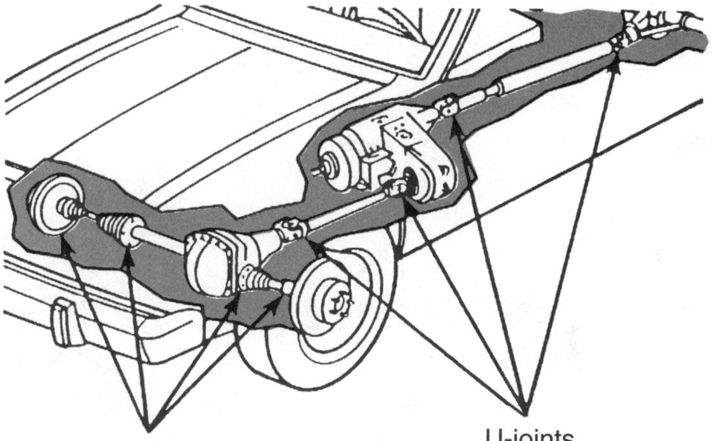

Figure 8-7 Location of CV- and U-joints on a typical 4WD vehicle.

either an automatic or manual transmission; front and rear drive shafts; front and rear drive axle assemblies; and a transfer case (Figure 8-5).

The transfer case is usually mounted to the side or rear of the transmission. When a drive shaft is not used to connect the transmission to the transfer case, a chain or gear drive (Figure 8-6) within the transfer case receives the engine's power from the transmission and transfers it to the drive shafts leading to the front and rear drive axles.

Classroom Manual
Chapter 6, page 115

The drive shafts from the transfer case shafts connect to differentials at the front and rear drive axles. As on 2WD vehicles, these differentials are used to compensate for road and operating conditions by altering the speed of the wheels connected to the axles. This is important when the vehicle is turning a corner and the outside wheel must travel more distance than the inner wheel.

2WD is an acronym commonly used for two-wheel-drive vehicles.

Universal or CV joints are used to connect the drive shafts to the differential and the transfer case. The rear axles are either connected directly to the hub of the wheels or are connected to the hubs by U-joints. Universal joints are also normally used to connect the front axles to the wheel hubs on heavy duty trucks. Light duty vehicles and 4WD passenger cars generally use half-shafts and CV-joints in their front drive axle assembly (Figure 8-7).

Classroom Manual
Chapter 5, page 96

The front wheel hubs on most 4WD vehicles must be locked or unlocked by the driver to provide for efficient operation in 2WD or 4WD. The rear wheel hubs are always engaged.

An electric switch or shift lever, located in the passenger compartment, controls the transfer case so that power is directed to the axles selected by the driver. Power can typically be directed to all four wheels, two wheels, or none of the wheels. On many vehicles, the driver can also select a low speed range for extra torque while traveling in very adverse conditions.

Full-time 4WD systems use a center differential that accommodates speed differences between the two axles, necessary for on-highway operation.

While most 4WD trucks and utility vehicles are design variations of basic RWD vehicles, most passenger cars equipped with 4WD are based on FWD designs. These modified FWD systems consist of a transaxle and differential to drive the front wheels, plus some type of mechanism for connecting the transaxle to a rear drive line. In many cases this mechanism is a simple clutch or differential. On some models, a transfer case, which transfers power to the rear axle, is fitted to the transaxle (Figure 8-8).

CUSTOMER CARE: Customers should be made aware that wear on 4WD systems is much greater than on a 2WD transaxle or transmission. This is especially true if the driver leaves the vehicle engaged in 4WD on dry pavement. Some manufacturers will not warranty the parts of the 4WD system if there is evidence of abuse, such as operation of 4WD while driving on dry surfaces.

Classroom Manual
Chapter 4 page 64

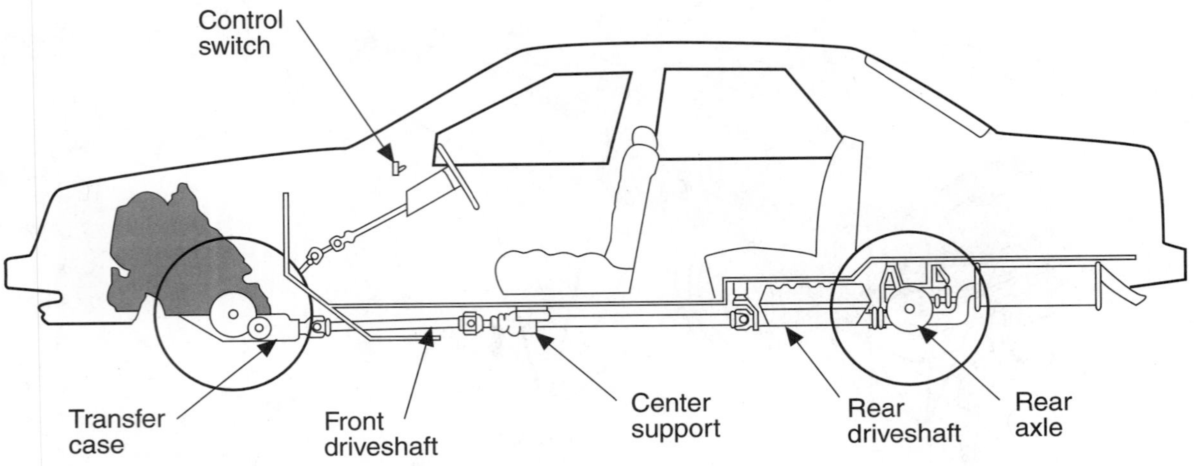

Figure 8-8 Location of 4WD drive line components on a typical FWD car. (Reprinted with the permission of Ford Motor Company)

Shop Manual
Chapter 8, page 271

Types of 4WD Systems

The rear drive axle of a 4WD vehicle is identical to those used in two-wheel-drive vehicles. The front drive axle is also like a conventional rear axle, except that it is modified to allow the front wheels to steer (Figure 8-9). Further modifications are also necessary to adapt the axle to the vehicle's suspension system. The differential units housed in the axle assemblies are similar to those found in a RWD vehicle.

The front differential is generally an open differential because of the great differences in wheel rotational speed when steering around corners. Some off-the-road vehicles have lockable front differentials for extreme conditions.

A locking differential is one that under certain circumstances will allow no speed difference between two wheels or drive shafts. This can aid in traction on loose or slippery surfaces.

The rear differential is necessary to allow for speed differences when turning corners and to avoid stresses on the drive train. Open differentials split torque evenly between both axle or output shafts. This action can be modified by the inclusion of a limited-slip differential.

Although four wheel drive offers increased traction, it also has disadvantages. The additional axle, differential, drive shaft, and transfer case add weight to the vehicle and therefore decrease its fuel economy. Also, less horsepower is available to the wheels due to the power lost in turning the additional axle assembly.

Most four-wheel-drive units are equipped to allow the driver to select in and out of four wheel drive. The systems capable of operating in both two wheel and four wheel drive are

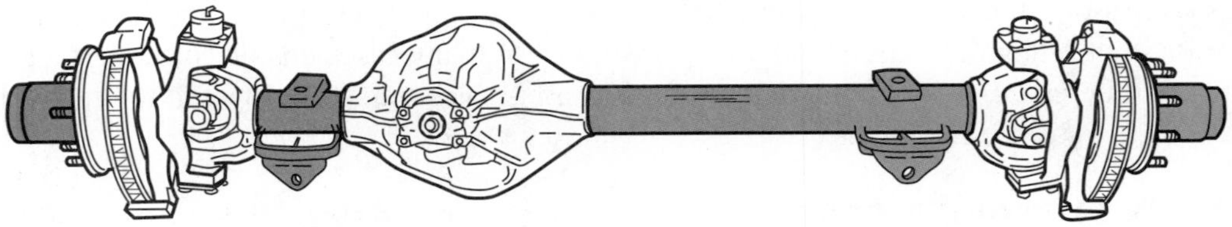

Figure 8-9 Typical front drive axle assembly. (Reprinted with the permission of Ford Motor Company)

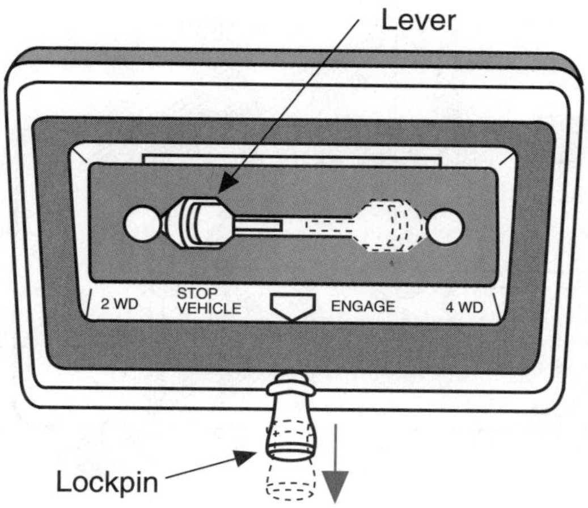

Figure 8-10 Typical 2WD-4WD selector switch. (Courtesy of Chrysler Corporation)

called part-time 4WD systems. Full-time 4WD systems cannot be selected out of four wheel drive. The selection of two wheel or four wheel drive is controlled by a shifter, an electric switch (Figure 8-10), or locking axle hubs (Figure 8-11).

Full-Time Systems

If a vehicle has full-time 4WD, a controllable differential is built into the transfer case. The purpose of this interaxle differential is to compensate for any difference in front wheel and rear wheel speeds and to allow the front and rear axles to operate at their own speeds.

A full-time system is ideal for use when driving through rapidly changing weather conditions or on surfaces that fluctuate between good and bad, such as when driving in and out of rain or on mostly clear roads spotted with patches of snow.

During turns, the front wheels travel a greater distance than the rear wheels. This is because the front wheels move through a wider arc than the rear wheels (Figure 8-12). With full-time 4WD, the center differential allows the front wheels to travel farther or turn faster than the rear wheels without slipping.

The use of an open differential in the transfer case does have a disadvantage. If the wheels on either axle lose traction and begin to spin, the differential continues to deliver maximum

Shop Manual
Chapter 8, page 272

Although most full-time systems cannot be selected out of 4WD, some, such as Jeep's full-time system, called Selec-Trac, can be disengaged for better fuel economy in dry conditions.

Integrated full-time 4WD systems use computer controls to enhance full-time operation, adjusting the torque split depending on which wheels have traction.

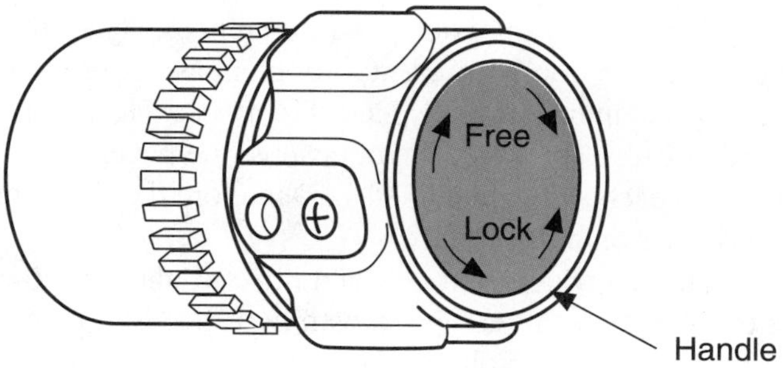

Figure 8-11 Typical locking hub.

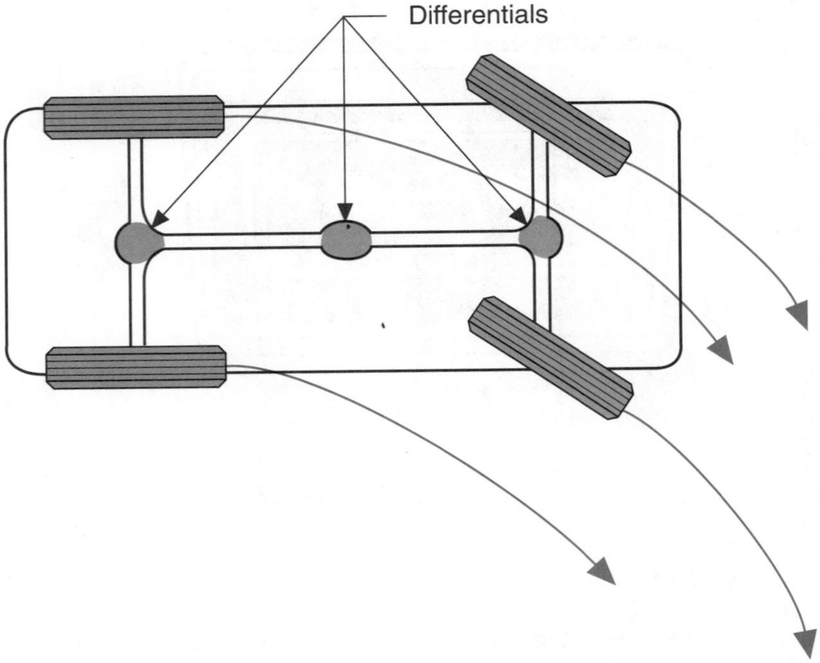

Figure 8-12 When turning a corner, the front wheels travel through a wider arc than the rear wheels.

torque to the axle with the minimum traction. As a result, insufficient torque to move the vehicle may be provided to the wheels that still have traction.

To overcome this problem, most full-time transfer cases are equipped with a limited-slip differential. These units may use a viscous clutch, cone clutch (Figure 8-13), or a multiple plate clutch assembly to control the differential action. By using a limited-slip differential, the amount of torque delivered to the axle with the least traction can be limited. This allows more torque to be delivered to the axle with the most traction.

On some models, when differential action is not wanted, such as for maximum engine braking or maximum wheel torque, the driver can move the transfer case shift lever either to the low range or to the LOCK position. This locks the output shaft to the front and rear axles together. This prevents differential action and allows equal torque to be delivered to both axles.

Part-Time Systems

Shop Manual
Chapter 8, page 273

Part-time 4WD systems can be shifted in and out of 4WD.

Part-time 4WD systems are designed to be used only when driving off the road or on slippery surfaces. When 4WD is engaged, a part-time system locks the front and rear axles together. Therefore the system does not allow for axle speed differences. When the vehicle is turning a corner on a slippery surface, this doesn't present a problem as the tires can easily skid or spin across the slick surface to accommodate the speed differences. However, on dry surfaces, the required speed differentials cause the tires to scrub against the pavement. A part-time 4WD system delivers equal power to each axle while in 4WD. Therefore the driver must shift the transfer case out of 4WD when driving on a dry surface.

A transfer case is equipped with a gear or chain drive to transmit power to one or both of the drive axles. Some transfer cases are equipped with two speeds: a high range for normal driving and a low range for especially difficult terrain, such as deep sand or a steep grade. The low range of the transfer case lowers all of the transmission's ratios, enabling the engine's power to be used more effectively in demanding situations. Most part-time systems and some full-time systems are equipped with a two-speed transfer case.

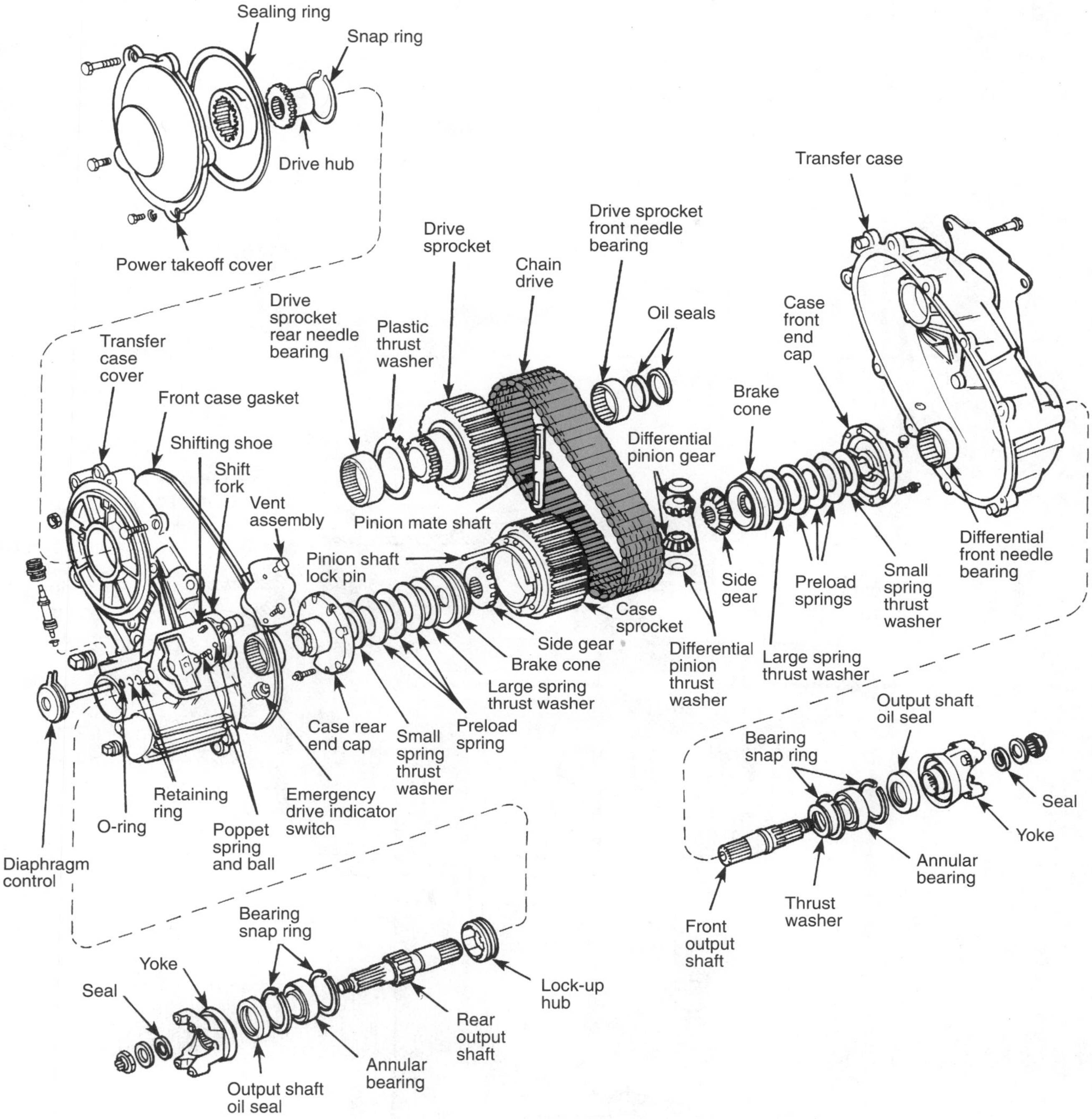

Figure 8-13 4WD transfer case equipped with a limited-slip differential. A commonly used early full-time 4WD system equipped with a limited-slip differential in the transfer case was called the Quadra-Trac. It was built by the Warner Gear Division of the Borg-Warner Corporation.

Two-speed transfer cases are controlled by a shift lever that typically has four positions (Figure 8-14): 2WD-High, which engages only the rear axle and is used for all dry-road driving; 4WD-High, which engages both axles and is used at any speed on slippery surfaces (Figure 8-15); neutral, which disengages both axles; and 4WD-Low, which engages both axles and lowers the ratios of the entire drive line. 4WD-Low should only be used at low speeds and on very demanding terrain.

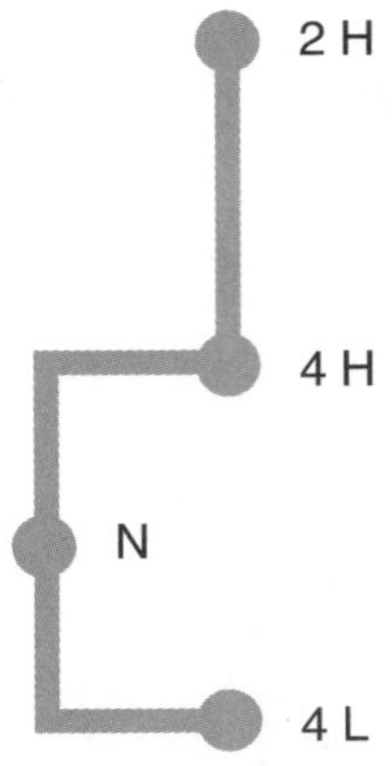

Figure 8-14 Typical shift lever positions.

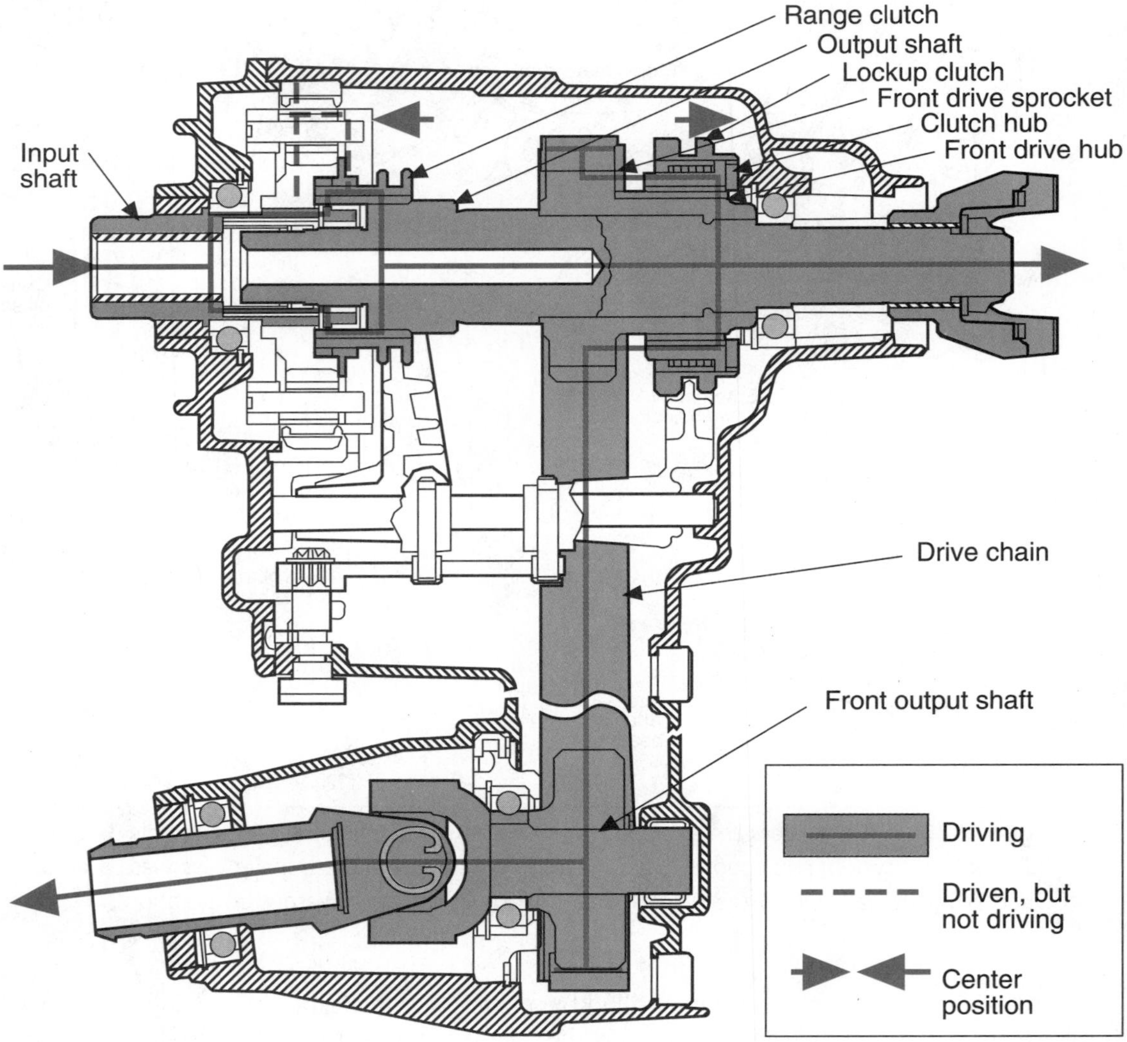

Figure 8-15 Power flow through a typical transfer case when it is in 4WD-High. (Reprinted with the permission of Ford Motor Company)

Other transfer cases are a single speed and only allow the driver to select between 2WD and 4WD modes. This switching is accomplished by an electric switch or a shift lever. The vehicle speed at which this changeover is permitted depends on the design of the system. Some systems require the vehicle to be at a stop before shifting into or out of 4WD, whereas others allow the change at any speed.

Some FWD vehicles are fitted with 4WD and use a compact transfer case bolted to the front drive transaxle. A drive shaft assembly carries the power to the rear differential. The driver can switch from 2WD to 4WD by pressing a dashboard switch. This switch activates a solenoid vacuum valve, which applies vacuum to a diaphragm unit in the transfer case. The linkage of the diaphragm unit linkage locks the output of the transaxle to the input shaft of the transfer case.

"Shift on the fly" 4WD is simply a system that can be shifted from two to four wheel drive while the vehicle is moving.

All-Wheel-Drive Systems

The driver cannot select between 2WD or 4WD in an all-wheel-drive system. These systems always drive four wheels. AWD vehicles are not designed for off-road operation; rather, they are designed to increase vehicle performance in poor traction situations, such as icy or snowy roads. AWD allows for maximum control by transferring a large portion of the engine's power to the axle with the most traction. Most AWD designs use a center differential to split the power between the front and rear axles (Figure 8-16). On some designs, the center differential locks automatically or the driver can manually lock it with a switch. AWD systems may also use a viscous coupling to allow variations in axle speeds.

Shop Manual
Chapter 7, page 191

On-demand 4WD systems power a second axle only after the first begins to slip.

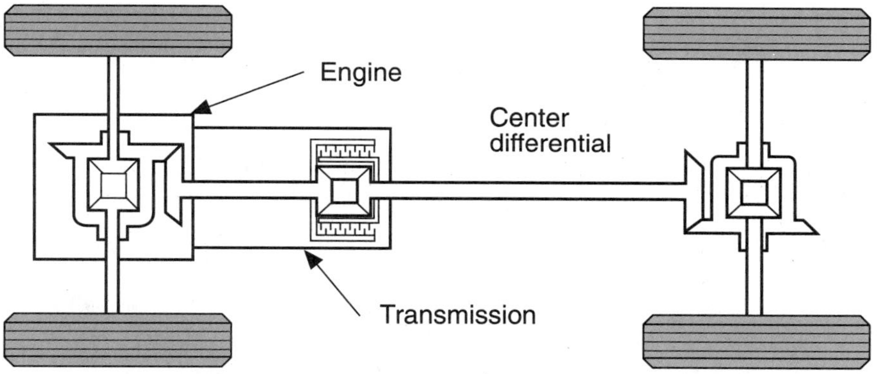

Figure 8-16 Typical AWD powertrain layout. (Courtesy of Chrysler Corporation)

Many automatic AWD systems are electronically controlled and are based on a FWD drivetrain. The rear drive shaft extends from the transaxle to the rear drive axle. To transfer the power to the rear, a multiple plate clutch is used. This clutch serves as an interaxle differential and permits a speed difference between the front and rear drive axles. Sensors monitor front- and rear-axle speeds, engine speed, and load on the engine and drive line. An electronic control unit receives information from the sensors and controls a solenoid that operates on a duty cycle to control the fluid flow that engages the transfer clutch (Figure 8-17). The duty solenoid pulses, cycling on and off very rapidly, which develops a controlled slip condition. As a result, the transfer clutch operates like an interaxle differential and allows for a power split from 95% FWD and 5% RWD to 50% FWD and 50% RWD. This power split takes place so rapidly that the driver is unaware of the traction problem.

The electronic control unit for AWD systems is called the transmission control unit or TCU.

The control solenoid in the transaxle is called a duty solenoid.

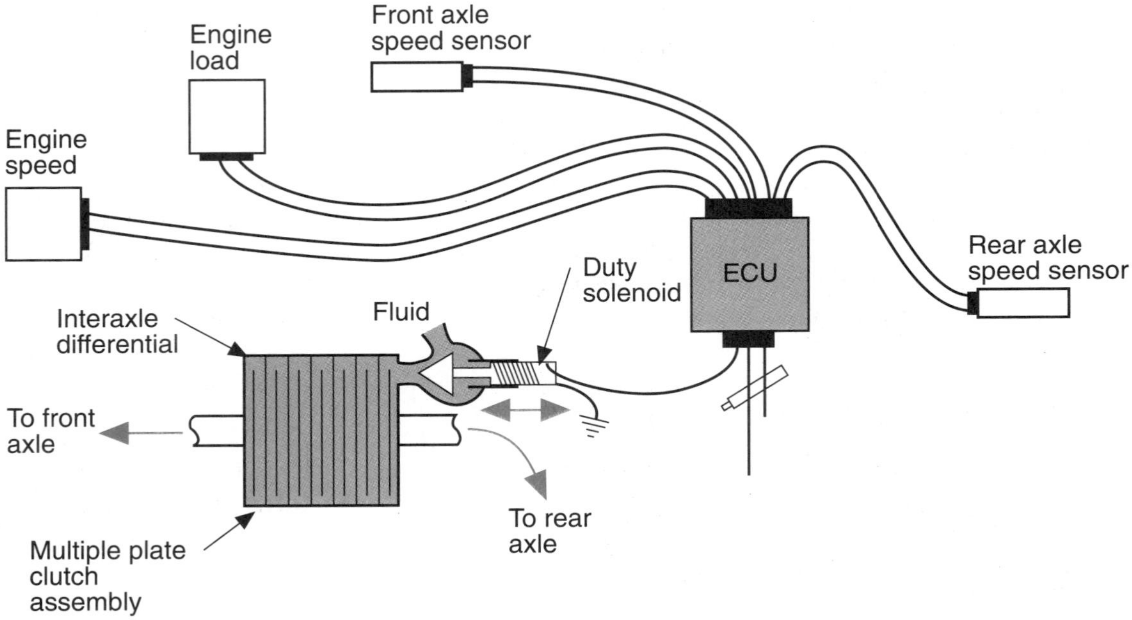

Figure 8-17 Simple schematic for an electronically controlled AWD system.

A duty cycle is also called a jitter cycle.

Transfer Cases

Shop Manual
Chapter 8, page 275

CAUTION: 4WD vehicles should not be operated in the 4WD high or low range on dry hard pavements. This will cause the tire to scrub and wear prematurely, and could cause damage to the drive line.

The transfer case delivers power to both the front and rear drive axle assemblies and is constructed much like a conventional transmission (Figure 8-18). It uses shift forks to select the operating mode, plus splines, gears, shims, bearings, and other components commonly found in transmissions. The outer case is made of magnesium, aluminum, or cast iron. It is filled with lubricant and seals are used to hold the lubricant in and prevent dirt from entering into the housing. Shims are used to maintain proper clearances between the parts of the transfer case.

The purpose of a transfer case is to transfer torque from the output of the transmission to the vehicle's front and rear axles. The transfer case is normally connected to each axle by two drive shafts, one between the transfer case and the front axle and the other from the transfer case to the rear axle. Each axle has its own differential unit that then turns the wheels. Torque is then multiplied by the gear reduction of the differential units and sent to the wheels.

Modes of Operation

Shop Manual
Chapter 8, page 274

A transfer case is an auxiliary transmission mounted to the side or in the back of the main transmission (Figure 8-19). Transfer cases are classified as full time or part time, depending on whether or not the front axle is engaged automatically as soon as the rear wheels begin to spin. With part-time 4WD, the transfer case must be manually shifted to engage or disengage the front axle.

In many vehicles, the transfer case also provides two different speed ranges, high and low. The change from the high speed range to the low speed range is made by moving the shift lever of the transfer case. The movement of the lever moves a gear on the transfer case's main drive shaft from the high speed drive gear to the low speed drive gear. High speed provides direct drive, and low speed usually produces a gear ratio of about 2:1.

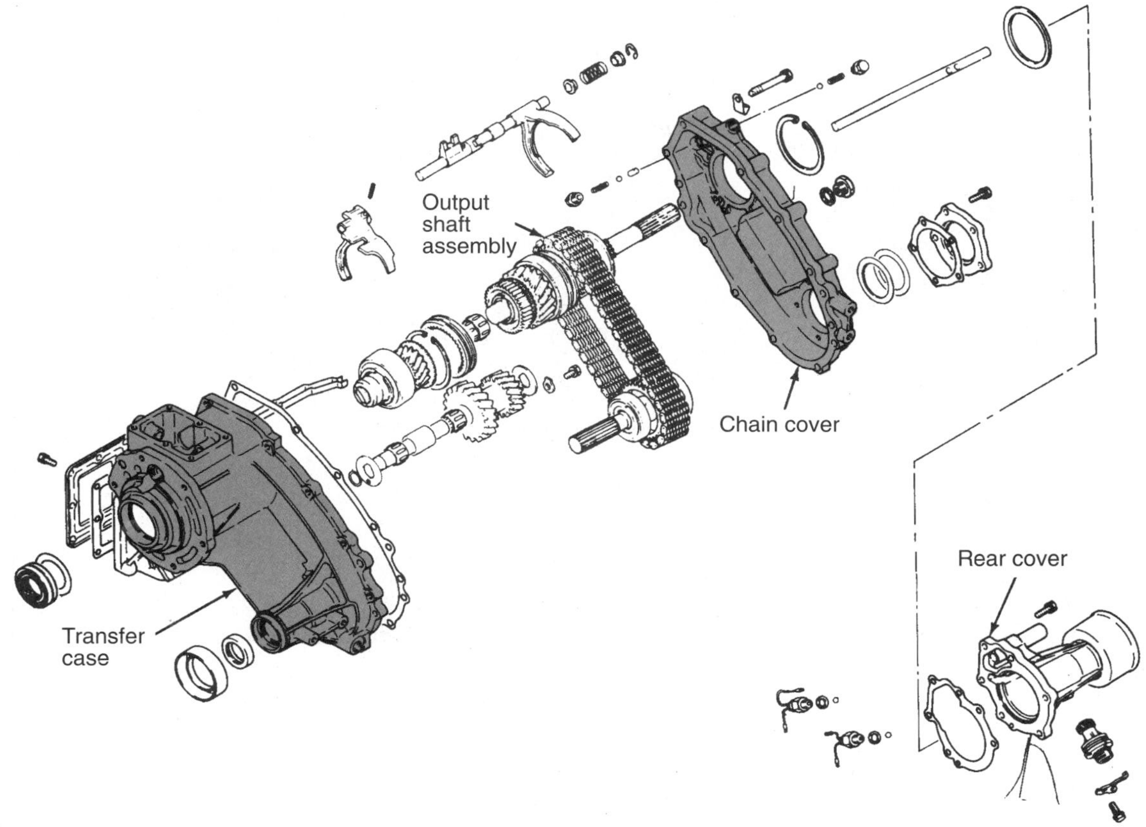

Figure 8-18 Typical transfer case. (Courtesy of Mitsubishi Motor Sales of America, Inc.)

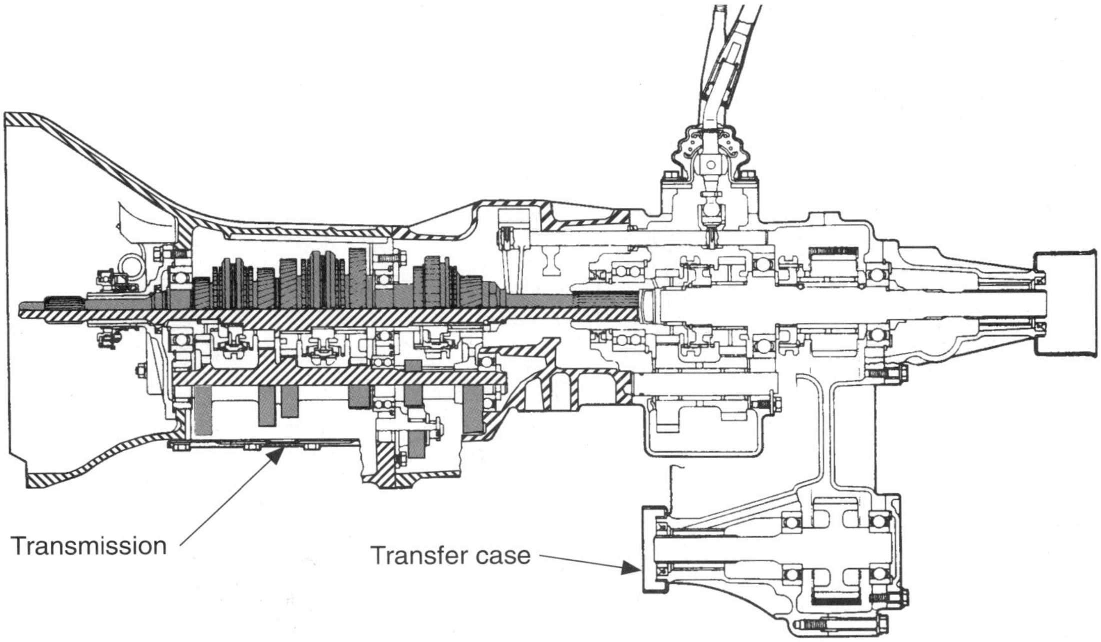

Figure 8-19 Location of transfer case on a transmission. (Courtesy of Chrysler Corporation)

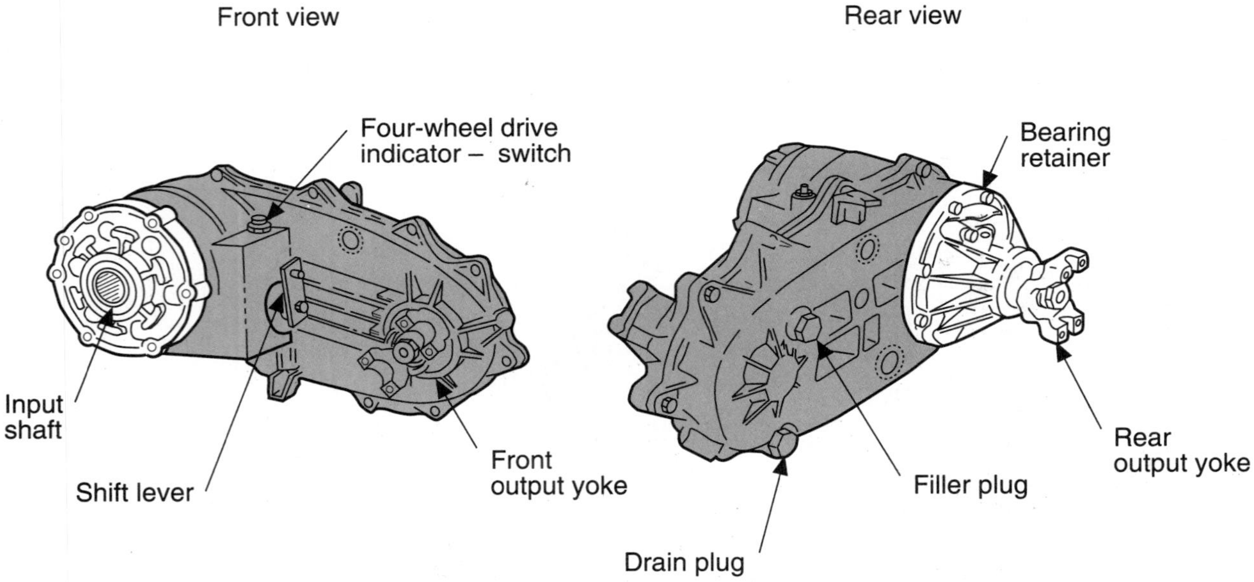

Figure 8-20 Location of front and rear drive shaft yokes on a typical transfer case. (Reprinted with the permission of Ford Motor Company)

Low speed is a torque multiplication gear that allows for increased power to the wheels. This torque multiplication is an addition to the gear reductions of the transmission and final drive gears. When the transfer case is in high speed, torque is only multiplied by the transmission and final drive gears. In addition to these two speeds, transfer cases may also have a neutral position. When the transfer case is in neutral, regardless of which speed gear the transmission is in, no power is applied to the wheels. Some 4WD vehicles are equipped with a single-speed transfer case and do not have a low range.

A part-time transfer case may also be equipped with two speeds, but in addition to the speed ranges, a part-time unit (Figure 8-20) has controls that send power to only the rear wheels (2WD) or to both the front and the rear wheels (4WD).

Most transfer cases use a planetary gearset (Figure 8-21) to provide for the different gear positions. Although the shifting mechanisms vary with the different models of transfer cases, the power flow through nearly all transfer cases is the same. The shift mechanism moves the planetary gear set carrier and the sun gear along a splined section of the input shaft. This movement engages and disengages various gears, which result in the different gear selections. A description of the power flow through a typical two-speed transfer case follows.

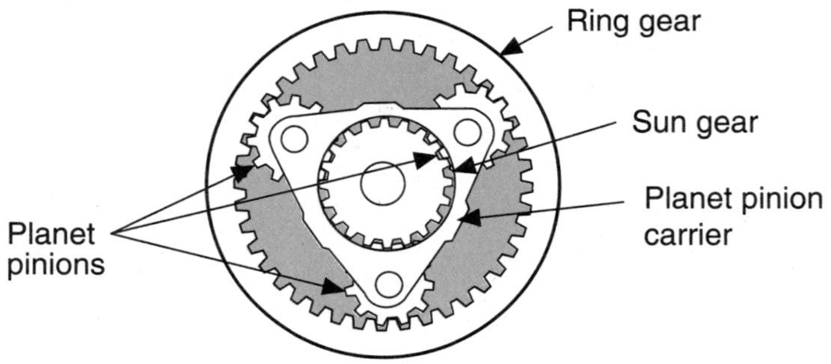

Figure 8-21 A simple planetary gearset.

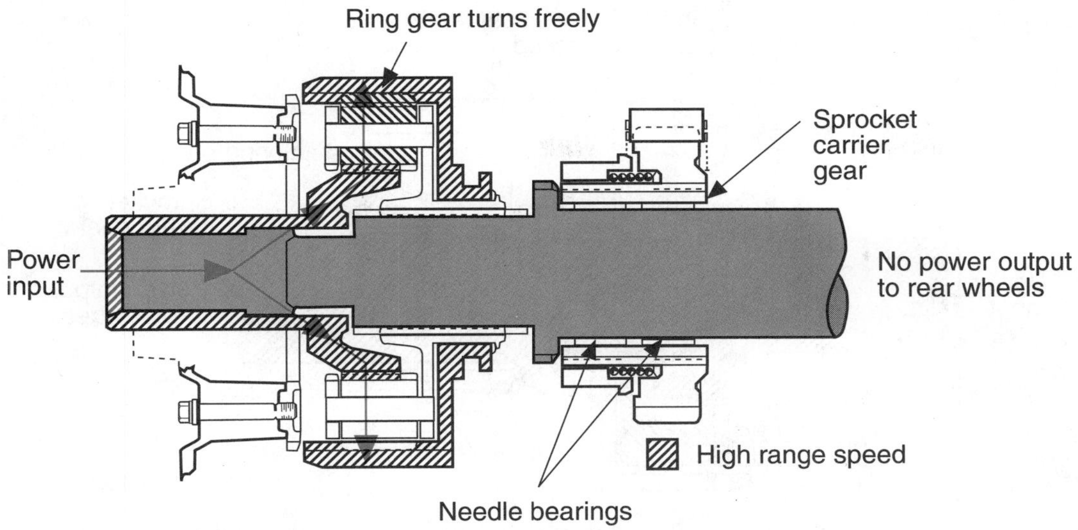

Figure 8-22 Power flow through a transfer case while it is in neutral. (Reprinted with the permission of Ford Motor Company)

While in neutral, the transfer case is driven directly by the main output shaft of the transmission. Power is not transmitted to the driving axles when the transfer case is in neutral, regardless of transmission gear position except Neutral (Figure 8-22). With the transfer case in neutral, the sun gear turns the planetary gears, which drive the ring gear. This gearset rotates with the input shaft. The planetary gear case, which is splined to the rear output shaft, remains stationary because the gearset is positioned away from and not engaged with the case. Therefore power is not transmitted to the rear output shaft.

The ring gear of a planetary gearset is often called the annulus gear.

While in 2WD-High, the planetary gearset moves rearward and the clutch shift fork holds the sliding clutch in the two-wheel-drive position. As the planetary gear set moves rearward, it locks into the planetary gear case. This prevents the rotation of the planetary gears on their axis causing the planetary gears, planetary case, and the ring gear to rotate as a single unit. As the gearset rotates, it drives the rear output shaft at the same speed as the input shaft (Figure 8-23). This provides for direct drive.

While in 4WD-High, the planetary gearset is in the same position as in 2WD-High, but the four-wheel-drive clutch shift fork is moved forward. This movement releases the sliding clutch, which is splined to the driving chain sprocket carrier gear (Figure 8-24). The clutch shift spring pushes the sliding clutch into engagement with the rear output shaft. This causes the chain to drive the front output shaft at the same speed as the rear output shaft, thus sending power to both axles and providing four wheel drive.

While in 4WD-Low, the four-wheel-drive lockup shift collar is positioned to provide for four wheel drive. However, the shifter moves the sun gear and planetary gearset assembly rearward, causing the ring gear to engage with a locking ring that is part of the bearing retainer assembly (Figure 8-25). This holds the ring gear stationary and allows the planetary gears to "walk around" the inside of the ring gear. This causes the planetary gears to drive the planetary case around slower than the input shaft speed. Because the planetary case hub is splined directly to the output shaft, the output shaft rotates at the slower speed. This speed reduction results in increased torque at the drive wheels.

Some transfer cases use a spur or helical gearset to provide for the speed ranges and for engagement and disengagement of 4WD. The front axle is engaged by shifting a sliding or clutching gear into engagement with the driven gear on the drive shaft for the front wheels inside the transfer case (Figure 8-26). The sliding gears and clutches are driven through splines on the shafts.

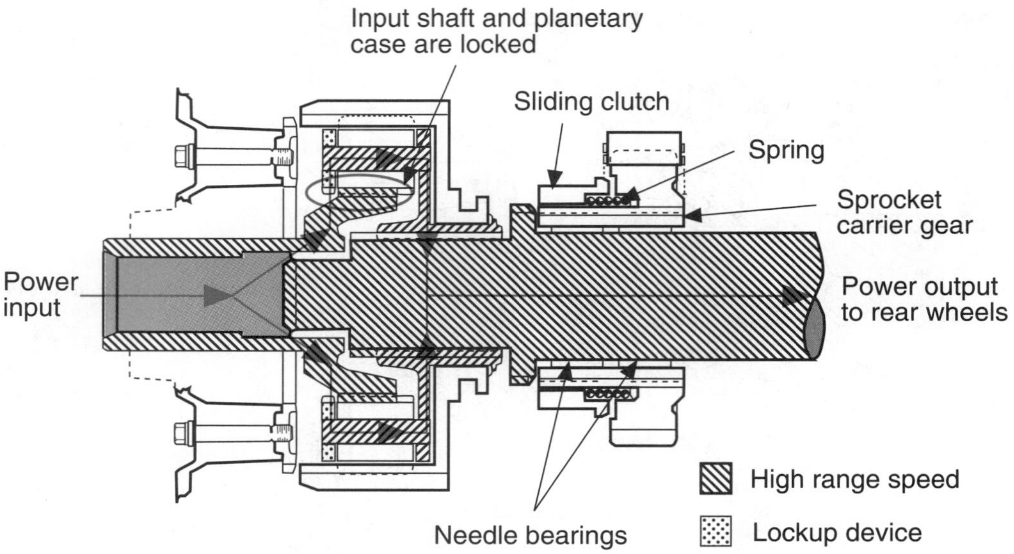

Figure 8-23 Power flow through a transfer case while it is in 2WD-High. (Reprinted with the permission of Ford Motor Company)

A sliding gear on the main shaft locks either the low speed gear or the high speed gear to the main shaft. In many transfer cases, this shift cannot be made unless the transmission is in neutral. Because the transfer case main shaft and the sliding gear, which is splined to it, will be turning at a different speed than the low or high speed gear, gear clash will occur.

To engage and disengage the front axle, another sliding gear or clutch is used to lock the front axle drive gear to the front axle drive shaft. On many transfer cases, the front axle can be engaged and disengaged while the vehicle is moving. This is done by releasing the accelerator pedal to remove the torque load through the gears and moving the shift lever. However, both the

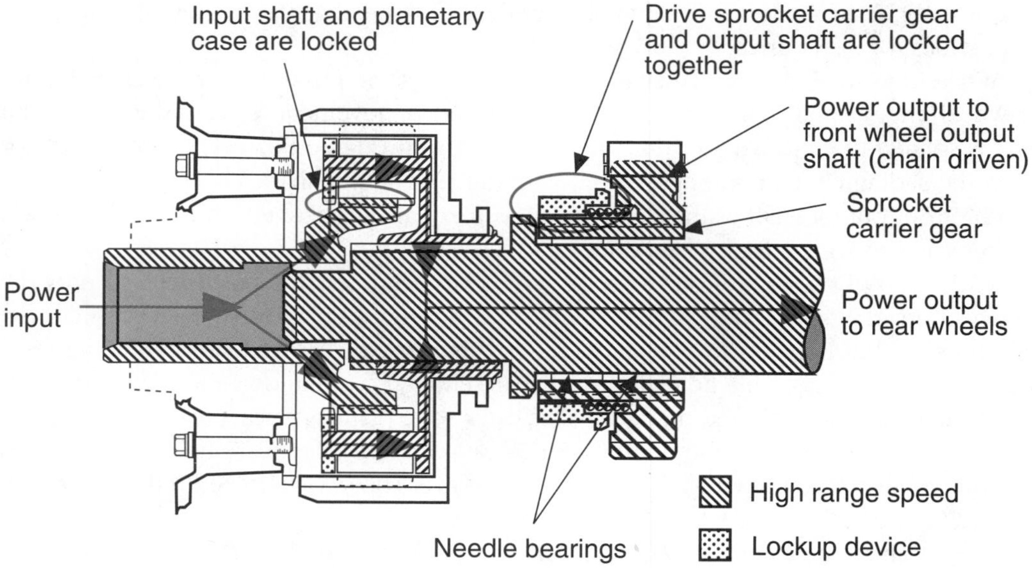

Figure 8-24 Power flow through a typical transfer case while it is in 4WD-High. (Reprinted with the permission of Ford Motor Company)

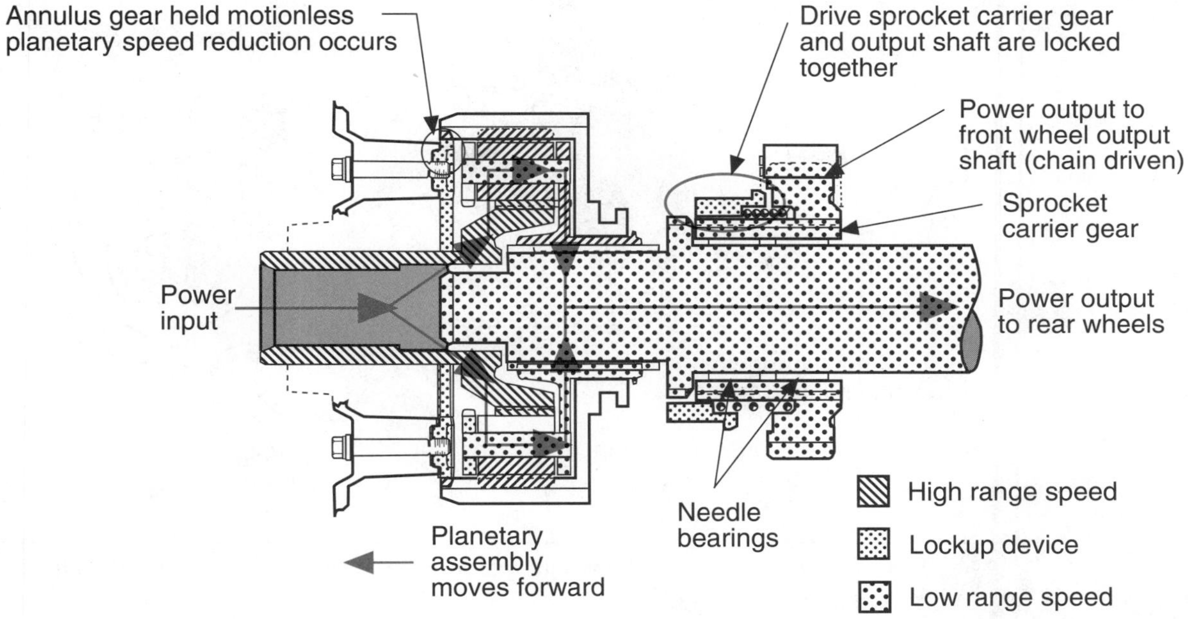

Figure 8-25 Power flow through a transfer case while it is in 4WD-Low. (Reprinted with the permission of Ford Motor Company)

front and the rear wheels must be turning at the same speed. If the rear wheels have lost traction and are spinning, or if the brakes are applied and either the front or the rear wheels are locked and sliding, gear clashing will occur when engagement of the front axle is attempted. On some 4WD systems, the vehicle must be stopped before the change can be made. Also the shift into 4WD-Low mode normally requires that the vehicle be stopped first.

Some transfer cases are single speed and only allow for a change between 2WD and 4WD. An integral main drive gear in the transfer case, which is driven by the output shaft of the trans-

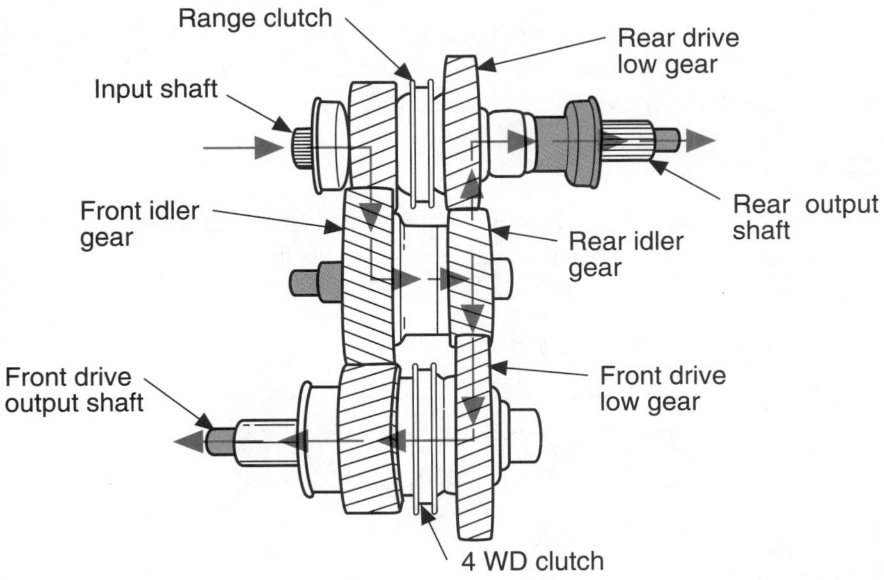

Figure 8-26 Power flow through a gear-driven transfer case while it is in 4WD-Low. (Reprinted with the permission of Ford Motor Company)

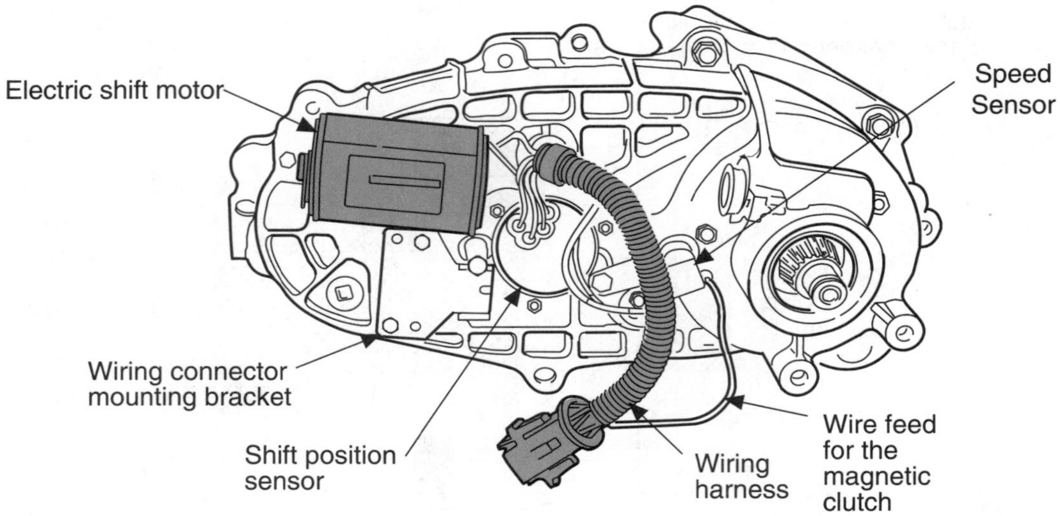

Figure 8-27 Location of electric shift motor on a transfer case. (Reprinted with the permission of Ford Motor Company)

mission, provides power to the rear drive line at all times. This main drive gear is in mesh with an idler gear. The idler gear, in turn, meshes with a front axle drive gear to rotate the front drive line when it is in 4WD. When the vehicle is in 2WD, the idler gear is moved out of mesh with the front axle drive gear.

Some of the newer "shift-on-the-fly" systems use a magnetic clutch in the transfer case to bring the front drive shaft, differential, and drive axles to the same speed as the transmission. When the speeds are synchronized, an electric motor in the transfer case (Figure 8-27) completes the shift. The system will not shift until the speeds are synchronized.

Transfer Case Designs

Some transfer cases rely entirely on gearsets to transfer power. Other designs use a combination of gears and a chain. Using a chain to link the drive axles instead of a gearset, reduces the weight of the transfer case, thereby improving fuel economy.

Drive Chains

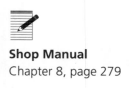

Shop Manual
Chapter 8, page 279

Drive chains are often used to link the input and output shafts in a transfer case. The chain only serves as a link and does not influence gear ratios. Chains are commonly used with planetary

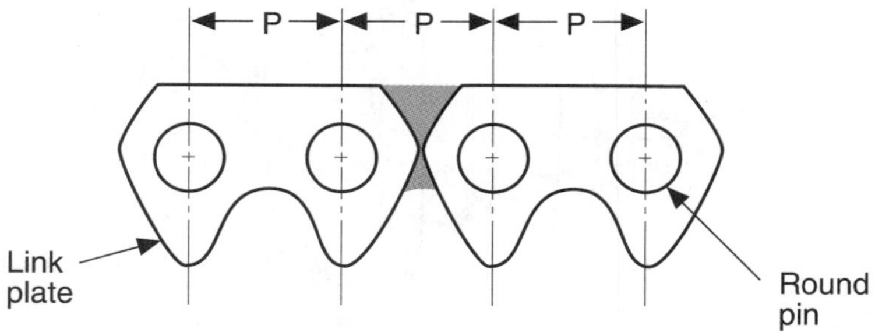

Figure 8-28 A round-pin design drive chain.

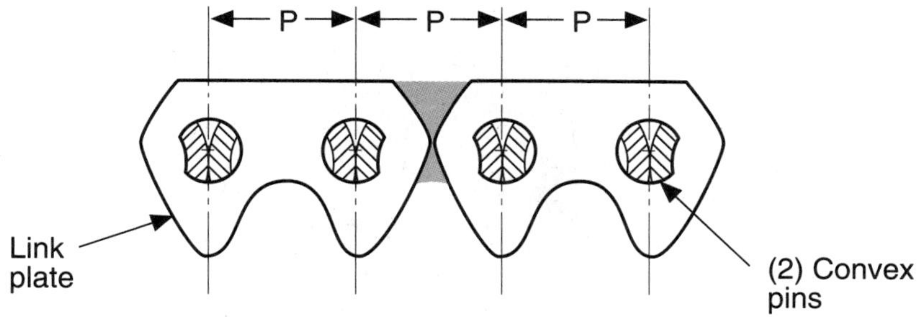

Figure 8-29 A pin-and-rocker-type drive chain.

gearsets. Chains are used because they are very efficient and quiet and they allow for flexible positioning of the transfer case's components. Two basic designs of chains are used: the round-pin style and the pin-and-rocker joint style.

In the round-pin design (Figure 8-28), a single pin is inserted into mating holes at each end of the link plate. Load is distributed over a large area allowing greater chain tensions and higher dynamic loads. The round-pin joint is widely used in transfer cases on part-time 4WD vehicles.

The pin-and-rocker joint design (Figure 8-29) uses two convex joints that roll against one another as the chain moves. This type of chain is very efficient at continuous high speeds and is used on full-time 4WD systems.

Planetary Gear Drives

A typical part-time transfer case uses an aluminum case, chain drive, and a planetary gearset for reduced weight and increased efficiency. Further efficiency is gained because the internal components of the transfer case do not rotate while the vehicle is operating in 2WD.

Shop Manual
Chapter 8, page 275

A simple planetary gearset is made up of three gears. In the center of the gearset is the sun gear. All other gears in the set revolve around the sun gear. Meshing with the sun gear are three or four planetary pinion gears. The pinions are held together by the planetary pinion carrier, or planetary carrier. The carrier holds the gears in place while allowing them to rotate around their shafts. On the outside of the planetary pinions is the ring gear. The ring gear has teeth around its inside circumference that mesh with the teeth of the planetary pinions. Each planetary pinion gear is mounted to the planetary carrier by a pin or shaft. The carrier assembly and the sun gear are mounted on their own shafts.

In any set of two or more gears, the smallest gear is often called the pinion gear.

When the transfer case is in neutral, the rotation of the input shaft spins the planetary pinion gears and the ring gear around them. With both the pinion gears and the ring gear spinning freely, no power is transmitted through the planetary gearset.

Low and high range speeds are available from the planetary gearset. Speed reduction provides the low range and direct drive provides the high range. Speed reduction is provided when the ring gear is held stationary (Figure 8-30). When the transfer case's shift lever is moved to low range, the planetary gear assembly slides forward on its shaft and engages with a locking plate. The locking plate is bolted to the case and engages with the teeth of the ring gear to hold it stationary.

The driving gear of the gearset is the sun gear, which is connected to the output of the transmission. When the ring gear is held stationary and the sun gear is driving, the planetary pinions rotate on their pins. As the pinions rotate, they must walk around the ring gear because they are in mesh with it. This action causes the pinion carrier to rotate in the same direction as the sun gear.

However, the planetary carrier turns slower than the sun gear because of the size difference between the pinions and the sun gear. As the pinions move around the inside of the ring gear, the shaft attached to the planet carrier is driven in the same direction as the sun gear, but at a

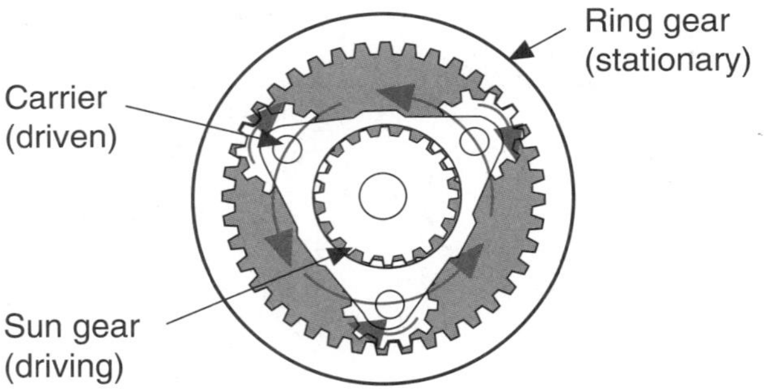

Figure 8-30 Speed reduction results from holding the ring gear and driving the sun gear.

lower speed. This action causes a speed reduction and torque multiplication, thereby providing the transfer case with low range.

The transfer case can be selected into the high range. In this operating mode, the ring gear and planetary carrier are locked together. As a result, the pinion gears cannot rotate on their pins and the entire assembly turns as a solid unit. Therefore the output shaft rotates at the same speed as the input shaft providing for direct drive.

Electronically Controlled Planetary Gearsets

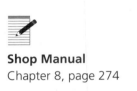

Shop Manual
Chapter 8, page 274

Some vehicles are equipped with an electronically controlled 4WD system, which couples the front and rear axles through a planetary gearset that allows each axle to run at its own speed (Figure 8-31). The planetary gearset is not used for speed reduction; rather, it is used as a differential. The transmission output shaft drives the planetary carrier. The ring gear is connected to the drive line for the rear wheels and the sun gear is connected by a chain to the front drive line.

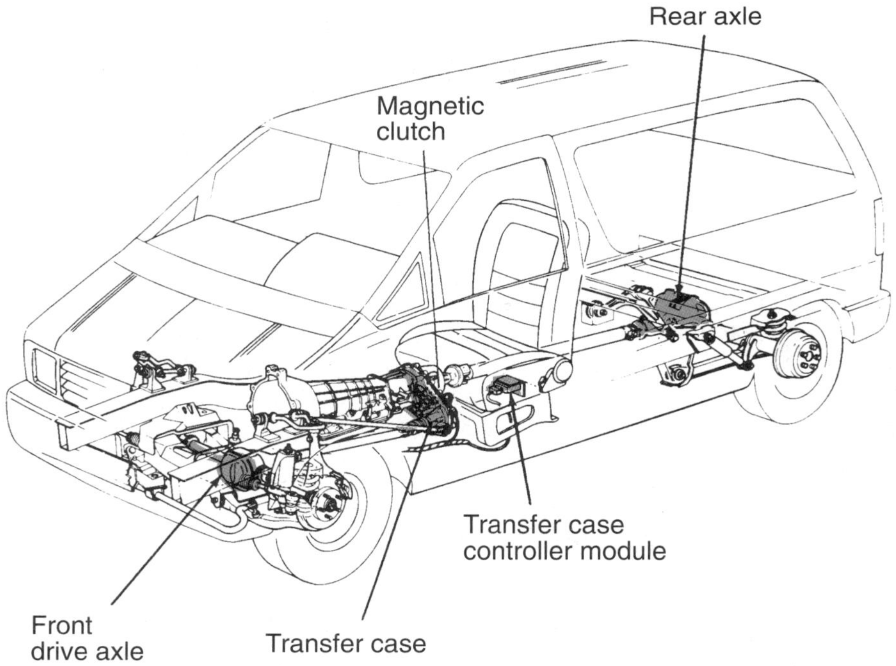

Figure 8-31 Typical electronic 4WD system. (Reprinted with the permission of Ford Motor Company)

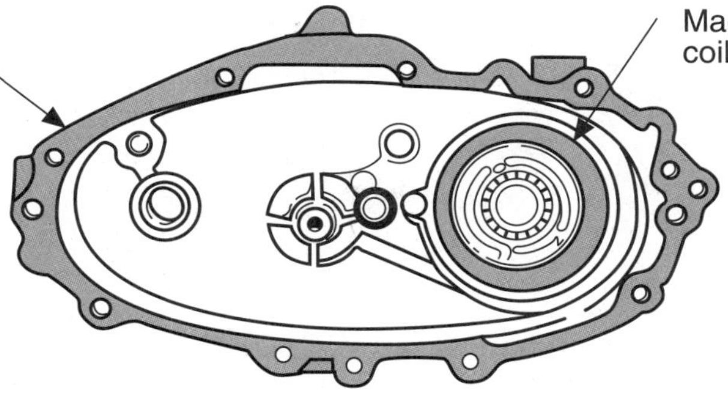

Figure 8-32 Magnetic clutch used in electronic 4WD systems. (Reprinted with the permission of Ford Motor Company)

This type of system uses a speed sensor at each side of the planetary gear, where power is split one third to the front axle and two thirds to the rear under normal driving conditions. When the sensors detect a difference between the speed of the front and rear wheels, a sign that one of the wheels has lost traction, an electronic controller sends current to an electromagnetic clutch that locks the planetary gear.

The electromagnetic clutch (Figure 8-32) is a multiple plate clutch, which has half of its plates splined to the sun gear and the other half to the ring gear. By locking the sun and ring gears together, the axles are locked together and the torque split is 50/50. The electronic controller holds the system in lockup for a few seconds, then unlocks the clutch and rechecks the axles' speeds. If slippage is still evident, the controller reengages the clutch.

This type of system is not intended for off-the-road use. It is intended only to improve traction, when necessary, under everyday driving conditions.

Locking Wheel Hubs

Shop Manual
Chapter 8, page 287

Most 4WD systems on trucks and utility vehicles use front wheel driving hubs that can be disengaged from the front axle when the vehicle is operating in the 2WD mode. When unlocked in 2WD, the front wheels still turn, but the entire front drivetrain, including the front axles, front differential, front drive shaft, and certain parts of the transfer case, stop turning. This helps reduce wear on these components. The rear wheels are being driven by the engine at all times, therefore the rear wheel hubs are permanently locked to the rear axle.

The front hubs must be locked in the 4WD mode. Some front hub designs automatically lock, whereas others require the driver to manually turn a lever or knob at the center of each front wheel to lock the hubs (Figure 8-33). Normally automatic hubs lock when the driver shifts into 4WD and automatically unlock when the vehicle is driven backward for a few feet. Automatic hubs are engaged by the rotational force of the axle shafts whenever the transfer case is in 4WD.

Locking hubs are also called freewheeling hubs because their purpose is to disengage the front axle and allow it to free wheel while the vehicle is operating in 2WD.

By locking the hubs, the wheels turn with the axle to provide the most traction. When unlocked, the wheels are free to spin at different rotational speeds, such as during cornering. The tires scrub the pavement if the hubs are locked, causing them to wear quicker. Therefore, they should only be locked when operating on slippery surfaces.

A locking hub is a type of clutch that disengages the outer ends of the axle shafts from the wheel hub. Then the axle shafts do not turn, or back-drive, the differential. Therefore, the differential does not turn the drive shaft from the transfer case to the front differential. The front wheels rotate on the hub's bearings and are not turned by the axle shafts. When the hubs are locked, the wheels and axles are locked so that they rotate together.

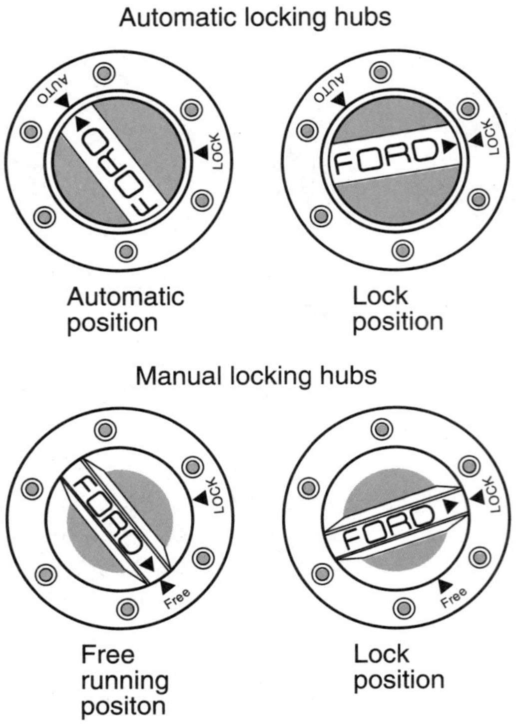

Figure 8-33 Knob positions for automatic and manual locking hubs. (Reprinted with the permission of Ford Motor Company)

A handle located in the center of manual hubs is turned to lock or unlock the hubs. This control handle applies or releases spring tension on the hub's clutch. When the hub is in the locked position, spring pressure causes the clutch to engage to the inner hub that is connected to the axle shaft (Figure 8-34). Because the clutch is connected to the outer hub, the engagement of the clutch connects the axle with the hub. In the unlocked position, the clutch does not engage with the inner hub and the wheel rotates freely on the bearing.

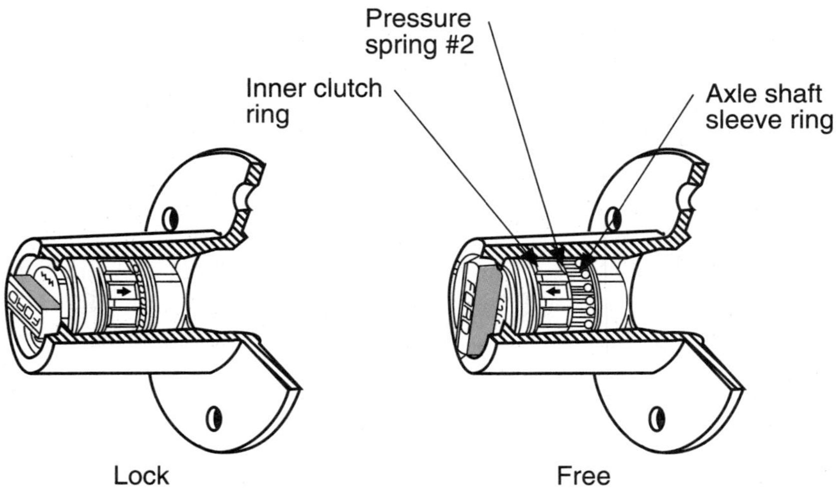

Figure 8-34 Action of a locking hub. (Reprinted with the permission of Ford Motor Company)

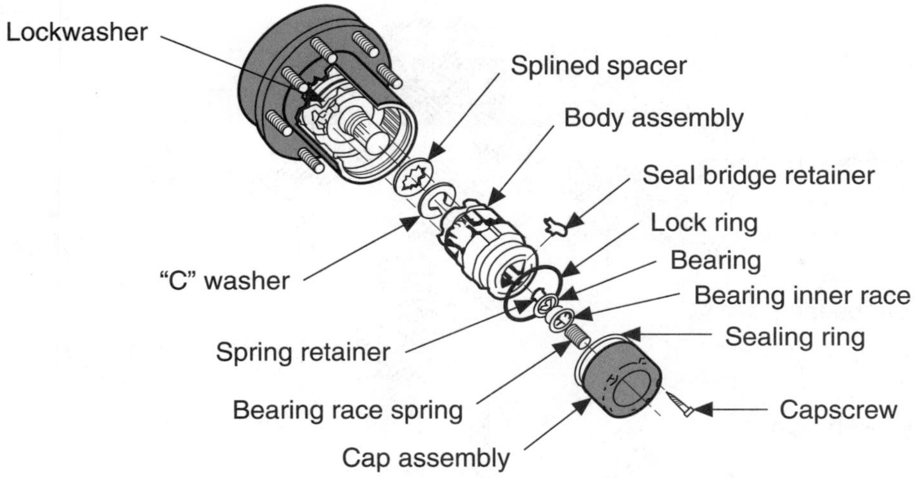

Figure 8-35 Disassembled automatic locking hub. (Reprinted with the permission of Ford Motor Company)

Although automatic hubs (Figure 8-35) are more convenient for the driver, they do have a disadvantage. Many self-locking hub designs are unlocked when the vehicle is moved in reverse. Therefore if the vehicle is stuck and needs to back out of the trouble spot, only rear wheel drive will be available to move it. Other automatic hubs unlock immediately when 4WD is disengaged without the need to back up. On these systems, the hubs are automatically locked, regardless of the direction the vehicle is moving.

Locking hubs are not needed with full-time 4WD. The wheels and hubs are always engaged with the axle shafts. The interaxle differential or transfer case prevents damage and undue wear of the parts of the powertrain.

Some 4WD vehicles are equipped with normal front wheel hubs and use a vacuum-operated axle disconnect shift mechanism. A vacuum motor moves a splined collar to connect or disconnect the front axle's drive shaft from the front differential. The axle shafts continue to turn, but the ring gear, pinion gear, and front drive shaft remain stationary.

Interaxle Differentials

Many 4WD vehicles have a different final drive ratio in each of the drive axles. This results in a push-pull action, which increases traction in very adverse conditions. It also results in a phenomenon called drive line windup. Drive line windup occurs when the drive axles are rotating at different speeds on dry pavements or when the vehicle is turning a corner. Without a means to compensate for these changes in speeds, the tires must skip and drag. This action may also cause damage to the transfer case. On slippery surfaces, the tires are able to slide and the potential for transfer case damage is lessened. To compensate for windup, most late model AWD vehicles are equipped with mechanisms that allow speed differential between the front and rear drive axles.

The most common method for speed differentiation or for dissipating drive line windup is to include a third differential in the transfer case.

The front and rear drive lines are connected to the interaxle differential (Figure 8-36) inside the transfer case. The interaxle differential allows for different front and rear drive line shaft speeds to prevent drive line windup but may also result in a loss of traction in very slippery conditions. This loss of traction will result from the use of an open differential assembly because it will tend to deliver more power to the wheels with the least traction. To counteract this problem, some interaxle differentials are designed to work as a limited-slip differential (Figure 8-37). They

Shop Manual
Chapter 8, page 271

The front axle must travel farther than the rear axle when the vehicle is making a turn.

The third differential is called an interaxle differential.

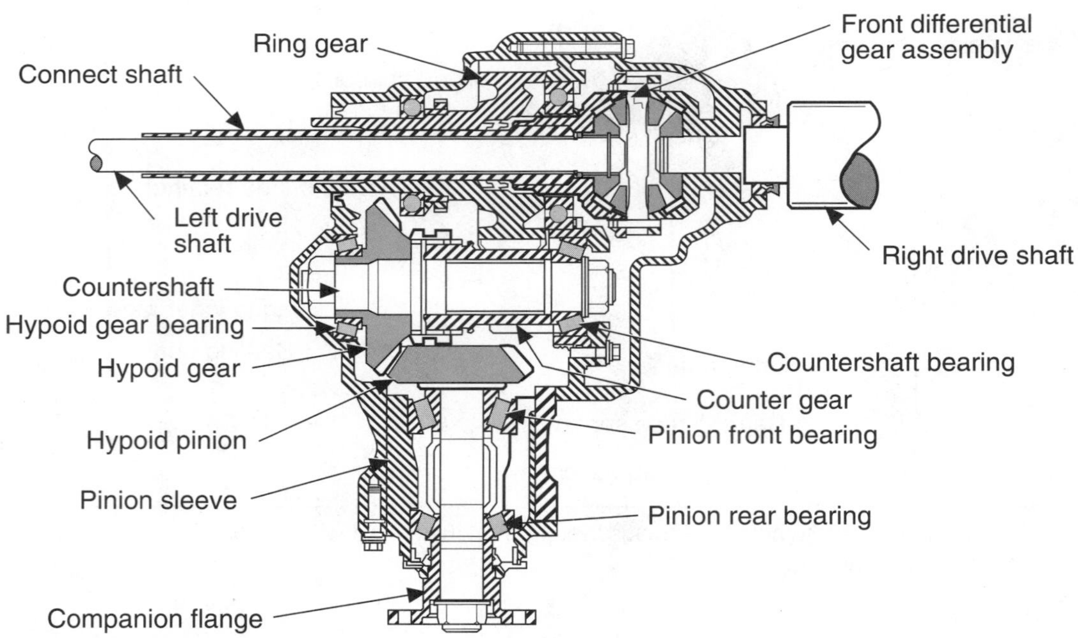

Figure 8-36 An interaxle differential. (Courtesy of Nissan Motors)

use a clutch assembly to maintain a predetermined amount of torque transfer before the differential action takes place. Therefore power is transferred to both axles, regardless of their traction.

Some high performance AWD cars use a vacuum system to allow the driver to lock the center and/or rear differential. This control allows the driver to select which set of wheels should receive the majority of the engine's torque.

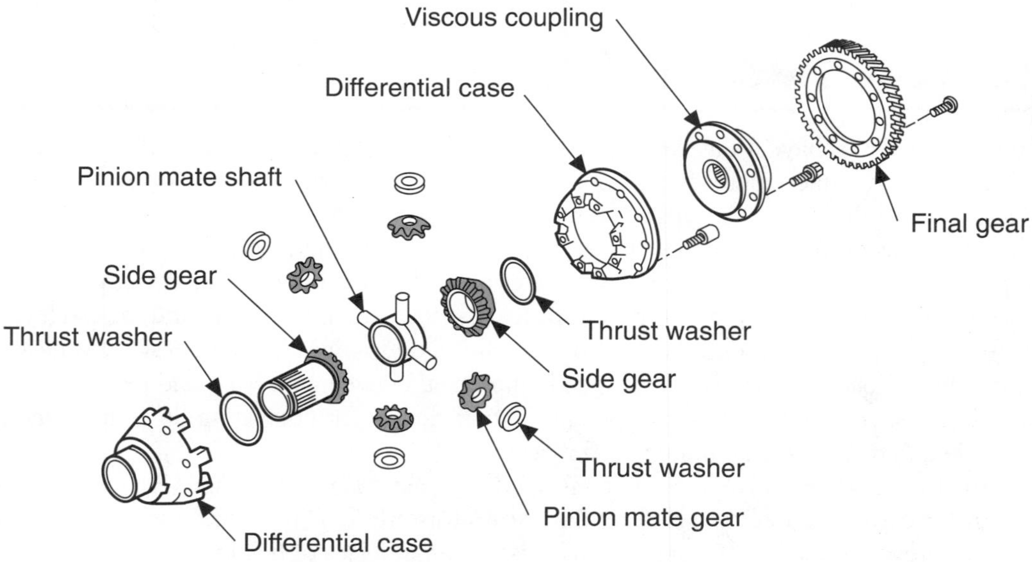

Figure 8-37 Disassembled interaxle differential with a viscous coupling. (Courtesy of Nissan Motors)

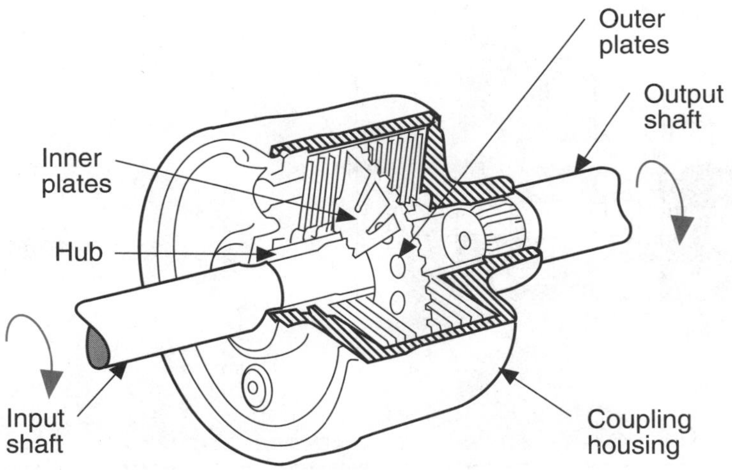

Figure 8-38 Typical viscous coupling. (Courtesy of Volkswagen of America)

Viscous Couplings

The handling characteristics of four-wheel-drive vehicles are excellent in all driving conditions except turning on a dry pavement. While turning a corner, the front wheels cannot rotate at the same speed as the rear wheels. This causes one set of wheels to scuff along the pavement. To overcome this tendency, many full-time 4WD and AWD vehicles use a viscous coupling between the front and rear axles.

A viscous coupling (Figure 8-38) is basically a drum filled with a thick fluid that houses several closely fitted, thin steel discs. One set of the discs is connected to the front wheels and the other to the rear (Figure 8-39). As the demand for more torque to one axle is evident, the fluid heats up and immediately changes viscosity. This change in viscosity reacts on the discs and torque is split according to the actual needs of the axles.

Shop Manual
Chapter 8, page 286

Viscous couplings are often referred to as viscous clutches because they engage and disengage power flow.

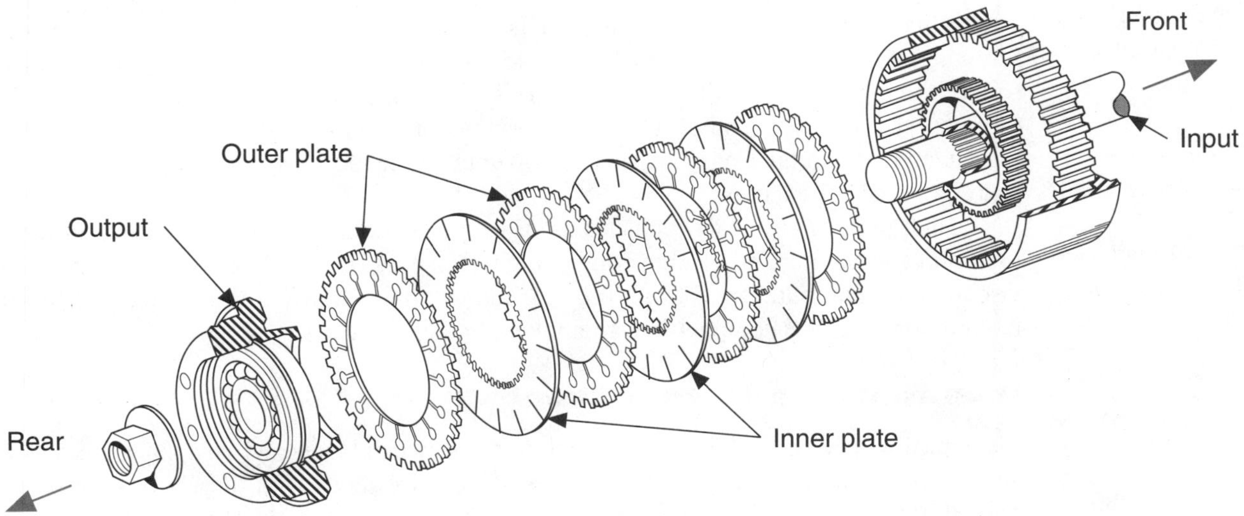

Figure 8-39 Disassembled view of a viscous coupling. (Courtesy of Nissan Motors)

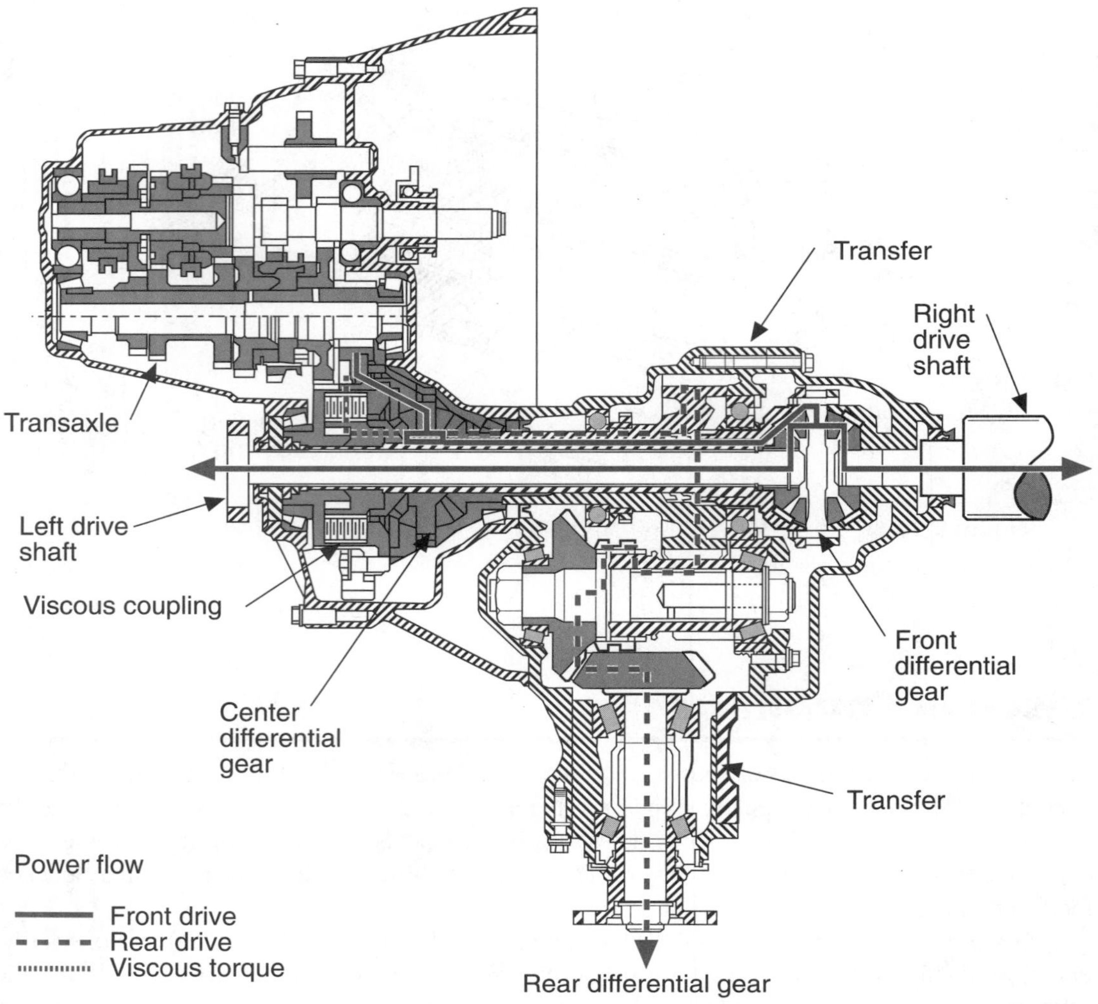

Figure 8-40 Power flow through a transfer case fitted with a viscous coupling and interaxle differential. (Courtesy of Nissan Motors)

When a viscous coupling provides holding ability for two parts of a planetary or bevel gearset, it is called a viscous control.

When a viscous coupling directly connects two shafts, it is called a viscous transmission.

Shop Manual
Chapter 8, page 286

Viscous couplings may also be used as a limited slip unit in the front and/or rear axle differentials (Figure 8-40). They provide a holding force between two shafts as they rotate under force. Like a limited-slip differential, a viscous coupling transfers torque to the drive wheel with more traction when the other wheel has less traction. A viscous clutch often takes the place of the interaxle differential. Viscous clutches operate automatically as soon as it becomes necessary to improve wheel traction.

High performance AWD vehicles use a viscous coupling in the center and rear differentials to improve cornering and maneuvering at higher speeds. The combination of a center differential viscous coupling with open front and rear differentials improves the vehicle's distribution of braking forces and is compatible with antilock braking systems.

Operation of a Viscous Coupling

In a typical viscous coupling, one of the two shafts, which have external splines, meshes with the viscous coupling housing's internal splines, which also mesh with the viscous coupling plates. The other shaft rotates on seals in the housing. The plates are made of steel with special slots cut into them. The inner plates have slots cut from its outer-diameter edge and the outer plates have

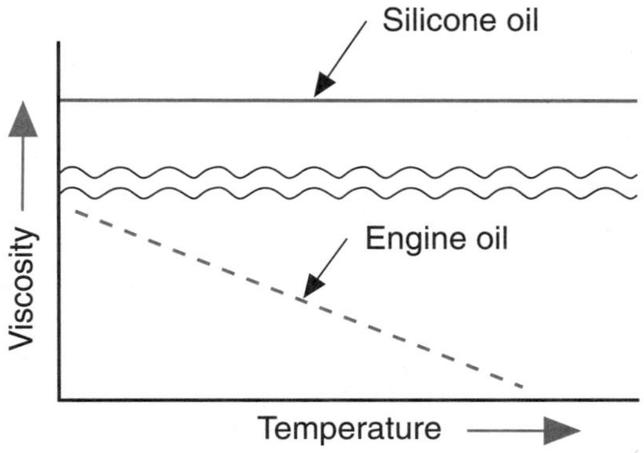

Figure 8-41 Chart comparing the viscosity of engine oil and silicone oil in response to heat. (Courtesy of Nissan Motors)

slots cut from its inner-diameter edge. The number and size of the plates depend on the torque-transfer ability planned for the viscous coupling.

The parts of the viscous clutch fit inside a sealed drum. The clutch pack is made up of alternating steel driving and driven plates. One set of steel plates is splined internally to the clutch assembly hub. The second set of clutch plates is splined externally to the clutch drum. The housing is filled with a small quantity of air and silicone fluid, which must have very specific properties, including controlled viscosity behavior as temperature and pressure change during operation. The amount of silicone fluid in the housing greatly determines coupling characteristics.

Silicone oil is used not only because its viscosity is less affected by temperature variations (Figure 8-41), but also because it features high shearing resistance and volumetric change at elevated temperatures.

Shearing means to cut through.

As the drivetrain operates on a road surface with good traction at both drive wheels attached to the axles, the coupling has no work to do. Both axles rotate at the same speed. As cornering occurs, the axle shafts have slightly different speeds. The viscous coupling plates rotate at different speeds and can shear the silicone fluid with ease. As one wheel loses traction, the drivetrain torque begins to flow to the axle of the slipping wheel. This moderate speed difference between the axles causes the plates to move together as they attempt to shear the silicone fluid. As a result, the fluid provides torque transfer to the wheel with good traction.

When a difference of speed exists between the input shaft driven by the axle with tractive effort, the clutch plates begin shearing the silicone fluid. This shearing action causes heat to build up within the housing, which causes the fluid to stiffen. This stiffening causes a locking action between the clutch plates to take place within one tenth of a second. This locking action results from the stiff silicone fluid becoming very hard for the plates to shear. The stiff silicone fluid transfers power flow from the driving to the driven plates. The driving shaft is then connected to the driven shaft through the clutch plates and stiff silicone fluid.

During acceleration, the viscous coupling transmits the driving torque to the rear wheels because of the relationship between the front and rear wheel rotating speeds. This reduces the front wheel driving torque. As acceleration increases, the rear wheel driving torque becomes greater than the front wheel driving torque, resulting in an increase in rear wheel slip. The center differential viscous coupling then activates to transfer additional driving torque, equivalent to that lost by the slipping of the rear wheel, to the front wheels. In this way, rear wheel spin is substantially restricted.

The unique principle of the viscous coupling is that it can vary its holding force without the need for control hydraulics or electronics. As a result of applying this principle to the center differential, any slipping wheel will not greatly reduce the torque transfer to the other wheel (Figure 8-42).

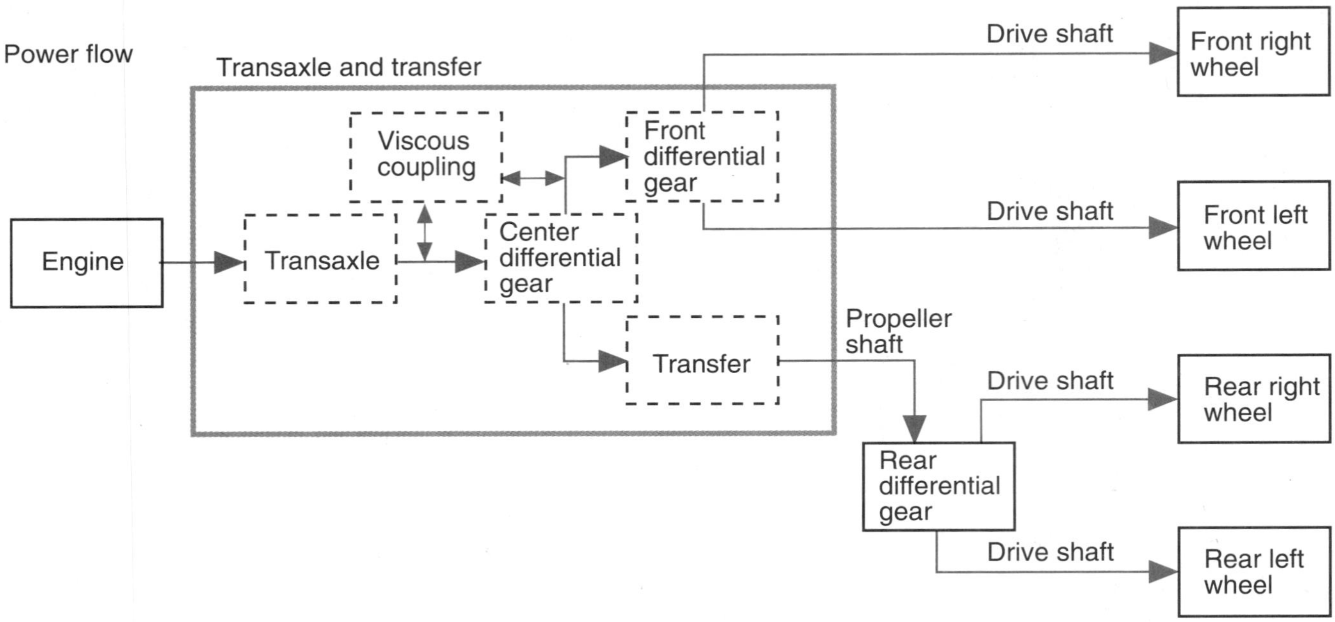

Figure 8-42 Chart showing power flow through a viscous clutch-type center differential. (Courtesy of Nissan Motors)

The viscous clutch has a self-regulating control. When the clutch assembly locks up, there is very little relative movement between the clutch plates. Because there is little movement, the temperature of the fluid drops, which reduces the pressure within the housing. But as speed fluctuates between the driving and driven members, heat increases, causing the fluid to stiffen. Speed differences between the driving and driven members regulate the amount of slip in a viscous clutch driveline.

Some systems use a viscous clutch across the planetary gear. This all-mechanical system bridges the gap between the front and rear wheels with a series of parallel plates alternatively connected to the front and rear wheels, and riding in a special silicone fluid. If all wheels have good traction, the plates rotate at the same speed, and the viscous coupling unit has no effect. But, if one wheel loses traction, the stiff liquid partially locks the front and rear wheels together.

The torque split is normally 35/65 front/rear, but the viscous clutch can transfer up to 100% to either axle, depending on the conditions and relative slip. An aluminum transfer case is attached to the rear of the transmission and houses the center differential, the viscous coupling, and a chain drive to transfer the front axle's power from the center differential to a drive shaft running alongside the engine.

The front and rear axles are coupled via a planetary-gear-type differential, which allows each axle to run at its own speed. Therefore, the 4WD system can be permanently engaged. Operating by itself, this would lead to one-wheel drive, because all of the torque would go to the wheel with the least traction. To prevent this, the set-up uses a viscous clutch-type limited-slip unit at the center differential and a conventional limited-slip unit at the rear axle.

Four-Wheel-Drive Suspensions

Shop Manual
Chapter 8, page 264

The suspension components necessary for 4WD vehicles is basically the same as for FWD and RWD vehicles. When the vehicle is based on a FWD set-up, the addition of the rear axle requires a rear suspension similar to those found on RWD models (Figure 8-43). The springs must be

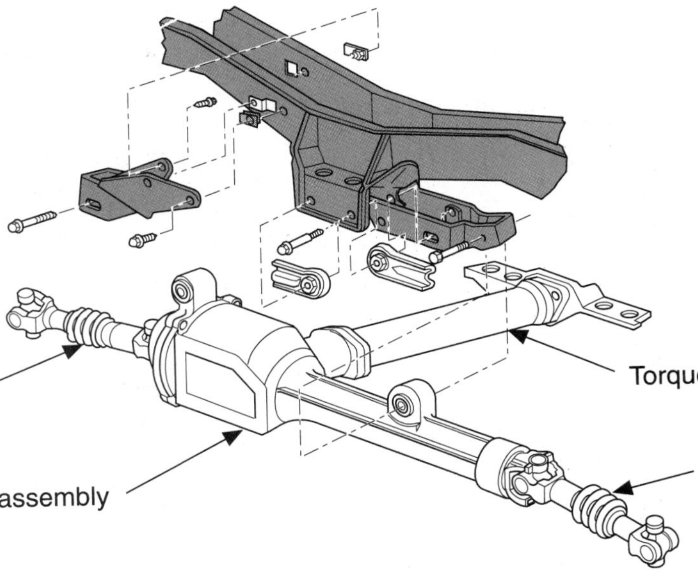

Figure 8-43 AWD rear drive axle assembly. (Reprinted with the permission of Ford Motor Company)

strong enough to support the axle housing and components must be used to accommodate the movement of the axle, such as control and trailing arms.

When the 4WD vehicle is based on a RWD platform, provisions for the front drive axle must be made. Basically there are two types of front drive axles used on 4WD vehicles: the solid axle and the independent front suspension axle assembly. The solid type is basically a rear axle turned so that the pinion gear faces the center of the vehicle. The entire axle assembly is suspended by springs. The front drive shaft is equipped with a slip yoke at the transfer case end to accommodate the movement of the axle. The ends of the axle housing are fitted with steering knuckles that allow for steering (Figure 8-44).

An independent front (IDF) suspension axle (Figure 8-45) has the differential housing mounted solidly to the vehicle's frame. Short drive shafts or half-shafts connect the differential to

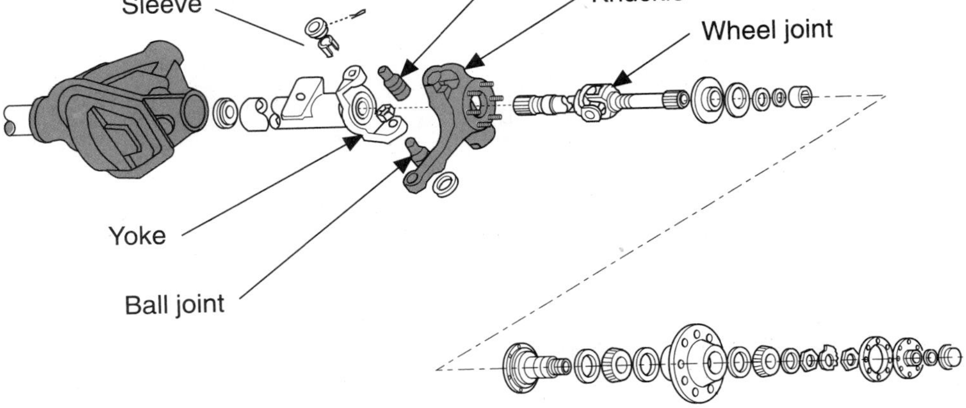

Figure 8-44 Exploded view of a front drive axle's steering knuckle and hub.

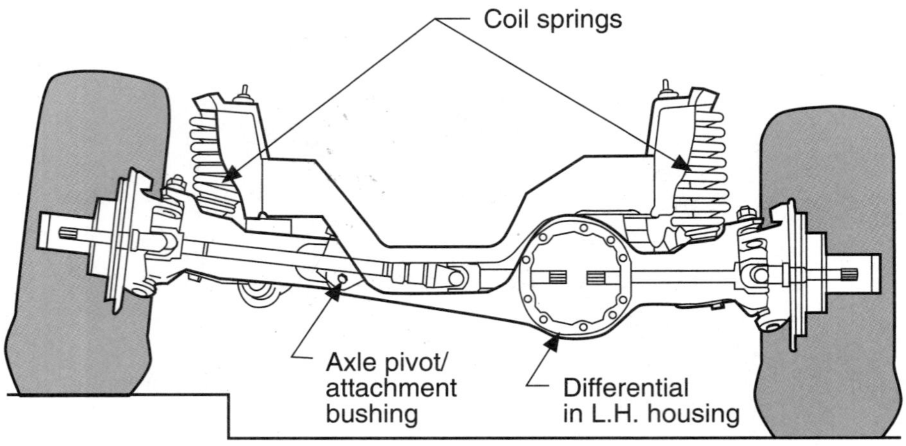

Figure 8-45 An independent suspension front drive axle assembly. (Reprinted with the permission of Ford Motor Company)

the wheels. These half-shafts are suspended by springs and are able to independently respond toroad surfaces. The half-shafts are normally equipped with CV-joints and U-joints to accommodate this movement. The ends of the shafts fit into the steering knuckles. Independent front suspension improves the vehicle's handling characteristics (Figure 8-46).

The most common independent front suspension on 4WD vehicles was introduced by Ford Motor Company in 1980. This system utilizes two steel axle carriers and a third universal joint in the right axle shaft next to the differential. The carrier and joints allow each front wheel to move up and down independently of the other. Instead of a one-piece tube to serve as the axle housing, the tube is split into two shorter sections and joined together at a pivot point. The axle shaft can flex near the pivot because of the centrally located third universal joint.

With independent front suspension, each wheel is able to react to the road conditions individually. Bumps are absorbed by each front wheel instead of some of the shock being transferred through the solid axle to the other wheel. Normally, these vehicles are equipped with coil springs, although some heavy duty models use leaf springs.

The wheel bearings in front and rear axle assemblies of 4WD vehicles are typically located so that they do not support the weight of the vehicle and are held in the hub assembly in the same way as other full-floating axle bearings.

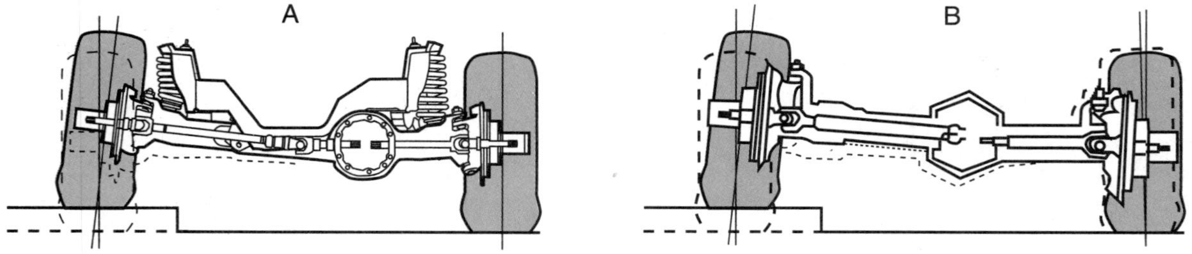

Figure 8-46 Comparison of an (A) independent front suspension and a (B) solid drive axle assembly in response to a bump.

Summary

- ❑ With four wheel drive, engine power can flow to all four wheels, which can greatly increase a vehicle's traction when traveling in adverse conditions and can also improve handling as side forces generated by the turning of a vehicle or by wind gusts will have less of an effect on a vehicle that has power applied to the road on four wheels.
- ❑ FWD cars are modified by adding a transfer case, a rear drive shaft, and a rear axle with a differential and some are equipped with a center differential.
- ❑ 4WD systems use a separate transfer case and allow the driver to select or feature automatic controls that select the transfer of the engine's power to either two or four wheels.
- ❑ Normally AWD systems use a center differential, viscous coupling, or transfer clutch assembly to transfer engine power to the front and rear axles.
- ❑ The transfer case is usually mounted to the side or rear of the transmission. A drive shaft is used to connect the transmission to the transfer case, or a chain or gear drive, within the transfer case, receives the engine's power from the transmission and transfers it to the drive shafts leading to the front and rear drive axles.
- ❑ The front wheel hubs on most 4WD vehicles must be locked or unlocked by the driver to provide for efficient operation in 2WD or 4WD. The rear wheel hubs are always engaged.
- ❑ Integrated full-time 4WD systems use computer controls to enhance full-time operation, adjusting the torque split depending on which wheels have traction.
- ❑ On-demand 4WD systems power a second axle only after the first begins to slip.
- ❑ Most four-wheel-drive units are equipped to allow the driver to select in and out of four wheel drive. The systems capable of operating in both two wheel and four wheel drive are called part-time 4WD systems. Full-time 4WD systems cannot be selected out of four wheel drive.
- ❑ During turns, the front wheels travel a greater distance than the rear wheels. This is because the front wheels move through a wider arc than the rear wheels.
- ❑ Part-time 4WD systems are designed to be used only when driving off-the-road or on slippery surfaces. When 4WD is engaged, a part-time system locks the front and rear axles together.
- ❑ A transfer case is equipped with a gear or chain drive to transfer power to one or both of the drive axles. Some transfer cases are equipped with two speeds: a high range for normal driving and a low range for especially difficult terrain.
- ❑ Two-speed transfer cases are controlled by a shift lever that typically has four positions: 2WD-High, which engages only the rear axle and is used for all dry road driving; 4WD-High, which engages both axles and is used at any speed on slippery surfaces; neutral, which disengages both axles; and 4WD-Low, which engages both axles and lowers the ratios of the entire drive line.
- ❑ AWD vehicles are not designed for off-road operation. Rather, they are designed to increase vehicle performance in poor traction situations, such as icy or snowy roads. AWD allows for maximum control by transferring a large portion of the engine's power to the axle with the most traction. Most AWD designs use a center differential to split the power between the front and rear axles.
- ❑ The purpose of a transfer case is to transfer torque from the output of the transmission to the vehicle's front and rear axles.
- ❑ Some transfer cases use a spur or helical gearset to provide for the speed ranges and for engagement and disengagement of 4WD.

Terms to Know

All wheel drive
Annulus gear
Center differential
Chain drive
Duty cycle
Duty solenoid
Drive line windup
Electromagnetic clutch
Free-wheeling hubs
Full-time 4WD
Integrated full-time 4WD
Interaxle differential
Locking hubs
On-demand 4WD
Part-time 4WD
Pin-and-rocker-joint-type chain
Planetary gearset
Ring gear
Round-pin-type chain
Shift on the fly
Silicone oil
Sprocket
Sun gear
Transfer case
Transmission control unit
Viscous couplings
Viscous control
Viscous transmission
Wheel hubs

- ❑ The front axle is engaged by shifting a sliding or clutching gear into engagement with the driven gear on the drive shaft for the front wheels inside the transfer case. The sliding gears and clutches are driven through splines on the shafts.
- ❑ Some "shift-on-the-fly" systems use a magnetic clutch in the transfer case to bring the front drive shaft, differential, and drive axles to the same speed as the transmission. When the speeds are synchronized, an electric motor in the transfer case completes the shift.
- ❑ Drive chains are often used to link the input and output shafts in a transfer case and serve as a link; they do not influence gear ratios.
- ❑ A simple planetary gearset is made up of three gears. In the center of the gearset is the sun gear. All other gears in the set revolve around the sun gear. Meshing with the sun gear are three or four planetary pinion gears. The pinions are held together by the planetary pinion carrier, or planetary carrier. The carrier holds the gears in place while allowing them to rotate around their shafts. On the outside of the planetary pinions is the ring gear. The ring gear has teeth around its inside circumference that mesh with the teeth of the planetary pinions. Each planetary pinion gear is mounted to the planetary carrier by a pin or shaft.
- ❑ Low and high range speeds are available from the planetary gearset. Speed reduction provides the low range and direct drive provides the high range. Speed reduction is provided when the ring gear is held stationary. When the transfer case's shift lever is moved to low range, the planetary gear assembly slides forward on its shaft and engages with a locking plate. The locking plate is bolted to the case and engages with the teeth of the ring gear to hold it stationary.
- ❑ Most 4WD systems on trucks and utility vehicles use front wheel driving hubs that can be disengaged from the front axle when the vehicle is operating in the 2WD mode. When unlocked in 2WD, the front wheels still turn, but the entire front drivetrain—including the front axles, front differential, front drive shaft, and certain parts of the transfer case—stop turning.
- ❑ Automatic hubs lock when the driver shifts into 4WD and normally automatically unlock when the vehicle is driven backward for a few feet. Automatic hubs are engaged by the rotational force of the axle shafts whenever the transfer case is in 4WD.
- ❑ A locking hub is a type of clutch that disengages the outer ends of the axle shafts from the wheel hub.
- ❑ A handle located in the center of manual hubs is turned to lock or unlock the hubs. This control handle applies or releases spring tension on the hub's clutch. When the hub is in the locked position, spring pressure causes the clutch to engage to the inner hub that is connected to the axle shaft.
- ❑ Locking hubs are not needed with full-time 4WD. The wheels and hubs are always engaged with the axle shafts. The interaxle differential or transfer case prevents damage and undue wear of the parts of the powertrain.
- ❑ Drive line windup occurs when the drive axles are rotating at different speeds on dry pavements or when the vehicle is turning a corner.
- ❑ To compensate for drive line windup, most late model AWD vehicles are equipped with mechanisms that allow speed differential between the front and rear drive axles.
- ❑ The most common method for speed differentiation or for dissipating drive line windup is to include a third differential, called an interaxle differential, in the transfer case.
- ❑ The interaxle differential allows for different front and rear drive line shaft speeds to prevent drive line windup but may also result in a loss of traction in very slippery conditions.
- ❑ A viscous coupling is basically a drum filled with a thick fluid that houses several closely fitted, thin steel disks. One set of the discs are connected to the front wheels and the other to the rear.

- ❑ In a typical viscous coupling, one of the two shafts that have external splines meshes with the viscous coupling housing's internal splines, which also mesh with the viscous coupling plates. The other shaft rotates on seals in the housing. The plates are made of steel with special slots cut into them. The inner plates have slots cut from its outer-diameter edge and the outer plates have slots cut from its inner-diameter edge.
- ❑ The viscous coupling plates rotate at different speeds and can shear the silicone fluid with ease.
- ❑ The suspension components necessary for 4WD vehicles is basically the same as for FWD and RWD vehicles.
- ❑ The solid front 4WD axle is basically a rear axle turned so that the pinion gear faces the center of the vehicle.
- ❑ An independent front (IDF) suspension axle has the differential housing mounted solidly to the vehicle's frame. Short drive shafts or half-shafts connect the differential to the wheels. These half-shafts are suspended by springs and are able to independently respond to road surfaces.
- ❑ The wheel bearings in front and rear axle assemblies of 4WD vehicles are typically located so that they do not support the weight of the vehicle and are held in the hub assembly in the same way as other full-floating axle bearings.

Review Questions

Short Answer Essays

1. What are the advantages of four wheel drive?
2. What are the main differences between four wheel drive and all wheel drive?
3. What is the primary purpose of a transfer case?
4. Describe the operation of front locking wheel hubs.
5. What are the differences between an integrated 4WD system and an on-demand 4WD system?
6. What are the primary differences between a full-time and part-time 4WD system?
7. Briefly explain how a viscous coupling works.
8. What is the primary purpose of an interaxle differential?
9. Why are chain drives used in many transfer cases?
10. Briefly explain the operation of a simple planetary gearset.

Fill-in-the-Blanks

1. Normally, AWD systems use a ________________ ________________, ________________ ________________, or ________________ ________________ assembly to transfer engine power to the front and rear axles.

2. A simple planetary gearset is made up of three gears. In the center of the gearset is the ________________ gear. Meshing with this gear are three or four

__________________ gears. These are held together by the __________________ . This holds the gears in place while allowing them to rotate around their shafts. On the outside of the planetary gearset is the __________________ gear. This gear has teeth around its inside circumference.

3. In a typical planetary gearset, speed reduction is provided when the __________________ gear is held stationary.

4. __________________ occurs when the drive axles are rotating at different speeds on dry pavements or when the vehicle is turning a corner.

5. A __________________ differential allows for different front and rear axle speeds.

6. Silicone oil is used in a viscous coupling because its __________________ is not affected very much by temperature and it has high __________________ resistance and __________________ changes at high temperatures.

7. When the plates of a viscous coupling rotate at different speeds, the plates __________________ the fluid.

8. A __________________ differential does not allow speed differences between the two shafts it connects.

9. __________________ __________________ are not used on full-time 4WD systems because the axles and hubs are always driven by the transfer case.

10. An independent front suspension 4WD axle has the __________________ housing mounted solidly to the vehicle's frame and uses short __________________ __________________ to connect the differential to the wheels.

ASE Style Review Questions

1. While discussing the purpose of a transfer case:
   Technician A says it transfers power to an additional drive axle.
   Technician B says most have two speeds that affect the overall gear ratio of the vehicle.
   Who is correct?
   A. A only
   B. B only
   C. Both A and B
   D. Neither A nor B

2. Who is correct?
   Technician A says power can be locked into all four wheels by turning the locking hubs into the locked position.
   Technician B says some transfer cases use an electric switch to engage both drive axles.
   A. A only
   B. B only
   C. Both A and B
   D. Neither A nor B

3. While discussing four-wheel-drive systems used on cars:
Technician A says cars are equipped the same as a truck would be.
Technician B says some cars do not use a transfer case, but they are equipped with a third differential unit.
Who is correct?
A. A only
B. B only
C. Both A and B
D. Neither A nor B

4. While discussing the gear sets of a transfer case:
Technician A says most use a planetary gear set to provide for the different gear selections.
Technician B says normally a chain and sprocket assembly is used instead of gears to connect the input shaft with the output shaft.
Who is correct?
A. A only
B. B only
C. Both A and B
D. Neither A nor B

5. While discussing the various gear positions of a transfer case:
Technician A says that when the transfer case is in low, the overall gear ratio is numerically increased.
Technician B says that when the transfer case is in the high position, the vehicle operates in an overdrive mode due to the decrease in torque multiplication.
Who is correct?
A. A only
B. B only
C. Both A and B
D. Neither A nor B

6. While discussing the different shift positions of a transfer case:
Technician A says 4WD-High engages both axles and is used at any speed on slippery surfaces.
Technician B says 4WD-Low engages both axles and lowers the ratios of the entire drive line.
Who is correct?
A. A only
B. B only
C. Both A and B
D. Neither A nor B

7. Who is correct?
Technician A says the systems capable of operating in both two wheel and four wheel drive are called part-time 4WD systems.
Technician B says full-time 4WD systems cannot be selected out of four wheel drive.
A. A only
B. B only
C. Both A and B
D. Neither A nor B

8. *Technician A* says when a vehicle is turning a corner, the rear wheels travel a greater distance than the front wheels.
Technician B says the front wheels move through a wider arc than the rear wheels when the vehicle is making a turn.
Who is correct?
A. A only
B. B only
C. Both A and B
D. Neither A nor B

9. *Technician A* says part-time 4WD systems are designed to be used only when driving off the road or on slippery surfaces.
Technician B says AWD systems are intended for off-the-road use only.
Who is correct?
A. A only
B. B only
C. Both A and B
D. Neither A nor B

10. While discussing the characteristics of full-time four-wheel-drive vehicles:
Technician A says that when they turn a corner, the front wheels cannot rotate at the same speed as the rear wheels, causing one set of wheels to scuff along the payment.
Technician B says most full-time four-wheel-drive vehicles use a viscous clutch between the front and rear axles.
Who is correct?
A. A only
B. B only
C. Both A and B
D. Neither A nor B

GLOSSARY

> *Note: Terms are highlighted in color, followed by Spanish translation in bold.*

Abrasion Wearing or rubbing away of a part.
Abrasión Desgaste o rozamiento de una pieza.

Acceleration An increase in velocity or speed.
Aceleración Aumento de velocidad o celeridad.

Accident Something that happens unintentionally and is a consequence of doing something else.
Accidente Suceso imprevisto que ocurre como consecuencia de otro.

Adhesives Chemicals used to hold gaskets in place during the assembly of an engine. They also aid the gasket in maintaining a tight seal by filling in the small irregularities on the surfaces and by preventing the gasket from shifting due to engine vibration.
Adhesivos Productos químicos que sirven para sujetar las guarniciones en su lugar durante el montaje de un motor. También ayudan a la guarnición a mantener una junta de estanqueidad hermética al rellenar las pequeñas irregularidades en las superficies y al impedir que la guarnición se desplace debido a la vibración del motor.

Alignment An adjustment to a line or to bring into a line.
Alineación Ajuste a una línea o poner en línea.

All wheel drive (AWD) Commonly refers to a full-time 4WD system.
Tracción en las cuatro ruedas Usualmente se refiere a un sistema en el cual la fuerza motriz se reparte a tiempo completo entre las cuatro ruedas.

Antifriction bearing A bearing designed to reduce friction. This type of bearing normally uses ball or roller inserts to reduce the friction.
Cojinete de antifricción Cojinete diseñado para disminuir la fricción. Normalmente a este tipo de cojinete se le insertan bolas o rodillos para disminuir la fricción.

Antiseize Thread compound designed to keep threaded connections from damage due to rust or corrosion.
Antiagarrotamiento Compuesto de filetes de un tornillo que sirve para mantener las conexiones fileteadas libres de averías causadas por el óxido o la corrosión.

Arbor press A small, hand-operated shop press used when only a light force is required against a bearing, shaft, or other part.
Prensa para calar Pequeña prensa de taller accionada a mano y utilizada solamente cuando se requiere una fuerza ligera contra un cojinete, un árbol u otra pieza.

Asbestos A material that was commonly used as a gasket material in places where temperatures are great. This material is being used less frequently today because of health hazards that are inherent to the material.
Asbesto Material que usualmente se utilizaba como material de guarnición en lugares donde las temperaturas eran muy elevadas. Hoy en día se utiliza dicho material con menos frecuencia debido a los riesgos para la salud inherentes al mismo.

Automatic locking/unlocking hubs Front wheel hubs that can engage or disengage themselves from the axles automatically.
Cubos automáticos de cierre/descerrar Cubos ubicados en las ruedas delanteras que pueden engranarse o desengranarse de los ejes automáticamente.

Automatic transmission A transmission in which gear or ratio changes are self-activated, eliminating the necessity of hand-shifting gears.
Transmisión automática Sistema en el cual los cambios de velocidades o de relación se activan automáticamente, eliminando así la necesidad de emplear la palanca del cambio de velocidades.

Axial Parallel to a shaft or bearing bore.
Axial Paralelo a un árbol o calibre de cojinete.

Axis The center line of a rotating part, a symmetrical part, or a circular bore.
Pivote Línea central de una pieza giratoria, una pieza simétrica, o un calibre circular.

Axle The shaft or shafts of a machine upon which the wheels are mounted.
Eje Árbol o árboles de una máquina sobre el cual se montan las ruedas.

Axle carrier assembly A cast-iron framework that can be removed from the rear axle housing for service and adjustment of the parts.
Conjunto portador del eje Armazón de hierro fundido que se puede remover del puente trasero para la reparación y el ajuste de las piezas.

Axle housing Designed in the removable carrier or integral carrier types to house the drive pinion, ring gear, differential, and axle shaft assemblies.
Puente trasero Diseñado en los tipos de portador desmontables o enterizos para alojar los conjuntos del piñón de mando, de la corona, del diferencial y del árbol motor.

Axle ratio The ratio between the rotational speed (rpm) of the drive shaft and that of the driven wheel; gear reduction through the differential, determined by dividing the number of teeth on the ring gear by the number of teeth on the drive pinion.
Relación del eje Relación entre las revoluciones por minuto (rpm) del árbol de mando y las de la velocidad de la rueda accionada; la desmultiplicación de engranajes por medio del diferencial, determinada al dividir el número de dientes en la corona por el número de dientes en el piñón de mando.

Axle shaft A shaft on which the road wheels are mounted.
Árbol motor Árbol sobre el cual se montan las ruedas de la carretera.

Axle shaft end thrust A force exerted on the end of an axle shaft that is most pronounced when the vehicle turns corners and curves.
Empuje longitudinal del árbol motor Fuerza ejercida sobre el extremo de un árbol motor. Dicha fuerza es más marcada cuando el vehículo dobla una esquina o una curva.

Axle shaft tubes These tubes are attached to the axle housing center section to surround the axle shaft and bearings.
Tubos del árbol motor Estos tubos se fijan a la sección central del puente trasero para rodear el árbol motor y los cojinetes.

Backlash The amount of clearance or play between two meshed gears.
Contragolpe Cantidad del espacio libre u holgura entre dos engranajes.

Balance Having equal weight distribution. The term is usually used to describe the weight distribution around the circumference and between the front and back sides of a wheel and tire assembly. Uneven weight distribution causes vibrations when the wheel is spun because the center of gravity of the wheel and tire assembly does not line up with the center line of the axle. A wheel and tire are balanced by adding weights to the wheel rim opposite the heavy section.
Equilibrio Que tiene igual distribución de peso. Usualmente este término se emplea para describir la distribución de peso alrededor de la circunferencia y entre las partes delantera y trasera de un conjunto de rueda y llanta. La distribución desigual de peso puede causar vibraciones cuando se gira la rueda porque el centro de gravedad del conjunto de rueda y llanta no está alineado con la línea central del eje. Una rueda y una llanta están en equilibrio cuando se le añade más peso a la llanta opuesta a la sección pesada.

Balanced resistance A situation where two objects, such as axle shafts, present the same resistance to driving rotation.
Resistencia equilibrada Situación en la que dos objetos, tales como los árboles motores, ofrecen la misma resistencia al mando de rotación.

Ball bearing An antifriction bearing that consists of a hardened inner and outer race with hardened steel balls that roll between the two races, and supports the load of the shaft.
Cojinete de bolas Cojinete de antifricción que consiste de un anillo interno y externo templado con bolas de acero templadas que giran entre los dos anillos, y apoya la carga del árbol.

Ball-and-trunnion universal joint A nonconstant velocity universal joint that combines the universal joint and slip joint. It uses a driveable housing connected to a shaft with a ball head through two other balls mounted on trunnions.
Junta universal de bola y muñequilla Junta universal de velocidad no constante que combina la junta universal y la junta deslizante. Utiliza un alojamiento accionable conectado a un árbol por medio de un péndulo a través de otras dos bolas montadas sobre las muñequillas.

Ball joint A suspension component that attaches the control arm to the steering knuckle and serves as the lower pivot point for the steering knuckle. The ball joint gets its name from its ball-and-socket design. It allows both up and down motion as well as rotation. In a MacPherson strut FWD suspension system, the two lower ball joints are nonload carrying.
Junta esférica Componente de la suspensión que fija el brazo de mando al muñón de dirección y sirve como punto de pivote inferior para el muñón de dirección. Así se le llama a la junta esférica por su diseño de rótula. Permite tanto el movimiento de ascenso y descenso, como el de rotación. En un sistema de suspensión de tracción delantera montante MacPherson, las dos juntas esféricas inferiores no tienen capacidad de carga.

Bearing The supporting part that reduces friction between a stationary and rotating part or between two moving parts.
Cojinete Pieza de soporte que disminuye la fricción entre una pieza fija y una pieza giratoria o entre dos piezas móviles.

Bearing cage A spacer that keeps the balls or rollers in a bearing in proper position between the inner and outer races.
Jaula de cojinete Espaciador que mantiene las bolas o los rodillos en posición correcta en el cojinete, entre los anillos interior y exterior.

Bearing caps In the differential, caps held in place by bolts or nuts that, in turn, hold bearings in place.
Casquillos de cojinete En el diferencial, los casquillos sujetados por tornillos o tuercas que a su vez sujetan los cojinetes en su lugar.

Bearing cone The inner race, rollers, and cage assembly of a tapered-roller bearing. Cones and cups must always be replaced in matched sets.
Cono de cojinete El anillo interior, los rodillos y el conjunto de jaula de un cojinete de rodillos cónicos. Se debe reemplazar siempre los conos y las rótulas en conjuntos que hagan juego.

Bearing cup The outer race of a tapered-roller bearing or ball bearing.
Rótula de cojinete Anillo exterior de un cojinete de rodillos cónicos o cojinete de bolas.

Bearing race The surface on which the rollers or balls of a bearing rotate. The outer race is the same thing as the cup, and the inner race is the one closest to the axle shaft.
Anillo de cojinete Superficie sobre la cual los rodillos o las bolas de un cojinete giran. El anillo exterior es lo mismo que la rótula, y el anillo interior es el que se encuentra más cerca del árbol motor.

Belleville spring A tempered spring steel cone-shaped plate used to aid the mechanical force in a pressure plate assembly.
Muelle de Belleville Lámina de muelle de acero templado en forma cónica utilizada para ayudar a la fuerza mecánica en un conjunto de placa de presión.

Bell housing A housing that fits over the clutch components and connects the engine and the transmission. A common term for a clutch housing.
Alojamiento de campana Alojamiento que se monta sobre los componentes del embrague y que conecta el motor y la transmisión. Término común para el alojaminto del embrague.

Bellows Rubber protective covers with accordionlike pleats used to contain lubricants and exclude contaminating dirt or water.
Fuelles Cubiertas protectivas de caucho con pliegues en forma de acordeón, utilizadas para contener los lubricantes y evitar la contaminación por polvo o agua.

Bevel spur gear Gear that has teeth with a straight center line cut on a cone.
Engranaje recto cónico Engranaje que tiene dientes con un corte de línea central recto, cortado en forma cónica.

Bolt torque The turning effort required to offset resistance as the bolt is being tightened.
Par de torsión del perno Esfuerzo giratorio requerido para neutralizar la resistencia al apretarse el perno.

Boots See bellows.
Botas Véase fuelles.

Brake horsepower (bhp) Power delivered by the engine and available for driving the vehicle; bhp = torque × rpm/5,252.
Potencia en caballos indicada al freno (bhp) Energía descargada por el motor y disponible para accionar el vehículo; bhp= par de torsión X rpm/5.252.

Brinnelling Rough lines worn across a bearing race or shaft due to impact loading, vibration, or inadequate lubrication.
Acción de Brinnell Líneas toscas a través de un anillo de cojinete o un árbol, causadas por la carga de un impacto, vibración o engrase inadecuado.

Bronze An alloy of copper and tin.
Bronce Aleación de cobre y estaño.

Burnish To smooth or polish by the use of a sliding tool under pressure.
Bruñir Suavizar o pulir con una herramienta deslizante bajo presión.

Burr A feather edge of metal left on a part being cut with a file or other cutting tool.
Rebaba Bisel de metal que queda en una pieza que ha sido cortada con una lima u otra herramienta de corte.

Bushing A cylindrical lining used as a bearing assembly made of steel, brass, bronze, nylon, or plastic.
Buje Forro en forma cilíndrica utilizado como un conjunto de cojinete, hecho de acero, latón, bronce, nilón o plástico.

C-clip A C-shaped clip used to retain the drive axles in some rear axle assemblies.
Grapa-C Grapa en forma de C utilizada para retener los ejes de mando en algunos conjuntos de eje trasero.

Cage A spacer used to keep the balls or rollers in proper relation to one another. In a constant velocity joint, the cage is an open metal framework that surrounds the balls to hold them in position.
Jaula Espaciador utilizado para sujetar las bolas o los rodillos en relación correcta entre sí. En una junta de velocidad constante, la jaula es un armazón abierto de metal que rodea las bolas para que se mantengan en su posición.

Camber A suspension alignment term used to define the amount that the centerline of a wheel is tilted inward or outward from the true vertical plane of the wheel. If the top of the wheel is tilted inward, the camber is negative. If the top of the wheel is tilted outward, the camber is positive.
Combadura Un término de alineamiento de suspensión que denota cuán inclinada está, hacia adentro o hacia afuera la línea central de una rueda del verdadero plano vertical de la rueda. Si la parte

superior de la rueda está inclinada hacia adentro, la combadura es negativa. Si la parte superior de la rueda está inclinada hacia afuera, la combadura es positiva.

Canceling angles Opposing operating angles of two universal joints cancel the vibrations developed by the individual universal joint.
Angulos de supresión Los ángulos de funcionamiento opuestos de dos juntas universales cancelan las vibraciones producidas por la junta universal individual.

Cardan universal joint A nonconstant velocity universal joint consisting of two yokes with their forked ends joined by a cross. The driven yoke changes speed twice in 360 degrees of rotation.
Junta de cardán Junta universal de velocidad no constante que consiste de dos horquillas con sus extremos unidos por una cruz. El yugo accionado cambia de velocidad dos veces en una rotación de 360°.

Carrier An object that bears, cradles, moves, or transports some other object or objects.
Portador Objeto que apoya, acojina, mueve o transporta otro u otros objetos.

Case-harden To harden the surface of steel. The carburizing method used on low carbon steel or other alloys to make the case or outer layer of the metal harder than its core.
Cementar Endurecer la superficie del acero. El método carburizante empleado en el acero de bajo contenido en carbón u otras aleaciones para hacer que la caja o capa exterior del metal sea más dura que el núcleo.

Castellate Formed to resemble a castle battlement, as in a castellated nut.
Entallar Formado para que se asemeje a la almena de un castillo, como una tuerca de corona.

Castellated nut A nut with six raised portions or notches through which a cotter pin can be inserted to secure the nut.
Tuerca de corona Tuerca con seis partes elevadas o muescas a través de las cuales se puede insertar un pasador de chaveta para sujetar la tuerca.

Caustic A material has the ability to destroy or eat through something. Caustic materials are considered extremely corrosive.
Cáustico Material que tiene la habilidad de destruir o corroer algo. A los materiales cáusticos se les considera extremadamente corrosivos.

Center hanger bearing Ball-type bearing mounted on a vehicle cross member to support the drive shaft and provide better installation angle to the rear axle.
Silleta de suspensión central Cojinete de tipo bola montado sobre la traviesa de un vehículo para apoyar el árbol de mando y proveerle un mejor ángulo de montaje al eje trasero.

Center section The middle of the integral axle housing containing the drive pinion, ring gear, and differential assembly.
Sección central Centro del puente trasero integral que contiene el piñón de mando, la corona y el conjunto del diferencial.

Centering joint Ball socket joint placed between two Cardan universal joints to ensure that the assembly rotates on center.
Junta centradora Junta de rótula montada entre dos juntas de cardán para asegurar que el montaje gire en el centro.

Centrifugal clutch A clutch that uses centrifugal force to apply a higher force against the friction disc as the clutch spins faster.
Embrague centrífugo Embrague que utiliza fuerza centrífuga para aplicar mayor fuerza contra el disco de fricción cuando el embrague gira con más rapidez.

Centrifugal force The force acting on a rotating body that tends to move it outward and away from the center of rotation. The force increases as rotational speed increases.
Fuerza centrífuga Fuerza que acciona un cuerpo giratorio y que tiende a moverlo hacia afuera y más lejos del centro de la rotación. La fuerza aumenta cuando aumenta la velocidad de rotación.

Chamfer A bevel or taper at the edge of a hole or a gear tooth.
Chaflán Bisel o cono al borde de un agujero o diente de engranaje.

Chamfer face A beveled surface on a shaft or part that allows for easier assembly. The ends of FWD drive shafts are often chamfered to make installation of the CV-joints easier.
Superficie achaflanada Superficie biselada en un árbol o pieza que facilita el montaje. Los extremos de los árboles de mando de tracción delantera se achaflanan para facilitar la instalación de las juntas CV.

Chase To straighten up or repair damaged threads.
Roscar Enderezar o reparar roscas averiadas.

Chasing To clean threads with a tap.
Filetear Limpiar las roscas con un macho de roscar.

Chassis The vehicle frame, suspension, and running gear. On FWD cars, it includes the control arms, struts, springs, trailing arms, sway bars, shocks, steering knuckles, and frame. The drive shafts, constant velocity joints, and transaxle are not part of the chassis or suspension.
Chasis El armazón del vehículo, la suspensión y el tren de ruedas. En vehículos de tracción delantera, incluye los brazos de mando, los montantes, los muelles, los brazos traseros, las barras de oscilación lateral, los amortiguadores, los muñones de dirección y el armazón. Los árboles de mando, las juntas de velocidad constante y el transeje no forman parte del chasis o de la suspensión.

Circlip A split steel snap ring that fits into a groove to hold various parts in place. Circlips are often used on the ends of FWD drive shafts to retain the constant velocity joints.
Grapa circular Anillo de resorte hendido, en acero, que se inserta en una ranura para sujetar varias piezas en su lugar. Con frecuencia se utilizan las grapas circulares en los extremos de árboles de mando de tracción delantera para retener las juntas de velocidad constante.

Clashing Grinding sound heard when gear and shaft speeds are not the same during a gearshift operation.
Entrechoque Rechinamiento que se escucha cuando las velocidades del engranaje y del árbol no son iguales durante el cambio de velocidades.

Clearance The space allowed between two parts, such as between a journal and a bearing.
Espacio libre Espacio permitido entre dos piezas, por ejemplo, entre un muñón y un cojinete.

Close ratio A relative term for describing the gear ratios in a transmission. If the gears are numerically close, they are said to be close ratio. This design gives quicker acceleration at the expense of initial acceleration and fuel economy.
Relación próxima Término relativo que describe las relaciones de los engranajes en una transmisión. Si existe una proximidad numérica entre los engranajes, se dice que su relación es próxima. Este diseño permite una aceleración más rápida a costa de la aceleración inicial y del rendimiento de combustible.

Cluster assembly A manual transmission related term applied to a group of gears of different sizes machined from one steel casting.
Conjunto desplazable Término relacionado a la transmisión manual que se aplica a un grupo de engranajes de diferentes tamaños hechos a máquina, de una pieza fundida en acero.

Cluster gear A common term for the counter gear assembly.
Engranajes desplazables Término común que denomina el conjunto de contraengranaje.

Clutch A device for connecting and disconnecting the engine from the transmission or for a similar purpose in other units.
Embrague Dispositivo que sirve para engranar y desengranar el motor de la transmisión o para un propósito parecido en otras unidades.

Clutch control cable A cable assembly with a flexible outer housing anchored at the upper and lower ends. Moving back and forth inside the flexible housing is a braided wire cable that transfers clutch pedal movement to the clutch release lever.

Cable de mando del embrague Conjunto de cable con un alojamiento exterior flexible sujetado a los extremos superior e inferior. Un cable trenzado de alambre que transfiere el movimiento del pedal del embrague a la palanca de desembrague se mueve de atrás para adelante dentro del alojamiento.

Clutch cover A term used by some manufacturers to describe a pressure plate.
Tapa del embrague Término empleado por algunos fabricantes para describir una placa de presión.

Clutch (friction) disc The friction material part of the clutch assembly that fits between the flywheel and pressure plate.
Disco de embrague (fricción) La pieza material de fricción del conjunto de embrague que encaja entre el volante y la placa de presión.

Clutch fork In the clutch, a Y-shaped member into which the throw-out bearing is assembled.
Horquilla de embrague En el embrague, una pieza en forma de Y sobre la cual se monta el cojinete de desembrague.

Clutch housing A large aluminum or iron casting that surrounds the clutch assembly. Located between the engine and transmission, it is sometimes referred to as bell housing.
Alojamiento del embrague Pieza grande fundida en hierro o en aluminio que rodea al conjunto del embrague. Ubicado entre el motor y la transmisión, a veces se le llama alojamiento de campana.

Clutch linkage A combination of shafts, levers, or cables that transmits clutch pedal motion to the clutch assembly.
Articulación de embrague Combinación de árboles, palancas o cables que transmite el movimiento del pedal del embrague al conjunto del embrague.

Clutch packs A series of clutch discs and plates installed alternately in a housing to act as a driving or driven unit.
Paquetes del embrague Serie de discos y placas del embrague instalados por turno en un alojamiento para que funcionen como una unidad de accionamiento o accionada.

Clutch pedal A pedal in the driver's compartment that operates the clutch.
Pedal del embrague Pedal que hace funcionar el embrague; ubicado en el compartimiento del conductor.

Clutch pedal free-play The amount the pedal can move without applying pressure on the pressure plate.
Juego libre del pedal del embrague Amplitud de movimiento del pedal sin que éste aplique presión sobre la placa de presión.

Clutch pushrod A solid or hollow rod that transfers linear motion between movable parts; that is, the clutch release bearing and release plate.
Varilla de empuje del embrague Varilla sólida o hueca que transfiere un movimiento lineal entre piezas móviles; es decir, el cojinete de desembrague del embrague y la placa de desembrague.

Clutch safety switch See neutral start switch.
Interruptor de seguridad del embrague Véase interruptor de arranque neutro.

Clutch shaft Sometimes known as the transmission input shaft or main drive pinion. The clutch driven disc drives this shaft.
Árbol de embrague A veces llamado árbol impulsor de la transmisión o piñón principal de mando. El disco accionado del embrague acciona este árbol.

Clutch slippage Engine speed increases but increased torque is not transferred through to the driving wheels because of clutch slippage.
Deslizamiento del embrague La velocidad del motor aumenta pero el par de torsión no se transmite a las ruedas motrices a causa del deslizamiento del embrague.

Coefficient of friction The ratio of the force resisting motion between two surfaces in contact to the force holding the two surfaces in contact.
Coeficiente de fricción Relación de la fuerza que resiste el movimiento entre dos superficies en contacto a la fuerza que mantiene el contacto de las dos superficies.

Coil spring A heavy wirelike steel coil used to support the vehicle weight while allowing for suspension motions. On FWD cars, the front coil springs are mounted around the MacPherson struts. On the rear suspension, they may be mounted to the rear axle, to trailing arms, or around rear struts.
Muelle helicoidal Espiral grueso de acero parecido al alambre, que sirve para apoyar el peso del vehículo mientras permite el movimiento de suspensión. En vehículos de tracción delantera, los muelles helicoidales delanteros se suspenden alrededor de los montantes MacPherson. En la suspensión trasera, pueden suspenderse del eje trasero, del los brazos traseros, o alrededor de los montantes traseros.

Coil preload springs Coil springs are made of tempered steel rods formed into a spiral that resist compression; located in the pressure plate assembly.
Muelles helicoidales de carga previa Muelles helicoidales se fabrican de varillas de acero templado configuradas en forma de espiral que resisten la compresión; ubicados en el conjunto de placa de presión.

Coil spring clutch A clutch using coil springs to hold the pressure plate against the friction disc.
Embrague de muelle helicoidal Embrague que emplea muelles helicoidales para mantener la placa de presión contra el disco de fricción.

Companion flange A mounting flange that fixedly attaches a drive shaft to another drive train component.
Brida acompañante Una brida de montaje que fija un árbol de mando a otro componente del tren de mando.

Compound A mixture of two or more ingredients.
Compuesto Mezcla de dos o más ingredientes.

Concentric Two or more circles, having a common center.
Concéntrico Dos o más círculos que tienen un centro común.

Cone clutch The driving and driven parts conically shaped to connect and disconnect power flow. A clutch made from two cones, one fitting inside the other. Friction between the cones forces them to rotate together.
Embrague cónico Piezas de accionamiento y accionadas, en forma cónica, empleadas para conectar y desconectar el regulador de fuerza. Un embrague hecho de dos conos, uno se inserta dentro del otro. La fricción entre los conos les hace girar juntos.

Constant mesh Manual transmission design permits the gears to be constantly enmeshed regardless of vehicle operating circumstances.
Engrane constante Diseño de transmisión manual permite que los engranajes permanezcan siempre engranados a pesar de las condiciones del funcionamiento del vehículo.

Constant mesh transmission A transmission in which the gears are engaged at all times, and shifts are made by sliding collars, clutches, or other means to connect the gears to the output shaft.
Transmisión de engrane constante Transmisión en la que los engranajes están siempre engranados, y los cambios se llevan a cabo a través de chavetas deslizantes, embragues u otros medios para conectar los engranajes al árbol de rendimiento.

Constant velocity joint (also called CV-joint) A flexible coupling between two shafts that permits each shaft to maintain the same driving or driven speed regardless of operating angle, allowing for a smooth transfer of power. The constant velocity joint consists of an inner and outer housing with balls in between, or a tripod and yoke assembly.
Junta de velocidad constante (llamada también junta-CV) Unión flexible entre dos árboles que permite que cada árbol mantenga la misma velocidad de accionamiento o accionada, a pesar del ángulo

de funcionamiento, y permite que la transferencia de fuerza sea una suave. La junta de velocidad constante consiste de un alojamiento interior y exterior entre el cual se insertan bolas, o un conjunto de trípode y yugo.

Contraction A reduction in mass or dimension; the opposite of expansion.

Contracción Disminución en masa o dimensión; lo opuesto de expansión.

Control arm A suspension component that links the vehicle frame to the steering knuckle or axle housing and acts as a hinge to allow up and down wheel motions. The front control arms are attached to the frame with bushings and bolts and are connected to the steering knuckles with ball joints. The rear control arms attach to the frame with bushings and bolts and are welded or bolted to the rear axle or wheel hubs.

Brazo de mando Componente de suspensión que une el armazón del vehículo al muñón de dirección o al puente trasero y funciona como una bisagra para permitir el movimiento de ascenso y descenso de la rueda. Los brazos de mando delanteros se fijan al armazón con bujes y pernos y se conectan a los muñones de dirección con juntas esféricas. Se fijan los brazos de mando traseros al armazón con bujes y pernos y se soldan o se empernan al eje trasero o a los cubos de la rueda.

Control cable An assembly with a flexible outer housing anchored at the upper and lower ends. Moving back and forth inside the housing is a braided stainless steel wire cable that transfers pedal movement to the release lever.

Cable de mando Conjunto con un alojamiento exterior flexible sujetado a los extremos superiores e inferiores. Un cable trenzado de acero inoxidable transfiere el movimimiento del pedal a la palanca de desembrague y se mueve hacia atrás y hacia adelante dentro del alojamiento.

Corrode To eat away gradually as if by gnawing, especially by chemical action.

Corroerse Carcomer gradualmente como al roer, especialmente debido a una acción química.

Corrosion Chemical action, usually by an acid, that eats away (decomposes) a metal.

Corrosión Acción química, normalmente producida por un ácido, que carcome (descompone) un metal.

Corrosivity A statement defining how likely it is that a substance will destroy or eat away at other substances.

Corrosividad Enunciado que define la posibilidad de que una sustancia destruya o carcoma otras sustancias.

Cotter pin A type of fastener made from soft steel in the form of a split pin that can be inserted in a drilled hole. The split ends are spread to lock the pin in position.

Pasador de chaveta Tipo de aparato de fijación hecho de acero recocido en forma de pasador hendido que puede insertarse en un agujero barrenado. Se separan los extremos hendidos para fijar la chaveta en la posición correcta.

Counter clockwise rotation Rotating the opposite direction of the hands on a clock.

Rotación a la izquierda Rotación en el sentido inverso a la dirección de las agujas del reloj.

Counter gear assembly A cluster of gears designed on one casting with short shafts supported by antifriction bearings. Closely related to the cluster assembly.

Mecanismo contador Grupo de engranajes diseñados en una sola fundición con árboles cortos apoyados por cojinetes de antifricción. Estrechamente relacionado al conjunto desplazable.

Countershaft An intermediate shaft that receives motion from a main shaft and transmits it to a working part; sometimes called a lay shaft.

Árbol de retorno Árbol intermedio que recibe movimiento de un árbol primario y lo transmite a una pieza móvil; llamado también árbol secundario.

Coupling A connecting means for transferring movement from one part to another; may be mechanical, hydraulic, or electrical.

Acoplamiento Método de conexión para transferir movimiento de una pieza a otra; puede ser mecánico, hidráulico o eléctrico.

Coupling yoke A part of the double Cardan universal joint that connects the two universal joint assemblies.

Yugo de acoplamiento Pieza de la junta doble de cardán que conecta los dos conjuntos de junta universal.

Cover plate A stamped steel cover bolted over the service access to the manual transmission.

Cubreplaca Cubreplaca de acero estampada, empernada sobre el acceso de reparación a la transmisión manual.

Critical speed The rotational speed at which an object begins to vibrate as it turns. This is mostly caused by centrifugal forces.

Velocidad crítica Velocidad de rotación a la cual un objeto comienza a vibrar mientras gira. Esto se debe en gran parte a fuerzas centrífugas.

Cross member A steel part of the frame structure that transverses the vehicle body to connect the longitudinal frame rails. Cross members can be welded into place or removed from the vehicle.

Traviesa Pieza de acero de la estructura del armazón que atraviesa la carrocería para conectar las barras longitudinales del armazón. Las traviesas se pueden soldar o remover del vehículo.

Crush sleeve A commonly used term for the collapsible spacer in a differential assembly.

Manguito de quiebra Término utilizado comunamente para denominar el espaciador desmontable en un conjunto de diferencial.

Cushioning springs A common name for a clutch disc's wave springs.

Muelles de acojinamiento Nombre común para los muelles ondulares del disco del embrague.

CV-joints Constant velocity joints that allow the angle of the axle shafts to change with no loss in rotational speed.

Juntas CV Juntas de velocidad constante que permiten que el ángulo de los árboles motores cambie sin que disminuya la velocidad de rotación.

Dead axle An axle that only supports the vehicle and does not transmit power.

Eje portante Eje que sirve sólo para apoyar el vehículo y que no transmite fuerza motriz.

Deflection Bending or movement away from normal due to loading.

Desviación Flexión o movimiento fuera de lo normal debido a la carga.

Degree A unit of measurement equal to 1/360th of a circle.

Grado Unidad de medida equivalente a cada una de las 360 partes de un círculo.

Density Compactness; relative mass of matter in a given volume.

Densidad Compacidad; masa relativa de materia en un volumen dado.

Detent A small depression in a shaft, rail, or rod into which a pawl or ball drops when the shaft, rail, or rod is moved. This provides a locking effect.

Retén Pequeña depresión en un árbol, una barra o una varilla sobre el/la cual cae un trinquete o una bola cuando se mueve el árbol, la barra o la varilla. Esto provee un efecto de blocaje.

Detent mechanism A shifting control designed to hold the manual transmission in the gear range selected.

Mecanismo de detención Control de cambio de velocidades diseñado para mantener la transmisión manual dentro del límite del engranaje elegido.

Diagnosis A systematic study of a machine or machine parts to determine the cause of improper performance or failure.
Diagnosis Estudio sistemático de una máquina o de piezas de un máquina para establecer la causa del mal funcionamiento o falla.

Dial indicator A measuring instrument with the readings indicated on a dial rather than on a thimble as on a micrometer.
Indicador de cuadrante Instrumento de medida que muestra las lecturas en un cuadrante en vez de en un tambor como en el caso de un micrómetro.

Diaphragm spring A circular disc shaped like a cone, with spring tension that allows it to flex forward or backward. Often referred to as a Belleville spring.
Muelle de diafragma Disco circular en forma de cono, con tensión en el muelle que le permite moverse hacia adelante o hacia atrás. Conocido también como muelle de Belleville.

Diaphragm spring clutch A clutch in which a diaphragm spring, rather than a coil spring, applies pressure against the friction disc.
Embrague de muelle de diafragma Embrague en el que un muelle de diafragma, en vez de un muelle helicoidal, ejerce presión contra el disco de fricción.

Differential A mechanism between drive axles that permits one wheel to run at a different speed than the other while turning.
Diferencial Mecansimo entre los ejes de mando que permite que una rueda gire a una velocidad diferente que la otra.

Differential action An operational situation where one driving wheel rotates at a slower speed than the opposite driving wheel.
Acción del diferencial Situación de funcionamiento donde una rueda motriz gira más despacio que la rueda motriz opuesta.

Differential case The metal unit that encases the differential side gears and pinion gears, and to which the ring gear is attached.
Caja del diferencial Unidad metálica que reviste los engranajes laterales del diferencial y los engranajes de piñón, y sobre la cual se monta la corona.

Differential drive gear A large circular helical gear driven by the transaxle pinion gear and shaft and drives the differential assembly.
Engranaje del diferencial Engranaje helicoidal circular grande accionado por el engranaje de piñón del transeje y el árbol, que acciona el conjunto del diferencial.

Differential housing Cast-iron assembly that houses the differential unit and the drive axles. This is also called the rear axle housing.
Alojamiento del diferencial Conjunto de hierro fundido que aloja a la unidad del diferencial y a los ejes de mando. Llamado también puente trasero.

Differential pinion gears Small beveled gears located on the differential pinion shaft.
Engranajes de piñón del diferencial Pequeños engranajes biselados ubicados en el árbol de piñón del diferencial.

Differential pinion shaft A short shaft locked to the differential case. This shaft supports the differential pinion gears.
Árbol de piñón del diferencial Árbol corto fijado a la caja del diferencial. Este árbol apoya los engranajes de piñón del diferencial.

Differential ring gear A large circular hypoid-type gear enmeshed with the hypoid drive pinion gear.
Corona del diferencial Engranaje hipoide circular grande engranado con el engranaje de piñón hipoide.

Differential side gears The gears inside the differential case that are internally splined to the axle shafts, and are driven by the differential pinion gears.
Engranajes laterales del diferencial Engranajes dentro de la caja del diferencial que son ranurados internamente a los árboles motores, y accionados por los engranajes de piñon del diferencial.

Direct drive One turn of the input driving member compared to one complete turn of the driven member, such as when there is direct engagement between the engine and drive shaft where the engine crankshaft and the drive shaft turn at the same rpm.
Toma directa Una vuelta de la pieza de accionamiento comparada a una vuelta completa de la pieza accionada, como por ejemplo, cuando hay un engrane directo entre el motor y el árbol de mando donde el cigüeñal del motor y el árbol mator giran a las mismas rpm.

Disengage When the operator moves the clutch pedal toward the floor to disconnect the driven clutch disc from the driving flywheel and pressure plate assembly.
Desengranar Cuando el conductor hunde el pedal del embrague para desconectar el disco de embrague accionado del volante motor y del conjunto de la placa de presión.

Distortion A warpage or change in form from the original shape.
Deformación Abarquillamiento o cambio en la forma original de la configuración original.

Dog tooth A series of gear teeth that are part of the dog clutching action in a transmission synchronizer operation.
Diente de sierra Serie de dientes de engranaje que forman parte de la acción del embrague de garras durante una sincronización de transmisión.

Double Cardan universal joint A near constant velocity universal joint that consists of two Cardan universal joints connected by a coupling yoke.
Junta doble de cardán Junta universal de velocidad casi constante compuesta de dos juntas de cardán conectadas por un yugo de acoplamiento.

Double-offset constant velocity joint Another name for the type of plunging, inner CV-joint found on many GM, Ford, and Japanese FWD cars.
Junta de velocidad constante de desviación doble Otro nombre para el tipo de junta de velocidad constante interior de pistón tubular, instalada en muchos automóviles de tracción delantera de la GM, de la Ford y japoneses.

Double reduction axle A drive axle construction in which two sets of reduction gears are used for extreme reduction of the gear ratio.
Eje de reducción doble Eje de mando en el que se utilizan dos juegos de reductores para lograr una mayor reducción de la relación de engranajes.

Dowel A metal pin attached to one object that when inserted into a hole in another object ensures proper alignment.
Pasador Chaveta metálica fijada a un objeto que, cuando se inserta dentro de un agujero en otro objeto, asegura una alineación correcta.

Dowel pin A pin inserted in matching holes in two parts to maintain those parts in fixed relation one to another.
Espiga de madera Chaveta insertada en agujeros parejos en dos piezas para mantener dichas piezas en una relación fija entre sí.

Downshift To shift a transmission into a lower gear.
Cambio de alta a baja velocidad Cambiar la transmisión a un engranaje de menos velocidad.

Drive line The universal joints, drive shaft, and other parts connecting the transmission with the driving axles.
Línea de transmisión Las juntas universales, el árbol de mando y otras piezas que conectan la transmisión a los ejes motores.

Drive line torque Relates to rear wheel drive line and is the transfer of torque between the transmission and the driving axle assembly.
Par de torsión de la línea de transmisión Relacionado a la línea de transmisión de las ruedas traseras y es la transferencia del par de torsión entre la transmisión y el conjunto del eje motor.

Drive line wrap-up A condition where axles, gears, U-joints, and other components can bind or fail if the 4WD mode is used on pavement where 2WD is more suitable.

Falla de la línea de transmisión Condición que ocurre cuando los ejes, los engranajes, las juntas universales, y otros componentes se traban o fallan si se emplea la tracción a las cuatro ruedas sobre un pavimiento donde la tracción a las dos ruedas es más adecuada.

Drive pinion The gear that takes its power directly from the drive shaft or transmission and drives the ring gear.

Piñón de mando Engranajes que obtienen su fuerza motriz directamente del árbol de mando o de la transmisión y que accionan la corona.

Drive pinion flange A rim used to connect the rear of the drive shaft to the rear axle drive pinion.

Brida de piñón de mando Corona utilizada para conectar la parte trasera del árbol de mando al piñón de mando del eje trasero.

Drive pinion gear One of the two main driving gears located within the transaxle or rear driving axle housing. Together the two gears multiply engine torque.

Engranaje del piñón de mando Uno de los dos mecanismos de accionamiento principales ubicados dentro del transeje o el puente trasero. Los dos engranajes actúan juntos para multiplicar el par de torsión del motor.

Drive shaft An assembly of one or two universal joints connected to a shaft or tube; used to transmit power from the transmission to the differential. Also called the propeller shaft.

Árbol de mando Conjunto de una o dos juntas universales conectadas a un árbol o tubo; utilizado para transmitir la fuerza motriz desde la transmisión hasta el diferencial. Llamado también árbol transmisor.

Drive shaft installation angle The angle the drive shaft is mounted off the true horizontal line measured in degrees.

Ángulo de montaje del árbol de mando El ángulo al que se monta el árbol de mando fuera de la línea horizontal verdadera, medido en grados.

Driven disc The part of the clutch assembly that receives driving motion from the flywheel and pressure plate assemblies.

Disco accionado Pieza del conjunto del embrague que recibe su fuerza motriz del conjunto del volante y del conjunto de la placa de presión.

Driven gear The gear meshed directly with the driving gear to provide torque multiplication, reduction, or a change of direction.

Engranaje accionado Engranaje engranado directamente al mecanimo de accionamiento para proveer multiplicación de par de torsión, reducción o un cambio de dirección.

Driving axle A term related collectively to the rear driving axle assembly where the drive pinion, ring gear, and differential assembly are located within the driving axle housing.

Eje motor Término relacionado colectivamente al conjunto del eje motor trasero donde el piñón de mando, la corona, y el conjunto del diferencial están ubicados dentro del puente trasero.

Drop forging A piece of steel shaped between dies while hot.

Estampado Pieza de acero conformada entre troqueles mientras está caliente.

Dry-disc clutch A clutch in which the friction faces of the friction disc are dry, as opposed to a wet-disc clutch, which runs submerged in oil. The conventional type of automobile clutch.

Embrague de disco seco Embrague en el que las placas de fricción del disco de fricción están secas, lo opuesto de un disco mojado, que funciona sumergido en aceite. Tipo convencional de embrague de automóviles.

Dry friction The friction between two dry solids.

Fricción seca Fricción entre dos sólidos secos.

Dual reduction axle A drive axle construction with two sets of pinions and gears, either of which can be used.

Eje de reducción doble Conjunto de eje de mando que tiene dos juegos de piñones y engranajes. Se puede utilizar cualquiera de los dos.

Dummy shaft A shaft, shorter than the countershaft, used during disassembly and reassembly in place of the countershaft.

Árbol falso Árbol, más corto que el árbol de retorno, empleado durante el desmontaje y el remonte en vez del árbol de retorno.

Dynamic In motion.

Dinámico En movimiento.

Dynamic balance The balance of an object when it is in motion; for example, the dynamic balance of a rotating drive shaft.

Equilibrio dinámico Equilibrio de un objeto cuando está en movimiento; por ejemplo, el equilibrio dinámico de un árbol de mando giratorio.

Eccentric One circle within another circle wherein both circles do not have the same center or a circle mounted off center. On FWD cars, front-end camber adjustments are accomplished by turning an eccentric cam bolt that mounts the strut to the steering knuckle.

Excéntrico Un círculo dentro de otro donde los dos círculos no comparten el mismo centro o un círculo colocado fuera del centro. En automóviles de tracción delantera, se realizan ajustes a la combadura del tren delantero girando un perno de leva excéntrica que levanta el montante al muñón de dirección.

Efficiency The ratio between the power of an effect and the power expended to produce the effect; the ratio between an actual result and the theoretically possible result.

Rendimiento La relación entre la fuerza de un efecto y la fuerza rendida para producir tal efecto; la relación entre un resultado verdadero y un resultado teóricamente posible.

Elastomer Any rubber-like plastic or synthetic material used to make bellows, bushings, and seals.

Elastómero Cualquier material plástico o sintético parecido al caucho que se utiliza para fabricar fuelles, bujes y juntas de estanqueidad.

Electrical short An alternative path for the flow of electricity.

Corto eléctrico Camino alternativo para el flujo de electricidad.

Ellipse A compressed form of a circle.

Elipse Forma comprimida de un círculo.

Endplay The amount of axial or end-to-end movement in a shaft due to clearance in the bearings.

Holgadura Amplitud de movimiento axial o movimiento de extremo a extremo en un árbol debido al espacio libre en los cojinetes.

Engage When the vehicle operator moves the clutch pedal up from the floor, this engages the driving flywheel and pressure plate to rotate and drive the driven disc.

Engranar Cuando el conductor del vehículo suelta el pedal del embrague, el volante y la placa de presión se engranan para girar y accionar el disco accionado.

Engagement chatter A shaking, shuddering action that takes place as the driven disc makes contact with the driving members. Chatter is caused by a rapid grip and slip action.

Vibración de acoplamiento Movimiento de agitación y estremecimiento que ocurre cuando el disco accionado entra en contacto con las piezas de accionamiento. La causa de esta vibración es un movimiento rápido de garra y de deslizamiento.

Engine The source of a power for most vehicles. It converts burned fuel energy into mechanical force.

Motor Fuente de fuerza para la mayoría de los vehículos. Convierte el combustible quemado en fuerza mecánica.

Engine torque A turning or twisting action developed by the engine measured in foot-pounds or kilogram meters.

Esfuerzo de rotación del motor Movimiento de giro o torcedura que produce el motor, medido en libras-pies o en kilográmetros.

Extension housing An aluminum or iron casting of various lengths that encloses the transmission output shaft and supporting bearings.

Alojamiento de extensión Pieza fundida en aluminio o en hierro de longitudes variadas que encubre el árbol de rendimiento de la transmisión y los cojinetes soporte.

External cone clutch The external surface of one part has a tapered surface to mate with an internally tapered surface to form a cone clutch.

Embrague cónico externo La superficie externa de una pieza tiene una superficie cónica para hacer juego con una superficie internamente cónica y así formar un embrague de cono.

External gear A gear with teeth across the outside surface.

Engranaje externo Engranaje que tiene dientes a través de la superficie exterior.

Externally tabbed clutch plates Clutch plates are designed with tabs around the outside periphery to fit into grooves in a housing or drum.

Placas de embrague con orejetas externas Se diseñan placas de embrague con orejetas alrededor de la periferia exterior para que puedan ajustarse a las acanaladuras de un alojamiento o de un tambor.

Extreme pressure lubricant A special lubricant for use in hypoid gear differentials; needed because of the heavy wiping loads imposed on the gear teeth.

Lubrificante para presión extrema Lubrificante especial que se utiliza en diferenciales de engranaje hipoide; necesario a causa del intenso esfuerzo al que los dientes de los engranajes están sometidos.

Face The front surface of an object.

Frente Superficie frontal de un objeto.

Fatigue The buildup of natural stress forces in a metal part that eventually causes it to break. Stress results from bending and loading the material.

Fatiga Acumulación de tensiones naturales en una pieza metálica que finalmente ocasiona una ruptura. La tensión es una consecuencia de la flexión y de la carga a las cuales el material está expuesto.

Feeler gauge A metal strip or blade finished accurately with regard to thickness, and used for measuring the clearance between two parts; such gauges ordinarily come in a set of different blades graduated in thickness by increments of 0.001 inch.

Calibrador de espesores Lámina metálica o cuchilla acabada con precisión de acuerdo al espesor que se utiliza para medir el espacio libre entre dos piezas. Dichos calibradores normalmente están disponibles en juegos de cuchillas diferentes graduadas según el espesor, en incrementos de 0,001 pulgadas.

Fiber composites A mixture of metallic threads along with a resin form a composite offering weight and cost reduction, long-term durability, and fatigue life. Fiberglass is a fiber composite.

Compuestos de fibra Una mezcla de hilos metálicos y resina forman un compuesto que ofrece una disminución de peso y costo, mayor durabilidad y resistencia a la fatiga. La fibra de vidrio es un compuesto de fibra.

Final drive The final set of reduction gears the engine's power passes through on its way to the drive wheels.

Transmisión final Ultimo juego de reductores por el cual pasa la fuerza del motor en camino a las ruedas motrices.

Final drive gears Main driving gears located in the axle area of the transaxle housing.

Engranajes de la transmisión final Mecanismos de accionamiento principales ubicados en la región del eje del alojamiento del transeje.

Final drive ratio The ratio between the drive pinion and ring gear.

Relación de la transmisión final Relación entre el piñón de mando y la corona.

First gear A small diameter driving helical- or spur-type gear located on the cluster gear assembly. First gear provides torque multiplication to get the vehicle moving.

Engranaje de primera velocidad Mecanismo de mando helicoide o recto de diámetro pequeño ubicado en el conjunto del tren desplazable. El engranaje de primera velocidad inicia la multiplicación de par de torsión para impulsar el vehículo.

Fit The contact between two machined surfaces.

Conexión Contacto entre dos superficies maquinadas.

Fixed-type constant velocity joint A joint that cannot telescope or plunge to compensate for suspension travel. Fixed joints are always found on the outer ends of the drive shafts of FWD cars. A fixed joint may be of either Rzeppa or tripod type.

Junta de velocidad constante de tipo fijo Junta que no puede extenderse o hundirse para compensar el movimiento de la suspensión. Siempre se encuentran juntas fijas en los extremos exteriores de los árboles de mando de vehículos de tracción delantera. Una junta fija puede ser de tipo Rzeppa o trípode.

Flammability A statement of how well the substance supports combustion.

Inflamabilidad Enunciado que formula cuán bien sustancia soporta la combustión.

Flange A projecting rim or collar on an object for keeping it in place.

Brida Corona proyectado o collar en un objeto que lo mantiene en su lugar.

Flange yoke The part of the rear universal joint attached to the drive pinion.

Yugo de brida Pieza de la junta universal trasera fijada al piñón de mando.

Flexplate A lightweight flywheel used only on engines equipped with an automatic transmission. The flexplate is equipped with a starter ring gear around its outside diameter and also serves as the attachment point for the torque converter.

Placa flexible Volante liviano empleado solamente en motores equipados con transmisión automática. La placa flexible está equipada de una corona de arranque alrededor de su diámetro exterior y sirve también de punto de fijación para el convertidor del par motor.

Fluid coupling A device in the powertrain consisting of two rotating members; transmits power from the engine, through a fluid, to the transmission.

Acoplamiento fluido Mecanismo en el tren transmisor de potencia que consiste de dos piezas giratorias; transmite la fuerza desde el motor, por medio de un fluido, hasta la transmisión.

Fluid drive A drive in which there is no mechanical connection between the input and output shafts, and power is transmitted by moving oil.

Transmisión hidráulica Transmisión en la que no existe conexión mecánica alguna entre el árbol impulsor y el árbol de rendimiento. La fuerza se transmite a través del aceite motor.

Flywheel A heavy metal wheel that is attached to the crankshaft and rotates with it; helps smooth out the power surges from the engine power strokes; also serves as part of the clutch and engine-cranking system.

Volante Rueda pesada de metal que se fija al cigüeñal y que gira con él; ayuda a neutralizar las sacudidas de fuerza de las carreras motrices del motor; sirve también como parte del sistema de embrague y de arranque del motor.

Flywheel ring gear A gear, fitted around the flywheel, that is engaged by teeth on the starting motor drive to crank the engine.

Corona del volante Engranaje ajustado alrededor del volante, engranado por dientes en el mando del motor de arranque para hacer arrancar el motor.

Foot-pound (or ft. lb.) This is a measure of the amount of energy or work required to lift 1 pound a distance of 1 foot.

Libra-pie Medida de la cantidad de energía o fuerza que se requiere para levantar una libra a una distancia de un pie.

Force Any push or pull exerted on an object; measured in pounds and ounces, or in newtons (N) in the metric system.

Fuerza Cualquier empuje o tirón que se ejerce sobre un objeto; medido en libras y onzas, o en newtons (N) en el sistema métrico.

Forward coast side The side of the ring gear tooth the drive pinion contacts when the vehicle is decelerating.

Cara de cabotaje delantera Cara del diente de la corona con el cual el piñón de mando entra en contacto mientras el vehículo deacelera.

Forward drive side The side of the ring gear tooth that the drive pinion contacts when accelerating or on the drive.

Cara de mando delantero Cara del diente de la corona con el cual el piñón de mando entra en contacto mientras acelera o está en marcha.

Four wheel drive (4WD) On a vehicle, driving axles at both front and rear, so that all four wheels can be driven. 4WD is the standard abbreviation for four wheel drive.

Tracción a las cuatro ruedas En un vehículo, los ejes motores se encuentran ubicados en las partes delantera y trasera, para que las cuatro ruedas se puedan accionar. 4WD es la abreviatura común para tracción a las cuatro ruedas.

Four wheel high A transfer case shift position where both front and rear drive shafts receive power and rotate at the speed of the transmission output shaft.

Alto de cuatro ruedas Posición en la caja de cambios donde los árboles de mando delantero y trasero reciben fuerza y giran a la mimsa velocidad que el árbol de rendimiento de la transmisión.

Frame The main understructure of the vehicle to which everything else is attached. Most FWD cars have only a subframe for the front suspension and drive train. The body serves as the frame for the rear suspension.

Armazón Chasis principal del vehículo al cual se fijan todas las demás piezas. La mayoría de los vehículos de tracción delantera sólo tiene un chasis que soporta la suspensión delantera y el tren de mando. La carrocería sirve como armazón para la suspensión trasera.

Free-running gears Gears that rotate independently on their shaft.

Engranajes de funcionamiento libre Engranajes que giran de manera independiente en su árbol.

Free-wheeling clutch A mechanical device that will engage the driving member to impart motion to a driven member in one direction but not the other. Also known as an "overrunning clutch."

Embrague de marcha en rueda libre Mecanismo que engranará a la pieza de accionamiento para impulsar movimiento a una pieza accionada en una dirección pero no en la otra. Conocido también como "embrague de giro libre".

Friction The resistance to motion between two bodies in contact with each other.

Fricción Resistencia al movimiento entre dos cuerpos en contacto el uno con el otro.

Friction bearing A bearing in which there is sliding contact between the moving surfaces. Sleeve bearings, such as those used in connecting rods, are friction bearings.

Cojinete de fricción Cojinete en el cual existe un contacto deslizante entre las superficies en movimiento. Los cojinetes de manguito, como los que se utilizan en las bielas, son cojinetes de fricción.

Friction disc In the clutch a flat disc, faced on both sides with friction material and splined to the clutch shaft. It is positioned between the clutch pressure plate and the engine flywheel. Also called the clutch disc or driven disc.

Disco de fricción Disco plano del embrague, revestido en las dos caras con material de fricción y ranurado al árbol de embrague. Está ubicado entre la placa de presión del embrague y el volante de la máquina. Llamado también disco de embrague o disco accionado.

Friction facings A hard-molded or woven asbestos or paper material that is riveted or bonded to the clutch driven disc.

Revestimiento de fricción Material de moldeado duro, de asbesto tejido o de papel que es remachado o adherido al disco accionado del embrague.

Front bearing retainer An iron or aluminum circular casting fastened to the front of a transmission housing to retain the front transmission bearing assembly.

Retenedor del cojinete de rueda delantero Pieza circular fundida en hierro o en aluminio fijada a la parte delantera de un alojamiento de la transmisión para sujetar el conjunto del cojinete de transmisión delantero.

Front differential/axle assembly Like a conventional rear axle but having steerable wheels.

Conjunto de diferencial/eje delantero Igual que el eje trasero convencional, pero con ruedas orientables.

Front wheel drive (FWD) The vehicle has all drive train components located at the front.

Tracción delantera Todos los componentes del tren de mando en el vehículo se encuentran en la parte delantera.

Fulcrum The support that provides a pivoting point for a lever.

Fulcro Soporte que le provee punto de apoyo a una palanca.

Fulcrum rings A circular ring over which the pressure plate diaphragm spring pivots.

Anillos de fulcro Anillo circular sobre el cual gira el muelle del diafragma de la placa de presión.

Full-floating rear axle An axle that only transmits driving force to the rear wheels. The weight of the vehicle (including payload) is supported by the axle housing.

Eje trasero enteramente flotante Eje que solamente transmite la fuerza motriz a las ruedas traseras. El puente trasero soporta el peso del vehículo (incluyendo la carga útil).

Full-time 4WD Systems that use a center differential that accommodates speed differences between the two axles; necessary for on-highway operation.

Tracción a las cuatro ruedas a tiempo completo Sistemas que utilizan un diferencial central que instala, acomoda diferentes velocidades entre los dos ejes; necesaria para el funcionamiento en la carretera.

Fully synchronized transmission A transmission in which all of its forward gears are equipped with a synchronizer assembly. In a manual transmission, the synchronizer assembly operates to improve the shift quality in all forward gears.

Transmisión enteramente sincronizada Transmisión en la que todos los engranajes delanteros han sido equipados con un conjunto sincronizador. En una transmisión manual, el conjunto sincronizador funciona para mejorar la calidad del desplazamiento en todos los engranajes delanteros.

Galling Wear caused by metal-to-metal contact in the absence of adequate lubrication. Metal is transferred from one surface to the other, leaving behind a pitted or scaled appearance.

Corrosión por rozamiento Desgaste causado por el contacto de un metal con otro metal debido a la ausencia de lubrificación adecuada. Se transfiere el metal de una superficie a la otra, lo cual deja un aspecto corroído o raspado.

Gasket A layer of material, usually made of cork, paper, plastic, composition, or metal, or a combination of these, placed between two parts to make a tight seal.

Guarnición Capa de un material, normalmente hecho de corcho, papel, plástico, pasta o metal, o una combinación de éstos, ubicada entre dos piezas para crear una junta de estanqueidad hermética.

Gasket cement A liquid adhesive material, or sealer, used to install gaskets.

Cemento de guarnición Material líquido adhesivo, o de juntura, utilizado para instalar guarniciones.

Gear A wheel with external or internal teeth that serves to transmit or change motion.

Engranaje Rueda con dientes externos o internos que sirve para transmitir o cambiar movimiento.

Gear lubricant A type of grease or oil blended especially to lubricate gears.

Lubrificante de engranaje Tipo de grasa o aceite mezclado especialmente para lubrificar engranajes.

Gear ratio The number of revolutions of a driving gear required to turn a driven gear through one complete revolution. For a pair of gears, the ratio is found by dividing the number of teeth on the driven gear by the number of teeth on the driving gear.

Relación de engranajes Número de revoluciones de un mecanismo de accionamiento requiridas para hacer girar un engranaje accionado una revolución completa. Para un par de engranajes, se obtiene la relación al dividir el número de dientes en el engranaje accionado por el número de dientes en el mecanismo de accionamiento.

Gear reduction When a small gear drives a large gear, there is an output speed reduction and a torque increase that results in a gear reduction.

Reducción de engranajes Cuando un engranaje pequeño acciona un engranaje grande, se produce una reducción de velocidad de rendimiento y un aumento de par de torsión que resulta en una reducción de engranajes.

Gear whine A high-pitched sound developed by some types of meshing gears.

Silbido del engranaje Sonido agudo producido por algunos tipos de engranajes.

Gearboxes A slang term for transmissions.

Cajas de engranajes Coloquialismo que significa transmisiones.

Gearshift A linkage-type mechanism by which the gears in an automobile transmission are engaged and disengaged.

Cambio de velocidades Mecanismo de tipo empalme a través del cual los engranajes en la transmisión de un vehículo se engranan y se desengranan.

Graphite Very fine carbon dust with a slippery texture used as a lubricant.

Grafito Polvo muy fino de carbón con una textura desbaladiza que se utiliza como lubrificante.

Grind To finish or polish a surface by means of an abrasive wheel.

Esmerilar Acabar o pulir una superficie con una rueda abrasiva.

Half-shaft Either of the two drive shafts that connect the transaxle to the wheel hubs in FWD cars. Half-shafts have constant velocity joints attached to each end to allow for suspension motions and steering. The shafts may be of solid or tubular steel and may be of different lengths.

Semieje Cualquiera de los dos árboles de mando que conecta el transeje a los cubos de rueda en automóviles de tracción delantera. Los semiejes tienen juntas de velocidad constante fijadas a cada extremo para permitir el movimiento de suspensión y la dirección. Los semiejes pueden ser de acero sólido o tubular y sus longitudes pueden variar.

Harshness A bumpy ride caused by a stiff suspension. Can be cured by installing softer springs or shock absorbers.

Aspereza Viaje de muchas sacudidas ocacionadas por una suspensión rígida. Puede remediarse con la instalación de muelles más flexbiles o amortigadores.

Heat sink A piece of material that absorbs heat to prevent the heat from settling on another component.

Fuente fría Pieza hecha de un material que absorbe el calor para impedir que el calor penetre otro componente.

Heat treatment Heating, followed by fast cooling, to harden metal.

Tratamiento térmico Calentamiento, seguido del enfriamiento rápido, para endurecer el metal.

Heel The outside, larger half of the gear tooth.

Talón Mitad exterior más grande del diente de engranaje.

Helical Shapes like a coil spring or a screw thread.

Helicoidal Formas parecidas a un muelle helicoidal o a un filete de tornillo.

Helical gear Gears with the teeth cut at an angle to the axis of the gear.

Engranaje helicoidal Engranajes que tienen los dientes cortados a un ángulo del pivote del engranaje.

Herringbone gear A pair of helical gears designed to operate together. The angle of the pair of gears forms a V.

Engranaje bihelocoidal Par de engranajes helicoidales diseñados para funcionar juntos. El ángulo del par de engranajes forma una V.

High gears Third, fourth, and fifth gears in a typical transmission.

Engranajes de alta multiplicación Engranajes de tercera, cuarta y quinta velocidad en una transmisión típica.

Horsepower A measure of mechanical power, or the rate at which work is done. One horsepower equals 33,000 ft.-lb. (foot-pounds) of work per minute. It is the power necessary to raise 33,000 pounds a distance of 1 foot in 1 minute.

Potencia en caballos Medida de fuerza mecánica, o velocidad a la que se realiza el trabajo. Un caballo de fuerza es equivalente a 33.000 libras-pies de trabajo por minuto. Es el esfuerzo necesario para levantar 33.000 libras a la distancia de un pie en un minuto.

Hotchkiss drive A type of rear suspension in which leaf springs absorb the rear axle housing torque.

Transmisión Hotchkiss Tipo de suspensión trasera en la cual muelles de láminas absorben el par de torsión del puente trasero.

Hub The center part of a wheel, to which the wheel is attached.

Cubo Parte central de una rueda, a la cual se fija la rueda.

Hydraulic clutch A clutch that is actuated by hydraulic pressure; used in cars and trucks when the engine is some distance from the driver's compartment so that it would be difficult to use mechanical linkages.

Embrague hidráulico Embrague accionado por presión hidráulica; utilizado en automóviles y camiones cuando el motor está lejos del compartimiento del conductor para dificultar la utilización de bielas motrices mecánicas.

Hydraulic fluid reservoir A part of a master cylinder assembly that holds reserve fluid.

Despósito de fluido hidráulico Parte del conjunto del cilindro primario que contiene el fluido de reserva.

Hydraulic press A piece of shop equipment that develops a heavy force by use of a hydraulic piston-and-jack assembly.

Prensa hidráulica Pieza del equipo de taller que desarrolla fuerza pesada por medio de un conjunto de gato de pistón hidráulico.

Hydraulic pressure Pressure exerted through the medium of a liquid.

Presión hidráulica Presión ejercida a través de un líquido.

Hydrocarbon A substance composed of hydrogen and carbon molecules.

Hidrocarburo Substancia compuesta de moléculas de carbono e hidrógeno.

Hypoid gear A gear that is similar in appearance to a spiral bevel gear, but the teeth are cut so that the gears match in a position where the shaft center lines do not meet; cut in a spiral form to allow the pinion to be set below the center line of the ring gear so that the car floor can be lower.

Engranaje hipoide Engranaje parecido a un engranaje cónico con dentado espiral, pero en el cual los dientes se cortan para que los engranajes se engranen en una posición donde las líneas centrales del árbol no se crucen; cortado en forma de espiral para permitir que el piñon sea colocado debajo de la línea central de la corona y que así el piso del vehículo sea más bajo.

Hypoid gear lubricant An extreme pressure lubricant designed for the severe operation of hypoid gears.
Lubrificante del engranaje hipoide Lubrificante de extrema presión diseñado para el funcionamiento riguroso de los engranajes hipoides.

ID Inside diameter.
ID Diámetro interior.

Idle Engine speed when the accelerator pedal is fully released and there is no load on the engine.
Marcha mínima Velocidad del motor cuando el pedal del acelerador se suelta completamente y no hay ninguna carga en el motor.

Ignitability A statement of how easily a substance can catch fire.
Inflamabilidad Afirmación sobre la facilidad con la cual una sustancia puede encenderse.

Impeller The pump or driving member in a torque converter.
Impulsor Bomba o miembro de accionamiento en un convertidor de torsión.

Inboard constant velocity joint The inner constant velocity joint, or the one closest to the transaxle. The inboard joint is usually a plunging-type joint that telescopes to compensate for suspension motions.
Junta de velocidad constante del interior Junta de velocidad constante interior, o la que está más cerca del transeje. La junta del interior es normalmente una junta de tipo sumergible que se extiende para compensar el movimiento de la suspensión.

Inclinometer Device designed with a spirit level and graduated scale to measure the inclination of a drive line assembly. The inclinometer connects to the drive shaft magnetically.
Inclinómetro Instrumento diseñado con nivel de burbuja de aire y escala graduada para medir la inclinación de un conjunto de la línea de transmisión. El inclinómetro se conecta magnéticamente al árbol de mando.

Increments Series of regular additions from small to large.
Incrementos Serie de aumentos regulares de lo pequeño a lo grande.

Independent rear suspension (IRS) The vehicle's rear wheels move up and down independently of each other.
Suspensión trasera independiente Las ruedas traseras del vehículo realizan un movimiento de ascenso y descenso de manera independiente la una de la otra.

Independent suspension A suspension system that allows one wheel to move up and down without affecting the opposite wheel. Provides superior handling and a smoother ride.
Suspensión independiente Sistema de suspensión que permite que una rueda realice un movimiento de ascenso y descenso sin afectar la rueda opuesta. Permite un manejo excelente y un viaje mucho más cómodo.

Index To orient two parts by marking them. During reassembly the parts are arranged so the index marks are next to each other. Used to preserve the orientation between balanced parts.
Alinear Orientar dos piezas marcándolas. Durante el remonte, se arreglan las piezas de modo que las indicaciones de alineación queden juntas. Utilizado para mantener la orientación entre piezas equilibradas.

Inner bearing race The inner part of a bearing assembly on which the rolling elements, ball or roller, rotate.
Anillo de cojinete interior Parte interior de un conjunto de cojinete sobre la cual los elementos rodantes, o sea, las bolas o los rodillos, giran.

Input shaft The shaft carrying the driving gear by which the power is applied, as to the transmission.
Árbol impulsor Árbol que soporta el mecanismo de accionamiento a través del cual se aplica la fuerza motriz; por ejemplo, a la transmisión.

Inserts One of several terms that could apply to the shift plates found in a synchronizer assembly.
Piezas insertas Uno de los varios téminos que puede aplicarse a las placas de cambio de velocidades encontradas en un conjunto sincronizador.

Insert springs Round wire springs that hold the inserts or shift plates in contact with the synchronizer sleeve. Located around the synchronizer hub.
Muelles insertos Muelles redondos de alambre que mantienen el contacto entre las piezas insertas o placas de cambio de velocidades y el manguito sincronizador. Ubicados alrededor del cubo sincronizador.

Inspection cover A removable cover that permits entrance for inspection and service work.
Cubierta de inspección Cubierta desmontable que permite la entrada para la inspección y la reparación.

Integral Built into, as part of the whole.
Integral Pieza incorporada, como parte del todo.

Integral axle housing A rear axle housing-type where the parts are serviced through an inspection cover and adjusted within and relative to the axle housing.
Puente trasero integral Tipo de puente trasero, donde se reparan las piezas a través de una cubierta de inspección y se ajustan dentro del y con relación al puente trasero.

Integrated full-time 4WD Systems that use computer controls to enhance full-time operation, adjusting the torque split depending on which wheels have traction.
Tracción a las cuatro ruedas a tiempo completo integral Sistemas que utilizan controles de computadoras para mejorar el funcionamiento a tiempo completo, ajustando el reparto de torsión dependiendo de cuales de las ruedas tienen tracción.

Interaxle differential The center differential in some 4WD systems.
Diferencial entre ejes Diferencial central en algunos sistemas de tracción a las cuatro ruedas.

Interlock mechanism A mechanism in the transmission shift linkage that prevents the selection of two gears at one time.
Mecanismo de enganche Mecanismo en la biela motriz del cambio de velocidades de la transmisión que impide la selección de dos velocidades a la vez.

Intermediate drive shaft Located between the left and right drive shafts, it equalizes drive shaft length.
Árbol de mando intermedio Ubicado entre los árboles de mando izquierdo y derecho, compensa la longitud del árbol de mando.

Intermediate plate A mechanism in the transmission shift linkage that prevents the selection of two gears at one time.
Placa intermedia Mecanismo en la biela motriz del cambio de velocidades de la transmisión que impide la selección de dos velocidades a la vez.

Internal gear A gear with teeth pointing inward, toward the hollow center of the gear.
Engranaje interno Engranaje con los dientes orientados hacia adentro, hacia el centro hueco del engranaje.

IRS A common abbreviation for independent rear suspension.
IRS Abreviatura común para suspensión trasera independiente.

Jam nut A second nut tightened against a primary nut to prevent it from working loose. Used on inner and outer tie-rod adjustment nuts and on many pinion bearing adjustment nuts.

Contratuerca Segunda tuerca apretada contra una tuerca principal para evitar que ésta se suelte. Utilizada en las tuercas de las barras de acoplamiento interiores y exteriores y en muchas tuercas de ajuste del cojinete del piñón.

Jitter cycle A common term for a duty cycle.
Ciclo jitter Término común para ciclo de trabajo.

Joint angle The angle formed by the input and output shafts of constant velocity joints. Outer joints can typically operate at angles up to 45 degrees, whereas inner joints have more restricted angles.
Ángulo de las juntas Ángulo formado por el árbol impulsor y el árbol de rendimiento de las juntas de velocidad constante. Las juntas exteriores pueden funcionar típicamente en ángulos de hasta 45 grados, mientras que las juntas interiores funcionan dentro de ángulos más limitados.

Jounce The up and down movement of the car in response to road surfaces.
Sacudida Movimiento de ascenso y descenso del vehículo debido a las superficies de la carretera.

Journal The area on a shaft that rides on the bearing.
Gorrón La sección en un árbol que se mueve sobre el cojinete.

Key A small block inserted between the shaft and hub to prevent circumferential movement.
Chaveta Pasador pequeño insertado entre el árbol y el cubo para impedir el movimiento circunferencial.

Keyway A groove or slot cut to permit the insertion of a key.
Chavetero Ranura o hendidura cortada para que pueda insertarse una chaveta.

Knock A heavy metallic sound usually caused by a loose or worn bearing.
Golpeteo Sondio metálico pesado normalmente causado por un cojinete suelto o desgastado.

Knuckle The part of the suspension that supports the wheel hub and serves as the steering pivot. The bottom of the knuckle is attached to the lower control arm with a ball joint, and the upper portion is usually bolted to the strut.
Muñón Parte de la suspensión que apoya al cubo de la rueda y que sirve como pivote de dirección. La parte inferior del muñón se fija al brazo de mando inferior por medio de una junta esférica, y la superior normalmente se emperna al montante.

Knurl To indent or roughen a finished surface.
Estriar Endentar o poner áspera una superficie acabada.

Lapping The process of fitting one surface to another by rubbing them together with an abrasive material between the two surfaces.
Pulido Proceso de ajustar una superficie contra otra rozando la una contra la otra con un material abrasivo colocado entre las dos superficies.

Lash The amount of free motion in a gear train, between gears, or in a mechanical assembly, such as the lash in a valve train.
Juego Cantidad de movimiento libre en un tren de engranajes, entre engranajes o en un conjunto mecánico, como por ejemplo, el juego en un tren de válvulas.

Leaf spring A spring made up of a single flat steel plate or of several plates of graduated lengths assembled one on top of another; used on vehicles to absorb road shocks by bending or flexing.
Muelle de láminas Muelle compuesto de una sola placa de acero plana o de varias placas de longitudes graduadas montadas la una sobre la otra; utilizado en vehículos para absorber la aspereza de la carretera a través de la flexión.

Limited-slip differential A differential designed so that when one wheel is slipping, a major portion of the drive torque is supplied to the wheel with the better traction; also called a nonslip differential.
Diferencial de deslizamiento limitado Diferencial diseñado para que cuando una rueda se deslice, una mayor parte del par de torsión de mando llegue a la rueda que tiene mejor tracción. Llamado también diferencial antideslizante.

Linkage Any series of rods, yokes, and levers, etc., used to transmit motion from one unit to another.
Biela motriz Cualquier serie de varillas, yugos y palancas, etc., utilizada para transmitir movimiento de una unidad a otra.

Live axle A shaft that transmits power from the differential to the wheels.
Eje motor Árbol que transmite fuerza motriz del diferencial a las ruedas.

Lock pin Used in some ball sockets (inner tie-rod end) to keep the connecting nuts from working loose. Also used on some lower ball joints to hold the tapered stud in the steering knuckle.
Pasador de cierre Utilizado en algunas rótulas para bolas (extremo interior de la barra de acoplamaiento) para que las tuercas de conexión no se suelten. Utilizado también en algunas juntas esféricas inferiores para mantener el espárrago cónico en el muñón de dirección.

Locked differential A differential with the side and pinion gears locked together.
Diferencial trabado Diferencial en el que el engranaje lateral y el de piñón están sujetos entre sí.

Locknut A second nut turned down on a holding nut to prevent loosening.
Contratuerca Una segunda tuerca montada boca abajo sobre una tuerca de ensamble para evitar que se suelte.

Lockplates Metal tabs bent around nuts or bolt heads.
Placa de cierre Orejetas metálicas dobladas alrededor de tuercas o cabezas de perno.

Lockwasher A type of washer that when placed under the head of a bolt or nut prevents the bolt or nut from working loose.
Arandela de muelle Tipo de arandela que cuando se coloca debajo de la cabeza de un perno o de una tuerca, evita que el perno o la tuerca se suelte.

Low gears First and second gears in a typical transmission.
Engranajes de baja velocidad En una transmisión típica, los engranajes de primera y segunda velocidad.

Low speed The gearing that produces the highest torque and lowest speed of the wheels.
Baja velocidad Engranajes que producen el par de torsión más alto y la velocidad más baja de las ruedas.

Lubricant Any material, usually a petroleum product such as grease or oil, that is placed between two moving parts to reduce friction.
Lubrificante Cualquier material, normalmente un derivado de petróleo, como la grasa o el aceite, que se coloca entre dos piezas móviles para disminuir la fricción.

Lug nut The nuts that fasten the wheels to the axle hub or brake rotor. Missing lug nuts should always be replaced. Overtightening can cause warpage of the brake rotor in some cases.
Tuerca de orejetas Tuercas que sujetan las ruedas al cubo del eje o el rotor de freno. Se deben reemplazar siempre las tuercas de orejetas que se hayan perdido. En algunos casos el apretar demasiado puede causar el torcimiento del rotor de freno.

Master cylinder The liquid-filled cylinder in the hydraulic brake system or clutch, where hydraulic pressure is developed when the driver depresses a foot pedal.
Cilindro primario Cilindro lleno de líquido en el sistema de freno hidráulico o en el embrague, donde se produce presión hidráulica cuando el conductor oprime el pedal.

Matched gear set code Identification marks on two gears that indicate they are matched. They should not be mismatched with another gear set and placed into operation.

Código del juego de engranaje emparejado Señales de identificación en dos engranajes que indican que ambos hacen pareja. No deben emparejarse con otro juego de engranaje y ponerse en funcionamiento.

Meshing The mating, or engaging, of the teeth of two gears.
Engranar Emparejar, o endentar, los dientes de dos engranajes.

Micrometer A precision measuring device used to measure small bores, diameters, and thicknesses. Also called a mike.
Micrómetro Instrumento de precisión utilizado para medir calibres, espesores y diámetros pequeños. Llamado también mic.

Misalignment When bearings are not on the same center line.
Mal alineamiento Cuando los cojinetes no se encuentran en la misma línea central.

Mounts Made of rubber to insulate vibrations and noise while they support a powertrain part, such as engine or transmission mounts.
Monturas Hechas de caucho para aislar vibraciones y ruido mientras apoyan una pieza del tren transmisor de potencia, como por ejemplo, las monturas del motor o de la transmisión.

Multiple disc A clutch with a number of driving and driven discs as compared to a single plate clutch.
Disco múltiple Embrague con muchos discos de accionamiento y accionados, comparado con un embrague de placa simple.

Needle bearing An antifriction bearing using a great number of long, small diameter rollers. Also known as a quill bearing.
Cojinete de agujas Cojinete de antifricción que utiliza una gran cantidad de rodillos largos con diámetros pequeños. Llamado también cojinete de manguito.

Needle deflection Distance of travel from zero of the needle on a dial gauge.
Desviación de aguja Distancia a la que viaja la aguja desde el cero en un indicador de cuadrante.

Neoprene A synthetic rubber not affected by the various chemicals that are harmful to natural rubber.
Neopreno Caucho sintético que no es afectado por las distintas sustancias químicas nocivas para el caucho natural.

Neutral In a transmission, the setting in which all gears are disengaged and the output shaft is disconnected from the drive wheels.
Neutral En una transmisión, la regulación a la cual se desengranan todos los engranajes y se desconecta el árbol de rendimiento de las ruedas motrices.

Neutral start switch A switch wired into the ignition switch to prevent engine cranking unless the transmission shift lever is in neutral or the clutch pedal is depressed.
Interruptor de encendido neutral Interruptor conectado al botón de encendido para impedir el arranque del motor a menos que la palanca de cambios esté en neutro o se oprima el pedal de embrague.

Newton-Meter (N•m) Metric measurement of torque or twisting force equal to foot-pounds multiplied by 1.355.
Metro-Newton (N•m) Medida métrica de par de torsión o fuerza de torsión equivalente a libras-pies multiplicado por 1,355.

Nominal shim A shim with a designated thickness.
Laminillas nominales Laminillas de un espesor específico.

Nonhardening A gasket sealer that never hardens.
Antiendurecedor Junta de estanqueidad de la guarnición que nunca se endurece.

Nut A removable fastener used with a bolt to lock pieces together; made by threading a hole through the center of a piece of metal that has been shaped to a standard size.
Tuerca Aparato de fijación desmontable utilizado con un perno para que las piezas queden sujetas entre sí; se hace abriendo un hueco en el centro de una pieza de metal conformada a un tamaño estándar.

O-ring A type of sealing ring, usually made of rubber or a rubberlike material. In use, the O-ring is compressed into a groove to provide the sealing action.
Anillo-O Tipo de anillo de estanqueidad, normalmente hecho de caucho o de un material parecido al caucho. Cuando se le utiliza, el anillo-O se comprime en una ranura para proveer estanqueidad.

OD Outside diameter.
OD Diámetro exterior.

Oil seal A seal placed around a rotating shaft or other moving part to prevent leakage of oil.
Junta de aceite Junta de estanqueidad colocada alrededor de un árbol giratorio u otra pieza móvil para evitar fugas de aceite.

On-demand 4WD Systems that power a second axle only after the first begins to slip.
Tracción a las cuatro ruedas por demanda Sistemas que proveen fuerza motriz a un segundo eje solamente cuando el primero comienza a patinar.

One-way clutch See sprag clutch.
Embrague de una sola dirección Véase embrague de horquilla.

Open differential A standard-type differential.
Diferencial abierto Diferencial de tipo estándar.

Operating angle The difference between the drive shaft and transmission installation angles is the operating angle.
Ángulo de funcionamiento El ángulo de funcionamiento es la diferencia entre los ángulos del montaje del árbol de mando y de la transmisión.

Out-of-round Wear of a round hole or shaft that when viewed from an end will appear egg shaped.
Con defecto de circularidad Desgaste de un agujero o árbol redondo que, vistos desde uno de los extremos, parecen ovalados.

Outboard constant velocity joint The outer constant velocity joint, or the one closest to the wheels. The outer joint is a fixed joint.
Junta de velocidad constante fuera de borde Junta de velocidad constante exterior, o la que está más cerca de las ruedas. La junta exterior es una junta fija.

Outer bearing race The outer part of a bearing assembly on which the balls or rollers rotate.
Anillo de cojinete exterior Parte exterior de un conjunto de cojinete sobre la cual giran las bolas o los rodillos.

Output shaft The shaft or gear that delivers the power from a device, such as a transmission.
Árbol de rendimiento Árbol o engranje que transmite la fuerza motriz desde un mecanismo, como por ejemplo, la transmisión.

Overcenter spring A heavy coil spring arrangement in the clutch linkage to assist the driver with disengaging the clutch and returning the clutch linkage to the full engagement position.
Muelle sobrecentro Distribución de muelles helicoidales gruesos en la biela motriz del embrague para ayudar al conductor a desengranar el embrague y devolver la biela motriz del embrague a la posición de enganche total.

Overall ratio The product of the transmission gear ratio multiplied by the final drive or rear axle ratio.
Relación total Producto de la relación del engranaje de la transmisión multiplicado por la relación de la transmisión final o del eje trasero.

Overdrive A gear ratio whereas the output shaft of the transmission rotates faster than the input shaft. Any arrangement of gearing that produces more revolutions of the driven shaft than of the driving shaft.
Sobremultiplicación Relación de engranajes donde el árbol de rendimiento de la transmisión gira de manera más rápida que el árbol impulsor. Cualquier distribución de engranajes que produce más revoluciones del árbol accionado que del árbol de accionamiento.

Overdrive ratio Identified by the decimal point indicating less than one driving input revolution compared to one output revolution of a shaft.

Relación de sobremultiplicación Identificada por el punto decimal, lo que indica menos de una revolución impulsora de mando comparada con una revolución de rendimiento de un árbol.

Overrun coupling A free-wheeling device to permit rotation in one direction but not in the other.

Acoplamiento de giro libre Mecanismo de marcha en rueda libre que permite que se lleve a cabo la rotación en una dirección pero no en la otra.

Overrunning clutch A device consisting of a shaft or housing linked together by rollers or sprags operating between movable and fixed races. As the shaft rotates, the rollers or sprags jam between the movable and fixed races. This jamming action locks together the shaft and housing. If the fixed race should be driven at a speed greater than the movable race, the rollers or sprags will disconnect the shaft.

Embrague de rueda libre Mecanismo que consiste de un árbol o un alojamiento conectados por rodillos u horquillas que funcionan entre anillos móviles o fijos. Mientras el árbol gira, los rodillos o las horquillas se acuñan entre los anillos móviles y los anillos fijos. Este acuñamiento sujeta al árbol y al alojamiento entre sí. Si el anillo fijo debe accionarse a una velocidad más alta que el anillo móvil, los rodillos o las horquillas desconectarán el árbol.

Oxidation Burning or combustion; the combining of a material with oxygen. Rusting is slow oxidation, and combustion is rapid oxidation.

Oxidación Quema o combustión; la combinación de oxígeno con otro elemento. La corrosión es una oxidación lenta, mientras que y la combustión es una oxidación rápida.

Parallel The quality of two items being the same distance from each other at all points; usually applied to lines and, in automotive work, to machined surfaces.

Paralelo Calidad de dos objetos que se encuentran a la misma distancia el uno del otro en todos los puntos; normalmente se aplica a líneas y, en la reparación de automóviles, a superficies maquinadas.

Part-time 4WD Systems that can be shifted in and out of 4WD.

Tracción a las cuatro ruedas a tiempo parcial Sistemas que pueden cambiarse a o de tracción a las cuatro ruedas según sea necesario.

Pawl A lever that pivots on a shaft. When lifted it swings freely and when lowered it locates in a detent or notch to hold a mechanism stationary.

Trinquete Palanca que gira sobre un árbol. Cuando se la levanta, se mueve libremente y cuando se la baja, se acuña en un retén o en una muesca para bloquear el movimiento de un mecanismo.

Pedal play The distance the clutch pedal and release bearing assembly move from the fully engaged position to the point where the release bearing contacts the pressure plate release levers.

Holgura del pedal Distancia a la que el pedal del embrague y el conjunto del cojinete de desembrague se mueven de una posición enteramente engranada a un punto donde el cojinete de desembrague entra en contacto con las palancas de desembrague de la placa de presión.

Peen To stretch or clinch over by pounding with the rounded end of a hammer.

Granallar Estirar o remachar golpeando con el extremo redondo de un martillo.

Phasing Rotational position of the universal joints on the drive shaft.

Fasaje Posición de rotación de las juntas universales sobre el árbol de mando.

Pilot bearing A small bearing, such as in the center of the flywheel end of the crankshaft, which carries the forward end of the clutch shaft.

Cojinete piloto Cojinete pequeño, como por ejemplo, el ubicado en el centro del extremo del volante del cigüeñal, que soporta el extremo delantero del árbol del embrague.

Pilot bushing A plain bearing fitted in the end of a crankshaft. The primary purpose is to support the input shaft of the transmission.

Buje piloto Cojinete sencillo insertado en el extremo de un cigüeñal. El propósito principal es apoyar el árbol impulsor de la transmisión.

Pilot shaft A shaft used to align parts and that is removed before final installation of the parts; a dummy shaft.

Árbol piloto Árbol que se utiliza para alinear piezas y que se remueve antes del montaje final de las mismas; árbol falso.

Pinion gear The smaller of two meshing gears.

Engranaje de piñón El más pequeño de los dos engranajes de engrane.

Pinion carrier The mounting or bracket that retains the bearings supporting a pinion shaft.

Portador de piñón Montaje o soporte que sujeta los cojinetes que apoyan un árbol del piñón.

Pitch The number of threads per inch on any threaded part.

Paso Número de filetes de un tornillo por pulgada en cualquier pieza fileteada.

Pivot A pin or shaft upon which another part rests or turns.

Pivote Chaveta o árbol sobre el cual se apoya o gira otra pieza.

Planet carrier In a planetary gear system, the carrier or bracket in a planetary system that contains the shafts upon which the pinions or planet gears turn.

Portador planetario En un sistema de engranaje planetario, portador o soporte en un sistema planetario que contiene los árboles sobre los cuales giran los piñones o los engranajes planetarios.

Planet gears The gears in a planetary gear set that connect the sun gear to the ring gear.

Engranajes planetarios Engranajes en un tren de engranaje planetario que conectan el engranaje principal a la corona.

Planet pinions In a planetary gear system, the gears that mesh with, and revolve about, the sun gear; they also mesh with the ring gear.

Piñones planetarios En un sistema de engranaje planetario, los engranajes que se engranan con y giran entorno del engranaje principal; se engranan también con la corona.

Planetary gear set A system of gearing modeled after the solar system. A pinion is surrounded by an internal ring gear, and planet gears are in mesh between the ring gear and pinion around which all revolves.

Tren de engranaje planetario Sistema de engranaje inspirado en el sistema solar. Un piñón está rodeado por una corona interna, y los engranajes planetarios se engranan entre la corona y el piñón, alrededor de los cuales todo gira.

Plate loading Force developed by the pressure plate assembly to hold the driven disc against the flywheel.

Carga de placa Fuerza producida por el conjunto de la placa de presión para sujetar el disco accionado contra el volante.

Plunging action Telescoping action of an inner front-wheel-drive universal joint.

Acción sumergible Acción telescópica de una junta universal interior de tracción delantera.

Plunging constant velocity joint Usually the inner constant velocity joint. The joint is designed so that it can telescope slightly to compensate for suspension motions.

Junta de velocidad constante sumergible Normalmente junta de velocidad constante interior. La junta está diseñada para que pueda extenderse ligeramente y así compensar el movimiento de la suspensión.

Powertrain The mechanisms that carry the power from the engine crankshaft to the drive wheels; these include the clutch, transmission, drive line, differential, and axles.
Tren transmisor de potencia Mecansimos que transmiten la potencia desde el cigüeñal del motor hasta las ruedas motrices; éstos incluyen el embrague, la transmisión, la línea de transmisión, el diferencial y los ejes.

Preload A fixed amount of pressure constantly applied to a component. Preload on bearings eliminates looseness. Preload on limited-slip differential clutches also provides torque transfer to the driven wheel with the least traction.
Carga previa Cantidad fija de presión aplicada continuamente a un componente. La carga previa sobre los cojinetes elimina el juego. La carga previa sobre los embragues del diferencial de deslizamiento limitado provee también transferencia de par de torsión a la rueda accionada de menor tracción.

Press fit Forcing a part into an opening that is slightly smaller than the part itself to make a solid fit.
Ajuste en prensa Forzar una pieza dentro de una apertura un poco más pequeña que la pieza misma para lograr un ajuste sólido.

Pressure Force per unit area, or force divided by area. Usually measured in pounds per square inch (psi) or in kilopascals (kPa) in the metric system.
Presión Fuerza por unidad de área, o fuerza dividida por área. Normalmente se mide en libras por pulgada cuadrada (lpc) o en kilopascales (kPa) en el sistema métrico.

Pressure plate That part of the clutch that exerts force against the friction disc; it is mounted on and rotates with the flywheel. A heavy steel ring pressed against the clutch disc by spring pressure.
Placa de presión Pieza del embrague que ejerce fuerza contra el disco de fricción; se monta sobre y gira con el volante. Anillo pesado de acero, comprimido contra el disco de embrague mediante presión elástica.

Propeller shaft A common term for a drive shaft.
Árbol transmisor Término común para árbol de mando.

Prussian blue A blue pigment; in solution, useful in determining the area of contact between two surfaces.
Azul de Prusia Pigmento azul; en una solución, sirve para determinar el área de contacto entre dos superficies.

psi Abbreviation for pounds per square inch, a measurement of pressure.
psi Abreviatura de libras por pulgada cuadrada; una medida de presión.

Puller Generally, a shop tool used to separate two closely fitted parts without damage. Often contains a screw, or several screws, which can be turned to apply a gradual force.
Tirador Generalmente, herramienta de taller utilizada para separar dos piezas fuertemente apretadas sin averiarlas. A menudo contiene uno o varios tornillos a los que se les puede dar vuelta para aplicar una fuerza gradual.

Pulsation To move or beat with rhythmic impulses.
Pulsación Mover o golpear con impulsos rítmicos.

Quadrant A section of a gear. A term sometimes used to identify the shift-lever selector mounted on the steering column.
Cuadrante Sección de un engranaje. Término utilizado en algunas ocasiones para identificar el selector de la palanca de cambio de velocidades montado sobre la columna de dirección.

Quill shaft The hollow shaft on the front of the front bearing retainer.
Árbol de manguito Árbol hueco en la parte frontal del retenedor del cojinete delantero.

Race A channel in the inner or outer ring of an antifriction bearing in which the balls or rollers roll.
Anillo Canal en el anillo interior o exterior de un cojinete de antifricción en el cual giran las bolas o los rodillos.

Raceway A groove or track designed into the races of a bearing or universal joint housing to guide and control the action of the balls or trunnions.
Anillo de rodadura Ranura o canal construido en el interior de os anillos de un cojinete o de un alojamiento de junta universal para guiar y controlar el movimiento de las bolas o de las muñequillas.

Radial The direction moving straight out from the center of a circle. Perpendicular to the shaft or bearing bore.
Radial Dirección que sale directamente del centro de un círculo. Perpendicular al árbol o al calibre de cojinete.

Radial clearance (radial displacement) Clearance within the bearing and between balls and races perpendicular to the shaft.
Espacio libre radial (desplazamiento radial) Dentro del cojinete y entre las bolas y los anillos, espacio libre perpendicular al árbol.

Radial load A force perpendicular to the axis of rotation. Loads applied at 90 degrees to an axis of rotation.
Carga radial Fuerza perpendicular al pivote de rotación. Cargas aplicadas a un pivote de rotación a 90°.

Ratcheting mechanism Uses a pawl and gear arrangement to transmit motion or to lock a particular mechanism by having the pawl drop between gear teeth.
Mecanismo de trinquete Utiliza un conjunto de retén y engranaje para transmitir movimiento o para bloquear un mecanismo específico haciendo que el retén caiga entre los dientes del engranaje.

Ratio The relation or proportion that one number bears to another.
Relación Razón o proporción que existe entre un número y otro.

Ravigneaux Designer of a planetary gear system with small and large sun gears, long and short planetary pinions, planetary carriers and ring gear.
Ravigneaux Fabricante de un sistema de engranaje planetario que consiste de engranajes prinipales pequeños y grandes, piñones planetarios largos y cortos, portadores planetarios y corona.

Reactivity A statement of how easily a substance can cause or be part of a chemical reaction.
Reactividad Enunciado que expresa cuán fácil una substancia puede provocar o ser parte de una reacción química.

Reamer A round metal cutting tool with a series of sharp cutting edges; enlarges a hole when turned inside it.
Escariador Herramienta redonda metálica de corte que tiene una serie de aristas agudas; ensancha un agujero cuando se le da vuelta dentro de éste.

Rear axle torque The torque received and multiplied by the rear driving axle assembly.
Torsión del eje trasero Par de torsión recibido y multiplicado por el conjunto del eje motor trasero.

Rear wheel drive (RWD) A term associated with a vehicle where the engine is mounted at the front and the driving axle and driving wheels at the rear of the vehicle.
Tracción trasera Término relacionado a un vehículo donde el motor se monta en la parte delantera, y el eje motor y las ruedas motrices en la parte trasera.

Rebound The movement of the suspension system as it attempts to bring the car back to normal heights after jounce.
Rebote Movimiento que se obsrva en la suspensión mientras intenta estabilizar el funcionamiento normal del vehículo después de una sacudida.

Release bearing A ball-type bearing moved by the clutch pedal linkage to contact the pressure plate release levers to either engage or disengage the driven disc with the clutch driving members.

Cojinete de desembrague Cojinete de tipo bola accionado por la biela motriz del pedal del embrague para entrar en contacto con las palancas de desembrague de la placa de presión o para engranar o desengranar el disco accionado con los mecanismos de accionamiento del embrague.

Release levers In the clutch, levers that are moved by throwout-bearing movement, causing clutch spring force to be relieved so that the clutch is disengaged, or uncoupled from the flywheel.
Palancas de desembrague En el embrague, las palancas accionadas por el movimiento del cojinete de desembrague, que hacen disminuir la fuerza del muelle del embrague para que el embrague se desengrane, o se desacople del volante.

Release plate Plate designed to release the clutch pressure plate's loading on the clutch driven disc.
Placa de desembrague Placa diseñada para desembragar la carga de la placa de presión del embrague en el disco accionado del embrague.

Removable carrier housing A type of rear axle housing from which the axle carrier assembly can be removed for parts service and adjustment.
Alojamiento portador desmontable Tipo de puente trasero del cual se puede desmontar el conjunto del portador del eje para la reparación y el ajuste de las piezas.

Retaining ring A removable fastener used as a shoulder to retain and position a round bearing in a hole.
Anillo de retención Aparato fijador desmontable utilizado como punto de apoyo para sujetar y colocar un cojinete redondo en un agujero.

Retractor clips Spring steel clips that connect the diaphragm's flexing action to the pressure plate.
Grapas retractoras Grapas de acero para muelles que conectan el movimiento flexible del diafragma a la placa de presión.

Reverse idler gear In a transmission, an additional gear that must be meshed to obtain reverse gear; a gear used only in reverse that does not transmit power when the transmission is in any other position.
Piñón de marcha atrás En una transmisión, engranaje adicional que debe engranarse para obtener un engranaje de marcha atrás; engranaje utilizado solamente durante la inversión de marcha, que no transmite fuerza cuando la transmisión se encuentra en cualquier otra posición.

Ring gear A gear that surrounds or rings the sun and planet gears in a planetary system. Also the name given to the spiral bevel gear in a differential.
Corona Engranaje que rodea los engranajes planetario y el principal en un sistema planetario. También es el nombre que se le da al engranaje cónico con dentado espiral en un diferencial.

Rivet A headed pin used for uniting two or more pieces by passing the shank through a hole in each piece, and securing it by forming a head on the opposite end.
Remanche Chaveta de cabeza utilizada para unir dos o más piezas insertando la espinilla en cada una de las piezas a través de un agujero. La espinella se asegura formando una cabeza en el extremo opuesto.

Roller bearing An inner and outer race upon which hardened steel rollers operate.
Cojinete de rodillos Anillo interior y exterior sobre el cual funcionan rodillos de acero templado.

Rollers Round steel bearings that can be used as the locking element in an overrunning clutch or as the rolling element in an antifriction bearing.
Rodillos Cojinetes redondos de acero que pueden utilizarse como el elemento de bloqueo en un embrague de rueda libre o como el elemento rodante en un cojinete de antifricción.

RPM Abbreviation for revolutions per minute, a measure of rotational speed.
RPM Abreviatura de revoluciones por minuto, una medida de velocidad de rotación.

RTV sealer Room temperature vulcanizing gasket material, which cures at room temperature; a plastic paste squeezed from a tube to form a gasket of any shape.
Junta de estanqueidad VTA Material vulcanizador de guarnición a temperatura ambiente, que se conserva a temperatura ambiente; pasta plástica que viene en tubo, utilizada para formar una guarnición de cualquier tamaño.

Rubber coupling Rubber-based disc used as a universal joint between the driving and driven shafts.
Acoplamiento de caucho Disco con base de caucho; utilizado como junta universal entre el árbol de accionamiento y el árbol accionado.

Runout Deviation of the specified normal travel of an object. The amount of deviation or wobble a shaft or wheel has as it rotates. Runout is measured with a dial indicator.
Desviación Desalineación del movimiento normal indicado de un objeto. Cantidad de desalineación o bamboleo que tiene un árbol o una rueda mientras gira. La desviación se mide con un indicador de cuadrante.

Rzeppa constant velocity joint The name given to the ball-type constant velocity joint (as opposed to the tripod-type constant velocity joint). Rzeppa joints are usually the outer joints on most FWD cars. Named after its inventor, Alfred Rzeppa, a Ford engineer.
Junta de velocidad constante Rzeppa Nombre que se le da a la junta de velocidad constante de tipo bola (en contraste con la junta de velocidad constante de tipo trípode). Las juntas Rzeppa normalmente son las juntas exteriores en la mayoría de automóviles de tracción delantera. Nombrada por su creador, Alfred Rzeppa, ingeniero de la Ford.

SAE Society of Automotive Engineers.
SAE Sociedad de Ingenieros Automotrices.

Score A scratch, ridge, or groove marring a finished surface.
Muesca Rayado, rotura o ranura que estropea una superficie acabada.

Scuffing A type of wear in which there is a transfer of material between parts moving against each other; shows up as pits or grooves in the mating surfaces.
Frotamiento Tipo de desgaste en el cual se transfiere material entre piezas que se mueven la una contra la otra; aparece en forma de hendiduras o ranuras en las superifices emparejadas.

Seal A material, shaped around a shaft, used to close off the operating compartment of the shaft, preventing oil leakage.
Junta de estanqueidad Material conformado alrededor de un árbol, que se utiliza para sellar el compartimiento de funcionamiento del árbol y así evitar la fuga de aceite.

Sealer A thick, tacky compound, usually spread with a brush, which may be used as a gasket or sealant to seal small openings or surface irregularities.
Líquido de estanqueidad Compuesto grueso y viscoso, normalmente esparcido con una brocha, que puede emplearse como guarnición o compuesto obturador para rellenar pequeñas aperturas o irregularidades en la superficie.

Seat A surface, usually machined, upon which another part rests or seats; for example, the surface upon which a valve face rests.
Asiento Superficie, normalmente maquinada, sobre la cual se coloca o sienta otra pieza; por ejemplo, la superficie sobre la cual se coloca una cara de válvula.

Self-adjusting clutch linkage Monitors clutch pedal play through a clutch control cable and ratcheting mechanism to automatically adjust clutch pedal play.
Biela motriz del embrague de ajuste automático Controla el juego del pedal del embrague mediante un cable de mando del embrague y un mecanimso de trinquete para ajustar automáticamente el juego del pedal del embrague.

Semicentrifugal pressure plate The release levers of this pressure plate are weighted to take advantage of centrifugal force to increase plate loading resulting in reduced driven disc slip.
Placa de presión semicentrífuga A las palancas de desembrague de esta placa de presión se les añade peso para aprovechar la fuerza centrífuga y hacer que ésta aumente la carga de la placa. El resultado será un deslizamiento menor del disco accionado.

Semifloating rear axle An axle that supports the weight of the vehicle on the axle shaft in addition to transmitting driving forces to the rear wheels.
Eje trasero semi-flotante Eje que apoya el peso del vehículo sobre el árbol motor además de transmitir las fuerzas motrices a las ruedas traseras.

Separators A component in an antifriction bearing that keeps the rolling components apart.
Separadores Componente en un cojinete de antifricción que mantiene los componentes de rodamiento separados.

Shank The portion of the shoe that protects the ball of your foot.
Arqueo Parte del zapato que protege el hueso de la planta del pie.

Shift-on-the-fly 4WD A system that can be shifted from two to four wheel drive while the vehicle is moving.
Tracción a las cuatro ruedas shift-on-the-fly Sistema que puede cambiarse de marcha a tracción a las dos ruedas a marcha a tracción a las cuatro ruedas mientras el vehículo está en movimiento.

Shift forks Mechanisms attached to shift rails that fit into synchronizer hub for change of gears.
Horquillas de cambio de velocidades Las ranuras en el anillo sincronizador del embrague cónico deben ser afiladas para lograr la sincronización.

Shift lever The lever used to change gears in a transmission. Also, the lever on the starting motor that moves the drive pinion into or out of mesh with the flywheel teeth.
Palanca de cambio de velocidades Palanca utilizada para cambiar las velocidades en una transmisión. También es la palanca en el motor de arranque que engrana o desengrana el piñón de mando con los dientes del volante.

Shift rails Rods placed within the transmission housing that are a part of the transmission gearshift linkage.
Barras de cambio de velocidades Varillas ubicadas dentro del alojamiento de transmisión que forman parte de la biela motriz del cambio de velocidades de la transmisión.

Shifter A common term for the shift lever of a transmission.
Cambiador Término común para la palanca de cambio de velocidades de una transmisión.

Shim Thin sheets used as spacers between two parts, such as the two halves of a journal bearing.
Chapa de relleno Láminas delgadas utilizadas como espaciadores entre dos piezas, como por ejemplo, las dos mitades de un cojinete liso.

Shim stock Sheets of metal of accurately known thickness that can be cut into strips and used to measure or correct clearances.
Material de chapa de relleno Láminas de metal de espesor preciso que puede cortarse en tiras y utilizarse para medir el espacio libre correcto.

Side clearance The clearance between the sides of moving parts when the sides do not serve as load-carrying surfaces.
Despojo lateral Espacio libre entre los dos lados de piezas móviles cuando éstos no sirven como superficies de carga.

Side gears Gears that are meshed with the differential pinions and splined to the axle shafts (RWD) or drive shafts (FWD).
Engranajes laterales Engranajes que se engranan con los piñones del diferencial y son ranurados a los árboles motores en vehículos de tracción trasera o a los árboles de mando en vehículos de tracción delantera.

Side thrust Longitudinal movement of two gears.
Empuje lateral Movimiento longitudinal de dos engranajes.

Slave cylinder Located at a lower part of the clutch housing. Receives fluid pressure from the master cylinder to engage or disengage the clutch.
Cilindro secundario Ubicado en la parte inferior del alojamiento del embrague. Recibe presión de fluido del cilindro primario para engranar o desengranar el embrague.

Sliding fit Where sufficient clearance has been allowed between the shaft and journal to allow free-running without overheating.
Ajuste deslizante Donde se ha permitido espacio suficiente entre el árbol y el gorrón para permitir un funcionamiento libre sin ocasionar un recalentimiento.

Sliding yoke Slides on internal and external splines to compensate for drive line length changes.
Yugo deslizante Se desliza sobre las lengüetas internas y externas para compensar los cambios de longitud de la línea de transmisión.

Sliding gear transmission A transmission in which gears are moved on their shafts to change gear ratios.
Transmisión por engranaje desplazable Transmisión en la cual los engranajes se mueven sobre sus árboles para cambiar la relación de los engranajes.

Slip fit Running or sliding fit.
Ajuste corredizo Ajuste deslizante o de marcha.

Slip joint In the powertrain, a variable length connection that permits the drive shaft to change its effective length.
Junta corrediza En el tren transmisor de potencia, una conexión de longitud variable que le permite al árbol de mando cambiar su longitud eficaz.

Snap ring Split spring-type ring located in an internal or external groove to retain a part.
Anillo de resorte Anillo hendido de tipo muelle ubicado en una ranura interna o externa para sujetar una pieza en su lugar.

Solid axle A rear axle design that places the final drive, axles, bearings, and hubs into one housing.
Eje sólido Diseño del eje trasero que coloca la transmisión final, los ejes, los cojinetes y los cubos dentro de un solo alojamiento.

Spalling A condition where the material of a bearing surface breaks away from the base metal.
Esquirla Condición en la que el material de la superficie de un cojinete se separa del metal de base.

Speed gears Driven gears located on the transmission output shaft. This term differentiates between the gears of the counter gear and cluster assemblies and gears on the transmission output shaft.
Engranajes de velocidades Engranajes accionados ubicados en el árbol de rendimiento de la transmisión. Este término distingue entre los engranajes de los conjuntos del mecanismo contador y de los engranajes desplazables y los engranajes sobre el árbol de rendimiento de la transmisión.

Spindle The shaft on which the wheels and wheel bearings mount.
Huso Árbol sobre el cual se montan las ruedas y los cojinetes de rueda.

Spiral bevel gear A ring gear and pinion wherein the mating teeth are curved and placed at an angle with the pinion shaft.
Engranaje cónico con dentado espiral Corona y piñón cuyos dientes emparejados son curvos y están montados en ángulo cor árbol de piñón.

Spiral gear A gear with teeth cut according to a mathemati on a cone. Spiral bevel gears that are not parallel have that intersect.
Engranaje helocoidal Engranaje con dientes corte según una curva matemática. Los engranajes có espiral que no son paralelos tienen líneas centr

Spline Slot or groove cut in a shaft or bore; a splined shaft onto which a hub, wheel, gear, etc., with matching splines in its bore is assembled so that the two must turn together.
Lengüeta Hendidura o ranura excavada en un árbol o un calibre; árbol ranurado sobre el que se montan un cubo, una rueda, un engranaje, etc., con lengüetas que hacen pareja en su calibre, para que ambos giren juntos.

Splined hub Several keys placed radially around the inside diameter of a circular part, such as a wheel or driven disc.
Cubo ranurado Varias chavetas ubicadas de manera radial alrededor del diámetro interior de una pieza circular, como por ejemplo, una rueda o un disco accionado.

Split lip seal Typically a rope seal sometimes used to denote any two-part oil seal.
Junta de estanqueidad de reborde hendido Típicamente una junta de estanqueidad de cable utilizada en algunas ocasiones para denominar cualquier junta de aceite de dos partes.

Split pin A round split spring steel tubular pin used for locking purposes; for example, locking a gear to a shaft.
Pasador hendido Pasador tubular redondo hendido de acero para muelles utilizado para sujetar; por ejemplo, para asegurar un engranaje a un árbol.

Sprag clutch A member of the overrunning clutch family using a sprag to jam between the inner and outer races used for holding or driving action.
Embrague de horquilla Miembro de la familia del embrague de rueda libre que se sirve de una horquilla para insertarse entre los anillos interior y exterior utilizados para la retención o el funcionamiento.

Spring A device that changes shape when it is stretched or compressed, but returns to its original shape when the force is removed; the component of the automotive suspension system that absorbs road shocks by flexing and twisting.
Muelle Pieza que cambia de forma cuando se estira o comprime, pero que recobra su forma original cuando se detiene la fuerza; componente del sistema de suspensión automotriz que absorbe las sacudidas de la carretera doblándose y torciéndose.

Spring retainer A steel plate designed to hold a coil or several coil springs in place.
Retenedor de muelle Placa de acero diseñada para sujetar uno o varios muelles helicoidales en su lugar.

Spur gear Gears cut on a cylinder with teeth that are straight and parallel to the axis.
Engranaje recto Engranajes cortados en un cilindro que tienen dientes rectos y paralelos al pivote.

Squeak A high-pitched noise of short duration.
Rechinamiento Sonido agudo de corta duración.

Squeal A continuous high-pitched noise.
Chirrido Sonido agudo continuo.

Standard shift A common name for a manual transmission.
Cambio de velocidades estándar Nombre común para transmisión manual.

Stick-shift A common name for a manual transmission.
Palanca de marcha Nombre común para transmisión manual.

Stress The force to which a material, mechanism, or component is subjected.
Esfuerzo Fuerza a la que se somete un material, mecanismo o componente.

trut assembly Refers to all the strut components, including the strut tube, shock absorber, coil spring, and upper bearing assembly.
njunto de montante Se refiere a todas las piezas del montante, clusive al tubo de montante, al amortigador, al muelle helicoidal, y conjunto del cojinete superior.

Stub axle A common name for the spindle shaft of an axle.
Muñón corto Nombre común para el árbol huso de un eje.

Stub shaft A very short shaft.
Árbol corto Árbol sumamente corto.

Sun gear The central gear in a planetary gear system around which the rest of the gears rotate. The innermost gear of the planetary gear set.
Engranaje principal Engranaje central en un sistema de engranaje planetario alrededor del cual giran los demás engranajes. Es el engranaje más interior del tren de engranaje planetario.

Synchromesh transmission Transmission gearing that aids the meshing of two gears or shift collars by matching their speed before engaging them.
Transmisión de engranaje sincronizado Engranaje transmisor que facilita el engrane de dos engranajes o collares de cambio de velocidades al igualar la velocidad de éstos antes de engranarlos.

Synchronize To cause two events to occur at the same time; for example, to bring two gears to the same speed before they are meshed to prevent gear clash.
Sincronizar Hacer que dos sucesos ocurran al mismo tiempo; por ejemplo, hacer que dos engranajes giren a la misma velocidad antes de que se engranen para evitar el choque de engranajes.

Synchronizer assemblies Device that uses cone clutches to bring two parts rotating at two speeds to the same speed. A synchronizer assembly operates between two gears; first and second gear, third and fourth gear.
Conjuntos sincronizadores Mecansimo que utiliza embragues cónicos para hacer que dos piezas que giran a dos velocidades giren a una misma velocidad. Un conjunto sincronizador funciona entre dos engranajes; engranjes de primera y segunda velocidad, y engranajes de tercera y cuarta velocidad.

Synchronizer blocker ring Usually a brass ring that acts as a clutch and causes driving and driven units to turn at the same speed before final engagement.
Anillo de bloque sincronizador Normalmente un anillo de latón que sirve de embrague y hace que las piezas de accionamiento y las accionadas giren a la misma velocidad antes del acoplamiento final.

Synchronizer hub Center part of the synchronizer assembly that is splined to the synchronizer sleeve and transmission output shaft.
Cubo sincronizador Pieza central del conjunto sincronizador ranurada al manguito sincronizador y al árbol de rendimiento de la transmisión.

Synchronizer sleeve The sliding sleeve that fits over the complete synchronizer assembly.
Manguito sincronizador Manguito deslizante que cubre todo el conjunto sincronizador.

Tap To cut threads in a hole with a tapered, fluted, threaded tool.
Aterrajar Cortar filetes de tornillo en un agujero con una herramienta cónica, estriada y fileteada.

Temper To change the physical characteristics of a metal by applying heat.
Templar Cambiar las características físicas de un metal aplicándole calor.

Tension Effort that elongates or "stretches" a material.
Tensión Fuerza que alarga o estira un material.

Thickness gauge Strips of metal made to an exact thickness, used to measure clearances between parts.
Calibrador de espesor Tiras de metal hechas a un espesor extacto, utilizadas para medir el espacio libre entre las piezas.

Thread chaser A device, similar to a die, that is used to clean threads.
Fileteadora de tornillo Utensilio, parecido a un troquel, utilizado para limpiar tornillos.

Threaded insert A threaded coil that is used to restore the original thread size to a hole with damaged threads.
Piezas insertas fileteadas Espiral fileteado utilizado para devolver el tamaño original del tornillo a un agujero que tiene tornillos averiados.

Three-quarter floating axle The axle housing carries the weight of the vehicle while the bearings support the wheels on the outer ends of the axle housing tubes.
Eje flotante de tres cuartos Puente trasero que soporta el peso del vehículo mientras los cojinetes soportan las ruedas en los extremos exteriores de los tubos del puente trasero.

Throwout bearing In the clutch, the bearing that can be moved inward to the release levers by clutch-pedal action to cause declutching, which disengages the engine crankshaft from the transmission. A common name for a clutch release bearing.
Cojinete de desembrague En el embrague, cojinete que puede moverse hacia adentro hasta las palancas de desembrague, por medio de la acción del pedal del embrague, para lograr el desembrague. Esta acción desengrana el cigüeñal del motor de la transmisión. Nombre común para cojinete de desembrague del embrague.

Thrust load A load that pushes or reacts through the bearing in a direction parallel to the shaft.
Carga de empuje Carga que empuja o reacciona mediante el cojinete en una dirección paralela al árbol.

Thrust washer A washer designed to take up end thrust and prevent excessive endplay.
Arandela de empuje Arandela diseñada para asegurar el empuje longitudinal y prevenir un juego longitudinal excesivo.

Tie rod The linkage between the steering rack and the steering knuckle arm. The tie rod is threaded into a tie-rod end or has a threaded split member for making toe adjustments.
Barra de acoplamiento Biela mortiz entre la cremallera y el brazo del muñón de dirección. La barra de acoplamiento se filetea en un extremo de la barra de acoplamiento o tiene una pieza fileteada hendida para ajustar el tope.

Tie-rod end The fittings on the ends of the tie rods. The outer tie-rod end connects to the steering arm, and the inner one connects to the steering rack. Both ends include ball sockets to allow pivotal action, as well as up and down flexing.
Extremo de la barra de acoplamiento Conexiones en los extremos de las barras de acoplamiento. El extremo exterior de la barra de acoplamiento se conecta al brazo de dirección, y el extremo interior a la cremallera. Ambos extremos incluyen juntas de rótula para permitir tanto el movimiento giratorio como el de ascenso y descenso.

Tolerance A permissible variation between the two extremes of a specification or dimension.
Tolerancia Variación permisible entre los dos extremos de una especificación o dimensión.

Torque A twisting motion, usually measured in ft.-lbs. (N•m).
Par de torsión Fuerza de torsión, normalmente medida en libras-pies (N•m).

Torque converter A turbine device utilizing a rotary pump, one or more reactors (stators) and a driven circular turbine or vane whereby power is transmitted from a driving to a driven member by hydraulic action. It provides varying drive ratios; with a speed reduction, it increases torque.
Convertidor de par de torsión Turbina que utiliza una bomba giratoria, uno o más reactores (estátores) y una turbina circular accionada o paleta; la fuerza se transmite del mecanismo de accionamiento al mecanismo accionado mediante acción hidráulica. Provee relaciones de accionamiento variadas; con una reducción de velocidad, aumenta el par de torsión.

Torque curve A line plotted on a chart to illustrate the torque personality of an engine. When the engine operates on its torque curve it is producing the most torque for the quantity of fuel being burned.
Curva de torsión Línea trazada en un gráfico para ilustrar las características de torsión de un motor. Cuando un motor funciona según su curva de torsión, produce mayor par de torsión por cantidad de combustible quemado.

Torque multiplication The result of meshing a small driving gear and a large driven gear to reduce speed and increase output torque.
Multiplicación de par de torsión Resultado de engranar un engranaje de accionamiento pequeño y un engranaje accionado grande para reducir la velocidad y aumentar el par de torsión de rendimiento.

Torque steer A self-induced steering condition in which the axles twist unevenly under engine torque. An action felt in the steering wheel as the result of increased torque.
Dirección de torsión Condición automática de dirección en la cual los ejes giran de manera irregular bajo el par de torsión del motor. Acción que se advierte en el volante de dirección como resultado de un aumento en el par de torsión.

Torque tube A fixed tube over the drive shaft on some cars. It helps locate the rear axle and takes torque reaction loads from the drive axle so the drive shaft will not sense them.
Tubo de eje cardán Tubo fijo sobre el árbol de mando en algunos automóviles. Ayuda a colocar el eje trasero y remueve las cargas de reacción de torsión del eje de mando para que el árbol de mando no las reciba.

Torsional springs Round, stiff coil springs placed in the driven disc to absorb the torsional disturbances between the driving flywheel and pressure plate and the driven transmission input shaft.
Muelles de torsión Muelles helicoidales redondos y rígidos ubicados en el disco accionado para absorber las alteraciones de torsión entre el volante motor y la placa de presión, y el árbol impulsor accionado de la transmisión.

Total travel Distance the clutch pedal and release bearing move from the fully engaged position until the clutch is fully disengaged.
Avance total Distancia a la que el pedal del embrague y el cojinete de desembrague se mueven de la posición enteramente engranada hasta que el embrague se desengrane por completo.

Toxicity A statement of how poisonous a substance is.
Toxicidad Enunciado que expresa la cualidad tóxica de una substancia.

Traction The gripping action between the tire tread and the road's surface.
Tracción Agarrotamiento entre la banda de la llanta y la superficie de la carretera.

Transaxle Type of construction in which the transmission and differential are combined in one unit.
Transeje Tipo de construcción en la que la transmisión y el diferencial se combinan en una sola unidad.

Transaxle assembly A compact housing most often used in front-wheel-drive vehicles that houses the manual transmission, final drive gears, and differential assembly.
Conjunto del transeje Alojamiento compacto que normalmente se utiliza en vehículos de tracción delantera y que aloja a la transmisión manual, a los engranajes de transmisión final y al conjunto del diferencial.

Transfer case An auxiliary transmission mounted behind the main transmission. Used to divide engine power and transfer it to both front and rear differentials, either full time or part time.
Caja de transferencia Transmisión secundaria montada detrás de la transmisión principal. Se utiliza para separar la energía del motor y transferirla a los diferenciales delantero y trasero, a tiempo completo o a tiempo parcial.

Transmission The device in the powertrain that provides different gear ratios between the engine and drive wheels as well as reverse

Transmisión Mecanismo en el tren transmisor de potencia que provee diferentes relaciones de engranajes tanto entre el motor y las ruedas motrices como en la marcha atrás.

Transmission case An aluminum or iron casting that encloses the manual transmission parts.
Caja de transmisión Pieza fundida en aluminio o en hierro que encubre las piezas de la transmisión manual.

Transverse Powertrain layout in a front-wheel-drive automobile extending from side to side.
Transversal Distribución del tren transmisor de potencia en un vehículo de tracción delantera, que se extiende de un lado al otro.

Tripod (also called tripot) A three-prong bearing that is the major component in tripod constant velocity joints. It has three arms (or trunnions) with needle bearings and rollers that ride in the grooves or yokes of a tulip assembly.
Trípode Cojinete de tres puntas; componente principal en juntas trípode de velocidad constante. Tiene tres brazos (o muñequillas) con cojinetes de agujas y rodillos que van montados sobre las ranuras o los yugos de un conjunto tulipán.

Tripod universal joints Universal joint consisting of a hub with three arm and roller assemblies that fit inside a casting called a tulip.
Juntas universales de trípode Juntas universales que consisten de un cubo con conjuntos de tres brazos y de rodillos que se insertan dentro de una pieza fundida llamada un tulipán.

Trunnion One of the projecting arms on a tripod or on the cross of a four-point universal joint. Each trunnion has a bearing surface that allows it to pivot within a joint or slide within a tulip assembly.
Muñequilla Uno de los brazos salientes en un trípode o en la cruz de una junta universal de cuatro puntas. La superficie del cojinete de cada muñequilla permite que la muñequilla gire dentro de una junta o se deslice dentro de un conjunto tulipán.

Tulip assembly The outer housing containing grooves or yokes in which trunnion bearings move in a tripod constant velocity joint.
Conjunto tulipán Alojamiento exterior que contiene ranuras o yugos en los cuales se mueven los cojinetes de muñequilla en una junta trípode de velocidad constante.

Two-disk clutch A clutch with two friction discs for additional holding power; used in heavy duty equipment.
Embrague de dos discos Embrague con dos discos de fricción para proporcionar más fuerza de retención; utilizado en equipo de gran potencia.

Two-speed rear axle See double-reduction differential.
Eje trasero de dos velocidades Véase diferencial de reducción doble.

U-bolt An iron rod with threads on both ends, bent into the shape of a U and fitted with a nut at each end.
Perno en U Varilla de hierro con filetes de tornillo en los dos extermos, acodada en forma de U y provista de una tuerca a cada extremo.

U-joint A four-point cross connected to two U-shaped yokes that serves as a flexible coupling between shafts.
Junta cardánica Cruz con cuatro puntas fijadas a dos yugos en forma de U, que sirve de acoplamiento flexible entre árboles.

Universal joint A mechanical device that transmits rotary motion from one shaft to another shaft at varying angles.
Junta universal Dispositivo mecánico que transmite movimiento giratorio de un árbol a otro a ángulos cambiantes.

Universal joint operating angle The difference in degrees between the drive shaft and transmission installation angles.
Ángulo de funcionamiento de la junta universal Diferencia en grados entre los ángulos del árbol de mando y del montaje de la transmisión.

Upshift To shift a transmission into a higher gear.
Cambio de velocidades ascendente Acción de cambiar la transmisión a un engranaje de alta multiplicación.

Vibration A quivering, trembling motion felt in the vehicle at different speed ranges.
Vibración Estremecimiento y temblor que se advierte en el vehículo a diferentes gamas de velocidades.

Vehicle identification number (VIN) The number assigned to each vehicle by its manufacturer, primarily for registration and identification purposes.
Número de identificación del vehículo Número asignado a cada vehículo por el fabricante, principalmente para su registración e identificación.

Viscosity The resistance to flow exhibited by a liquid. A thick oil has greater viscosity than a thin oil.
Viscosidad Resistencia de un fluido al movimiento relativo. Un aceite pesado tiene mayor viscosidad que un aceite fluido.

Viscous Thick; tending to resist flowing.
Viscoso Espeso; que tiende a resistir el movimiento.

Viscous friction The friction between layers of a liquid.
Fricción viscosa Fricción entre las capas de un fluido.

Volatility A statement of how easily the substance vaporizes or explodes.
Volatilidad Enunciado que expresa la facilidad de una substancia de vaporizarse o explotar.

Wet-disc clutch A clutch in which the friction disc (or discs) is operated in a bath of oil.
Embrague de disco húmedo Embrague en el que el disco de fricción (o discos) funciona bañado en aceite.

Wheel A disc or spokes with a hub at the center that revolves around an axle, and a rim around the outside for mounting the tire on.
Rueda Disco o rayos con un cubo en el centro que gira alrededor de un eje, y una llanta alrededor del exterior sobre la cual se monta la rueda.

Worm gear A gear with teeth that resemble a thread on a bolt. It is meshed with a gear that has teeth similar to a helical tooth except that it is dished to allow more contact.
Engranaje sinfín Engranaje con dientes parecidos a los filetes de tornillo en un perno. Se engrana con un engranaje cuyos dientes son parecidos a un diente helicoidal, pero se comba para permitir un mejor contacto.

Yoke In a universal joint, the driveable torque-and-motion input and output member, attached to a shaft or tube.
Yugo En una junta universal, el mecanismo accionable impulsor y de rendimiento de torsión y movimiento, que se fija a un árbol o un tubo.

Yoke bearing A U-shaped, spring-loaded bearing in the rack-and-pinion steering assembly that presses the pinion gear against the rack.
Cojinete de yugo En el conjunto de dirección de cremallera y piñón, cojinete en forma de U, con cierre automático, que sujeta el piñón contra la cremallera.

Zerk fitting A very small check valve that allows grease to be injected into a component part, but keeps the grease from squirting out again.
Conexión Zerk Válvula de retención sumamente pequeña que permite inyectar la grasa en un componente, y que a la vez impide que esa grasa se derrame nuevamente.

INDEX

Note: Page numbers in **BOLD** type reference nontext material.